An Introduction to
Mathematical Analysis

An Introduction to
Mathematical
Analysis

Jonathan Lewin
Kennesaw College

Myrtle Lewin
Agnes Scott College

The Random House/Birkhäuser Mathematics Series
Random House
New York

First Edition

9 8 7 6 5 4 3 2

Library of Congress Cataloging-in-Publication Data

Lewin, Jonathan.
    An introduction to mathematical analysis/Jonathan Lewin, Myrtle
Lewin.

    p.   cm.—(The Random House/Birkhäuser mathematics series)
    Includes indexes.
    ISBN 0–394–37262–X
    1. Mathematical analysis.   I. Lewin, Myrtle.   II. Title.
III. Series.
QA300.L52   1987
515—dc19   87–21476
                                            CIP

Preface

This book provides a rigorous course in the calculus of one real variable to students who have previously studied calculus at the elementary level and are possibly entering their first upper-level mathematics course. In many undergraduate programs, the first course in analysis is expected to provide students with their first solid training in mathematical thinking and writing and their first real appreciation of the nature and role of mathematical proof. Therefore, a beginning analysis text needs to be much more than just a sequence of rigorous definitions and proofs. The book must shoulder the responsibility of introducing its readers to a new culture, and it must encourage them to develop an aesthetic appreciation of this culture.

This book is meant to serve two functions (and two audiences): On the one hand it is intended to be a gateway to analysis for students of mathematics and for students majoring in the sciences, engineering, and computer science. It is also intended, however, for other groups of students, such as prospective high-school teachers, who will probably see their course in analysis as the hardest course they have ever taken and for whom the most important role of the course will be as an introduction to mathematical thinking.

At the same time this book is meant to be a recruiting agent. It should motivate talented students to develop their interest in mathematics and to continue their studies when the present course ends. Each topic is presented in a way that extends naturally to more advanced levels of study, and it should not be necessary for students to "unlearn" any of the material of this book when they enter more advanced courses in analysis and topology.

In common with most other introductory texts in real analysis, this book confines much of its attention to the analysis of functions of one variable. Our approach is particularly gentle in the first few chapters but gradually becomes more demanding. By the time one reaches the last few chapters, both the pace and the depth have been increased. We expect students who use this book for a second course to be generally stronger than those who take analysis for one term only. Even in the later chapters, however, we have preserved our commitment to strong motivation and clean, well-explained proofs, and these chapters contain a number of innovative sections on topics not usually covered in a book at this level.

This book omits the usual dreary opening chapter on sets and functions. While such an introduction may have been necessary in the analysis books of a few decades ago, almost all students who have studied elementary calculus (and possibly linear algebra) have at least some acquaintance with the notation of elementary set theory, including unions, intersections, functions, and domain and range. We therefore introduce set theoretic notation as needed in the body of the text, and we hope that this will save a considerable amount of classroom time.

For those who prefer a somewhat deeper approach to set theory, we have provided Appendix A as an optional addendum to Chapter 2. Appendix A discusses Russell's paradox, the Schroeder-Bernstein equivalence theorem, the concept of countability, and Cantor's inequality. Some instructors might elect to cover part of Appendix A—for example, Sections A.1 through A.8. Or one might cover these sections but omit the proof of the Schroeder-Bernstein theorem. The main body of the text does not require the material of Appendix A except in a few starred exercises, which are clearly marked as depending upon this appendix.

The Prelude gives an idea of what a course in analysis is meant to achieve and how it complements a typical elementary calculus course. The Prelude also provides a historical perspective of the development of calculus and of its rigorous formulation in the nineteenth century. Chapter 1 begins the training in the reading and writing of mathematical language. Both the Prelude and Chapter 1 are designed to be given as reading assignments. Many instructors will use them in this way and begin the main body of their courses with Chapter 2.

Chapter 2 introduces the real number system and the notion of completeness, which plays a prominent role throughout Chapters 2 through 7. In Chapter 3, we introduce some of the simple topological properties of R. These properties are used extensively in the subsequent theory of sequences (Chapter 4), limits of functions (Chapter 5), continuity (Chapter 6), derivatives (Chapter 7), and integrals (Chapter 8).

In Chapter 8 we have chosen to introduce the Riemann integral by extending the notion of integration of step functions. Although the opening discussion of the integration of step functions postpones the definition of the Riemann integral, we have found that the delay is well worthwhile. This discussion takes little classroom time and provides simple, clean, and precise notation in which to present the mainstream of Riemann integration. The notation allows us to give short, clean proofs of several theorems that look quite formidable in many texts, and it also facilitates the proof of the bounded convergence theorem, which is one of the high points of the chapter. Appendix B on sets of measure zero is an optional addendum to Chapter 8.

The availability of the bounded convergence theorem from Chapter 8 onward has a number of far-reaching consequences. This theorem is used in the later chapters to provide simple proofs of several interesting results that are considerably sharper than those usually found in a text at this level. Among these is Fichtenholz's theorem on the inversion of repeated Riemann integrals (Chapter 9). In addition, the role of uniform convergence has been modified. Since we no longer require it to justify the interchange of limits with derivatives or integrals, uniform convergence is able to take its proper place as a mode of convergence that is strong enough to guarantee that the limit of a sequence of continuous functions is continuous and that the limit of a sequence of Riemann integrable functions is Riemann integrable. The discussion of uniform convergence in Chapter 11 follows Chapter 9 on improper integrals and Chapter 10 on infinite series.

Following tradition, we have made extensive use of the standard transcendental functions in our illustrative examples throughout the text, even before these functions have been defined rigorously. The functions log and exp are defined in Chapter 9, and the trigonometric functions are defined in Chapter 11. Naturally, these functions are not used to develop the theory until after they have been defined properly.

Sections and exercises marked with a star (*) may be omitted without loss of continuity. They are sometimes, but not necessarily, harder than the regular material. The starred material digresses a little from the main theme of the chapter, to broaden and deepen the concepts discussed. This material is ideal for better students or as a source for special assignments and projects.

In a two-course sequence the material will probably be taught in the order in which it appears. For those who would use this book for a single course in analysis, we suggest the following options:

(1) Chapters 2 through 7.

(2) Chapters 2 through 6, the first ten sections of Chapter 7, and the first seventeen sections of Chapter 8.

(3) Chapters 2 through 6, the first ten sections of Chapter 7, and the first eleven sections from Chapter 10. Those selecting this option will naturally omit the parts of Chapter 10 that involve integration, and especially

the sections that make use of the bounded convergence theorem. On the whole, this will not greatly impair the continuity of Chapter 10. One should of course omit Example 10.2(7) and its more general analogue 10.4.2, and one should omit the integral test 10.6.2. In addition, one should discuss the convergence of the p-series by the alternative method given in Section 10.6.4 instead of using the integral test as in Section 10.6.3.

(4) For classes with an unusually strong calculus background: brief coverage of Chapters 1 through 7, dwelling on Appendix A and a few topics in Chapters 4 and 6, followed by the major part of Chapters 8, 9, 10, and 11.

At the back of the book we have provided answers, solutions, or hints to the solutions of selected exercises, particularly those exercises that play a role in future material. For some exercises we have provided complete solutions within the text itself, and in these cases the reader is told where to find the solution. For each exercise for which assistance is provided, either in the text or at the back of the book, the exercise is bulleted (■). Sometimes, a hint is given with the exercise itself. This is done when we expect the vast majority of readers to need it. For many of the exercises, including most of the starred ones, we have provided no hints at all. This should leave ample material for unaided homework assignments and special projects. The instructor's manual for this book contains further solutions to the exercises and is available from the publisher.

We would like to take this opportunity to express our appreciation to the reviewers of this text for the many hours they have spent in painstaking examination of this work. Their comments resulted in many significant improvements to the manuscript. We would particularly like to express our thanks to Duane Blumberg, University of Southwestern Louisiana; William Bray, University of Maine; Ezra Brown, Virginia Polytechnic Institute; Stephen Fisher, Northwestern University; Chaitan Gupta, Northern Illinois University; Alexander Kleiner, Drake University; Steven Krantz, Washington University; Thomas McCoy, Michigan State University; and William Row, University of Tennessee. We wish also to express our sincere thanks to the many students who have studied from preliminary versions of this book. The insight gained from their responses has been particularly valuable. We take this opportunity to express our appreciation for the support we have received from Agnes Scott College for the writing of this book. We would also like to mention that the original typed version of this work was prepared using a T^3 scientific word processing system (generously lent to us by Kennesaw College), and we would like to express our thanks to the staff of T.C.I. Software Research, Inc., of Las Cruces, New Mexico, the makers of T^3, for the help they gave us while we were working on the manuscript. Finally, we would like to express our profound appreciation to the staff of Random House, for the patience and care they displayed during the production of this work.

Contents

An Introduction to
Mathematical Analysis

Prelude

P.1 WHAT IS MATHEMATICAL ANALYSIS? WHAT IS THIS BOOK ABOUT?

In a nutshell, **mathematical analysis** is the critical study of calculus. As opposed to *discrete* mathematics, or *finite* mathematics, calculus can be thought of as being *infinite* mathematics; and as such, it must rank as one of the greatest, most powerful, and most profound creations of the human mind.

> The infinite! No other question has ever moved so profoundly the spirit of man.— David Hilbert (1921)

Now as you might expect, great, profound, and powerful thoughts don't come all that easily; and in the case of calculus, it took the best part of twenty-five hundred years from the time the first calculuslike problems tormented Pythagoras until the first really solid foundation of calculus was laid, in the nineteenth century. During the seventeenth and eighteenth centuries calculus blossomed, becoming an important branch of mathematics and at the same time a powerful tool, able to describe such physical phenomena as the motion of the planets, the stability of a spinning top, the behavior of a wave, and the laws of electrodynamics. This period saw the emergence of almost all the concepts which one might expect to see in an elementary calculus course today.

But if the blossoms of calculus were formed during the seventeenth and eighteenth centuries, then its roots were formed during the nineteenth. Calculus underwent a revolution during the nineteenth century, a revolution in which its fundamental ideas were revealed and in which its underlying theory was properly understood for the first time. In this revolution calculus was rewritten from its foundations by a small band of pioneers, among whom were Bernhard Bolzano, Augustin Cauchy, Karl Weierstrass, Richard Dedekind, and Georg Cantor. You will see their names repeatedly in this book, for largely as a result of their efforts, the subject we know as mathematical analysis was born. Their work enabled us to appreciate the nature of our number system and gave us our first solid understanding of the concepts of limit, continuity, derivative, and integral. This great and profound theory is the legacy to which you, our reader, are heir.

If you are a typical reader of this book, then you have already completed some courses in a first-calculus sequence; and the words *limit, continuity, derivative,* and *integral* are already familiar to you. But there is a great difference between the way you will see these concepts in this book and the way you saw them in your elementary calculus courses. There, the purpose was to get on with the material as quickly as possible so as to give you a bird's-eye view of the subject and to allow you to see some of its applications. Here, on the other hand, we shall strive for *understanding*. Now that you have your bird's-eye view, it is time to go back and make a careful and critical study of the central ideas of calculus.

It is not the purpose of this book to introduce you to any really new topics; as a matter of fact, the calculus you have already learned contains much more than we can possibly cover. All we are going to do is take a critical look at limits of sequences and at limits, continuity, derivatives, and integrals of functions of one variable. This may not seem like much. There will be almost nothing on functions of several variables, nothing on line and surface integrals, Fourier series, or differential equations. It may seem as though you will be covering very little, and yet you will be working hard. Very hard! But the fruits of your labors as you study this book will more than repay you for the efforts that you will make. Your reward will be the thrill of genuine understanding; and as your understanding of mathematics increases, so will your appreciation of its beauty. You will experience in mathematics the kind of stimulation and pleasure that we associate with the great masterpieces in art, literature, and music. Perhaps you will come to feel that mathematical analysis is the greatest masterpiece of them all.

P.2 A SHORT HISTORY OF THE ROLE OF RIGOR IN MATHEMATICS[1]

Somewhere around the year 500 B.C.E., someone in the school of Pythagoras noticed that according to the Greek concept of number, in which all numbers were

---

[1] For an additional source of reading on this topic, consult the book by Morris Kline [8].

1 unit

Figure P.1

rational, one cannot compare the length of a side of a square with the length of its diagonal. Suppose, for example, that we have a square of side one unit (see Figure P.1). According to Pythagoras's theorem, the length of the diagonal of this square must be $\sqrt{2}$, and it is well known that the number $\sqrt{2}$ is irrational.[2]

This Pythagorean reasoned that if one could compare the side of this square with its diagonal, then one could find some little measuring rod that fits an exact number of times into the side and also fits an exact number of times into the diagonal. But he realized with a shock that such a measuring rod cannot exist. He reasoned as follows: Suppose such a measuring rod can be found, and suppose that this rod fits m times into the diagonal and n times into the side, where m and n are natural numbers. Then we must have $\sqrt{2} = m/n$. But this is impossible, because $\sqrt{2}$ is not a rational number. From our standpoint today we can see that this discovery reveals the inadequacy of the rational number system and of the Greek concept of length; but to them, the discovery was a shock. Just how much of a shock it was can be gauged from the writings of the Greek philosopher Proclus, who tells us that the Pythagorean who made this terrible discovery suffered death by shipwreck as a result of it.

About a hundred years later, something even more devastating happened. In the fifth century B.C.E. Zeno of Elea invented four innocent-sounding paradoxes which brought the philosophers to their knees and kept them there until the time of Bolzano and Cauchy early in the nineteenth century. The first three of **Zeno's paradoxes** appear in Bell [3] as follows:

> **(1)** Motion is impossible, because whatever moves must reach the middle of its course *before* it reaches the end; but *before* it has reached the middle, it must have reached the quarter mark, and so on, *indefinitely*. Hence the motion can never start.
>
> **(2)** Achilles running to overtake a crawling tortoise ahead of him can never overtake it, because he must first reach the place from which the tortoise started; when Achilles reaches that place, the tortoise has departed and so is still ahead. Repeating the argument, we easily see that the tortoise will always be ahead.
>
> **(3)** A moving arrow at any instant is either at rest or not at rest, that is, moving. If the instant is indivisible, the arrow cannot move, for if it did, the instant would immediately be divided. But time is made up of instants. As the arrow cannot move in any one instant, it cannot move in any time. Hence it always remains at rest.

[2] You can find a proof of the irrationality of $\sqrt{2}$ in Section 1.8.

Much has been said about these paradoxes; and quite obviously, we are not going to do them justice here. But let's talk about paradox (3) for a moment. At any one instant of time the arrow does not move. Does that really mean that the arrow will not find its target? Would Zeno have been prepared to stand in front of the arrow? We think not. Then what was Zeno trying to tell us? Perhaps he was trying to warn us that velocity can only be meaningful in any physical sense as an *average velocity over a period of time*. While it is all very well to say that an arrow covers a distance of (say) 60 feet during the course of a second, and that the arrow therefore has an average velocity of 60 feet per second during this time, Zeno warns us that our senses can make nothing out of a notion of *velocity of the arrow at any one instant*.

Now let's look at what is essentially the same problem phrased in terms of *slope* (see Figure P.2). Suppose A is the point $(x_1, f(x_1))$ on the graph of a function f, and B is some other point $(x_1 + \Delta x, f(x_1 + \Delta x))$. Then while it is all very well to say that the slope of the line segment AB is $\Delta y/\Delta x$, where $\Delta y = f(x_1 + \Delta x) - f(x_1)$, there is no obvious physical meaning to the notion of *slope of the graph at the point A*.

"But," you may ask, "isn't this what calculus is all about? Are the paradoxes of Zeno trying to tell us to abandon the idea of a derivative?" They are not. But what we should learn from these paradoxes is that if we want to *define* the derivative of a function f at a point A, then that's just fine with Zeno. Only we can't blame Zeno if this derivative which we have *defined* doesn't measure how the function f increases at A, because as Zeno quite rightly tells us, the function f can't change its value at any one point. In other words, Zeno's paradoxes tell us that (referring to Figure P.2) we may speak of the slope $\Delta y/\Delta x$ of the line segment AB; and if we want to give to the derivative of f at A, which we have *defined*, the name dy/dx, we may do so. But we should not think of dy/dx as the ratio of two numbers dy and dx, the amounts by which y and x increase at the point A; because as Zeno tells us, there are *no* increases in y and x at the point A.

So there is a message for us in the paradoxes of Zeno. The concepts of infinite mathematics (i.e., calculus) must be given properly conceived *definitions*. All properties of mathematical concepts like length, area, slope of a curve, or velocity must be proved directly and logically from the definitions; and outside their definitions mathematical concepts will have no physical existence. For example, once the derivative of a function has been defined, it will not be an auto-

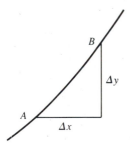

Figure P.2

matic truth that a function with a zero derivative has to be constant nor that a function with a positive derivative has to be increasing. These statements will be *theorems,* and we shall need to understand just what it is about the definition of a derivative and just what it is about the nature of our number system that makes these statements true.

It is precisely this kind of understanding that emerged during the mathematical revolution that took place during the nineteenth century. Although the calculus which had been developed by Newton, Leibniz, and others during the seventeenth century represented a magnificent contribution and gave us the notation for derivatives and integrals that we still use today, that calculus did not rest on a solid foundation. The idea of continuity and its role had not been discovered, the notion of a limit was very inadequate, and the theory did not include an analysis of the nature of the number system and its role. In a sense, the calculus of Newton and Leibniz did not pay sufficient heed to the paradoxes of Zeno. Although there is evidence that Newton and Leibniz themselves may have had considerable understanding of the fundamental ideas upon which the concepts of derivative and integral depend, many of those who followed them did not. Until the end of the eighteenth century the majority of mathematicians based their work upon an impossible mythology. During this time proofs of theorems in calculus commonly depended upon a notion of "infinitely small" numbers, numbers which were zero for some purposes yet not for others. These numbers were known as *evanescent numbers*, *differentials*, or **infinitesimals**; and undeniably, their use provided a beautiful, revealing, and elegant way of looking at many of the important theorems of calculus. Even today, using infinitesimals to motivate some calculus theorems is helpful. But it is one thing to use the idea of an infinitesimal to *motivate* a theory, and it is quite another matter to base virtually the entire theory upon them. Furthermore, the concept of an infinitesimal can be made precise today in a modern mathematical theory which we call **nonstandard analysis**, but there was no precision in the way infinitesimals were used in the eighteenth century. On the whole, the calculus of the seventeenth and eighteenth centuries was not mathematics as we know it today.

During the eighteenth century the voices of critics began to be heard. In 1733 Voltaire [15] described calculus as

> the art of numbering and measuring exactly a thing whose existence cannot be conceived.

Then in 1734 Bishop George Berkeley [4], the philosopher, wrote an essay in which he rebuked the mathematicians for the weak foundations upon which their calculus had been based. And he no doubt took great pleasure in asking

> whether the object, principles, and inferences of the modern analysis are more distinctly conceived, or more evidently deduced than religious mysteries and points of faith.

There is no need to ask you to guess how he answered this question. Some mathematicians composed weak answers to Berkeley's criticism, and others tried vainly to make sense of the idea of infinitely small numbers. But it was not until the early nineteenth century that any real progress was made, when the work of Bernhard Bolzano gave us the first coherent definition of limits and continuity and the first understanding of the need for a **complete number system**. Then came Cauchy, Weierstrass, Dedekind, and Cantor, who, between them, gave us a solid, understandable calculus, both differential and integral, and who settled many important questions about the nature of the real number system R. The work of these pioneers made possible the understanding which we have promised you.

However, before giving you the final go-ahead to start Chapter 1, we must make it clear that even after you have read this book, your understanding will be incomplete. In the first place, this book has been written for a typical two-term sequence of introductory mathematical analysis, and the time and space available to us do not permit the inclusion of all the important ideas which were developed during the nineteenth century. As a matter of fact, just about all the great work of Dedekind and Cantor, which provides us with much of our understanding of the real number system, has been omitted.

Furthermore, you should understand that great as it was, the work of the nineteenth century pioneers left a number of important stones unturned and that there are questions about the underlying theory which even today's mathematicians are unable to answer. In 1897 the Italian mathematician Burali-Forti discovered what is known today as the **Burali-Forti paradox,** which shows that there are serious flaws in Cantor's theory of sets, upon which our understanding of the real number system had been based. Then a few years later, Bertrand Russell posed his famous paradox, which asks:

> If A is the set of all those sets that are not members of themselves, is A a member of itself?

It turns out that whichever way you answer this question, the answer leads to a contradiction (see Appendix A for more details); and this paradox too demonstrated flaws in Cantor's set theory. To see just how much these paradoxes stunned the mathematical community, you might want to look at the book *Grundgesetze der Arithmetik* (*The Fundamental Laws of Arithmetic*), which was written by the German philosopher Gottlob Frege and published in two volumes, the first in 1893 and the second in 1903. This book was Frege's life's work and his pride and joy. He had bestowed upon the mathematical community the first sound analysis of the meaning of number and the laws of arithmetic; and although the book is quite technical in places, it is worth skimming through, if only to see the sarcastic way in which Frege speaks of the "stupidity" of those who had come before him. An example of this sarcasm is Frege's description of his attempt to induce other mathematicians to tell him what the number *one* means. "One object," would be the reply. "Very well," answered Frege, "I choose the *moon*! Now I ask you please to tell me: *Is one plus one still equal to two*??!!!" As things turned out, the second volume of Frege's book was published just after Russell had sent Frege his famous paradox. There was just space at the end of Frege's book for the following acknowledgment:

A scientist can hardly encounter anything more undesirable than to have the foundation collapse just as the work is finished. I was put in this position by a letter from Mr. Bertrand Russell when the work was almost through the press.

During the first few decades of the twentieth century, work was begun in an attempt to put the house of set theory in order, and this work has led to the fundamental **Zermelo-Fraenkel axioms** of set theory that form the basis of much of our mathematics today. Within the framework of the Zermelo-Fraenkel axioms we can once again make use of Frege's important work. What sort of shape are we in now? Perhaps the answer to this question can be summed up by the following words of James Pierpont in 1928:

> The notion of infinity is our greatest friend; it is also the greatest enemy of our peace of mind . . . . Weierstrass taught us to believe that we had at last thoroughly tamed and domesticated this unruly element. Such, however, is not the case; it has broken loose again. Hilbert and Brouwer have set out to tame it once more. For how long? We wonder.

Much progress has been made in the theory of foundations of mathematics since 1928. But the uncertainty remains, for we still have to ask: How good are the Zermelo-Fraenkel axioms? We must ask you, therefore, to look upon your studies in this book as just one step into the world of understanding. If, by the time you reach the last page, the desire for further study has been kindled in you, then we shall have succeeded in our purpose.

Chapter *1*

An Introduction to Mathematical Grammar

1.1 THE ROLE OF PROOFS IN MATHEMATICS

Why do we have to prove theorems in mathematics? Among all the questions that students ask, this one is probably heard most often, and it is a good question which deserves a good answer. But answers to this question aren't so easy to find; and rightly or wrongly, many students would identify the answers they have received in their elementary mathematics courses with one or more of the following:

Because I say what goes in this classroom and if you don't learn these proofs, you will fail.

You don't! I was just mentioning this proof in case anyone happened to be interested, and since it is obvious that nobody is, let's get on with the only thing that really matters, and that is my instructions on how to do the homework problems.

Look! I don't like this stuff any more than you do. But it's in the syllabus and so we have to do it.

If you don't prove a theorem in mathematics, you won't really be sure that it is true.

Well, as you may have guessed, none of these answers comes close to providing the message which this book is meant to give you; and in fact, none of them represents true mathematical spirit. This statement applies even to the fourth answer, for it is quite wrong to say that we prove theorems in mathematics simply to make sure that they are true. If this were our only reason for proving theorems, why should more than one person have to prove them? If one person with a Ph.D. and an honest face announces that a theorem is true, then why shouldn't the rest of us just get on with our lives in the blissful knowledge that if we ever have to use that theorem, it will be there? And this question brings us back to the question we started with: Why do we need to prove theorems? We have two answers to suggest to you; the first takes a somewhat practical point of view, and the second reminds us that, after all, mathematics is an art form:

(1) A mathematical theorem represents much more than just a single statement. It represents a host of many statements, all of which can be deduced by roughly the same method of proof. A theorem therefore represents a level of mathematical understanding, for whoever understands its proof will command all the statements that can be deduced by the same method.

(2) Every mathematical theorem has two parts. The first part, which is called the *hypothesis,* contains the information which is assumed (or given); the second part, called the *conclusion,* is the part we have to prove. What the theorem really says is that *if* the hypothesis is assumed, *then* the conclusion must follow. Thus it is not the truth of the conclusion that is so important to us but, rather, the existence of a bridge between the hypothesis and the conclusion. The proof of the theorem is that bridge; and therefore, the proof of a theorem is the most important thing about it.

Let us look at some examples to illustrate the points we have been making.

(1) The opposite sides of a parallelogram have the same length. (See Figure 1.1.) If ever there were an obvious statement, then this simple fact of high school geometry must be it. Who could doubt that the opposite sides of a parallelogram must have the same length? Now cast your mind back a few years. Do you

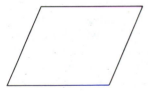

Figure 1.1

remember how this theorem is proved? We draw a diagonal which splits the parallelogram into two congruent triangles. Therefore, what the proof is really telling us is not so much that the opposite sides of a parallelogram must have the same length but, rather, that there is a bridge between this fact and the principles of congruent triangles. An understanding of this bridge enables us to deduce other theorems in geometry that are possibly less obvious.

$$\text{(2)} \qquad \sqrt[3]{\frac{9\sqrt{3} + 5\sqrt{11}}{6\sqrt{3}}} + \sqrt[3]{\frac{9\sqrt{3} - 5\sqrt{11}}{6\sqrt{3}}} = 1.$$

At first sight this statement probably looks unbelievable, but now that we have told you that it is true, you can believe it. Why would we lie? Do you feel an urge to put the left side of this identity into your calculator? Go ahead and do it, but you will be wasting your time. Even before you begin, you know that your calculator must come up with the number 1. What point can there be in putting this expression into your calculator if you already know what the result will be? And after you are done, will you have any more understanding of the ideas that might lead to an identity like this one? Perhaps you would like to cube both sides. If so, prepare for a long, hard battle. And if you succeed, will you then understand how this identity was derived? The fact is that if you were to learn how one might *arrive* at this identity, then you would also know how to make many more of the same type; and you would *understand* why the identity has to hold.

$$\text{(3)} \qquad \frac{1}{1^2} + \frac{1}{2^2} + \frac{1}{3^2} + \ldots = \frac{\pi^2}{6}.$$

This example, like example (2), looks a little unbelievable, and once again, you could use your calculator to verify that the identity is at least approximately true. But once again, using your calculator wouldn't give you the slightest idea how you might arrive at such an identity. Surely, we need to ask just what it is about the infinite series on the left side that has anything to do with the number π. A proof of this identity can be found in Chapter 11.

In this book you will find many proofs you will need to know. Do not try to memorize them or they will rise against you and dominate you. Strive, instead, to understand them, and they will be your faithful servants. When studying a proof, do not be content with the knowledge of how each individual step follows from the one before it. Every proof has a *theme*, a master plan, which suggests what the individual steps should be. You have understood a proof only when you have looked into it deeply enough to perceive that theme, and you have understood a proof only when you feel able to write it down or explain it to others. As you progress and as you begin to understand the proofs this book contains, you will begin to command the ideas which underlie calculus.

1.2 THE DENIAL (NEGATION) OF A MATHEMATICAL SENTENCE

The *denial* (sometimes called the *negation*) of a given sentence is the statement that the given sentence is false. In the following examples we illustrate this notion.

  (1) Suppose a given sentence says: *Everyone in this room speaks French.* In order for this statement to be false, there would have to be at least one person in this room who does not speak French. So the denial of the given sentence says: *At least one person in this room does not speak French.*

  (2) Suppose a given sentence says: *The diagonals of a rectangle have the same length.* What this sentence really means in high school geometry is that every rectangle has the property that its diagonals have the same length. This example is therefore very similar to example (1), and its denial says: *There is at least one rectangle whose diagonals do not have the same length.*

  (3) Suppose $f(x) = 32x/(1 + 4x^2)^2$ for $0 \le x \le 1$, and suppose that the given sentence says: *The maximum value of f is $3\sqrt{3}$.* This sentence really tells us two things. On the one hand, it tells us that $f(x) \le 3\sqrt{3}$ for every $x \in [0, 1]$; and on the other hand, it tells us that there is at least one number $x \in [0, 1]$ for which $f(x) = 3\sqrt{3}$. The denial of this sentence therefore says: *Either there must exist a number $x \in [0, 1]$ for which $f(x) > 3\sqrt{3}$, or there is no number $x \in [0, 1]$ for which $f(x) = 3\sqrt{3}$.*

As you study mathematics, you should cultivate the habit of writing the denial of any sentence that seems difficult to understand. The general principle is that whenever you ask yourself what it means to say that a given statement is false, you will understand more clearly what it means to say that the statement is true. So, for example, in Chapter 6 you will see the definition of continuity of a function at a point: A function *f* is said to be continuous at a point *a* when. . . . When you come to Chapter 6, you will help yourself understand the definition of continuity if you ask yourself what it means to say that a given function *fails* to be continuous. The examples of Section 1.3.4 will help you learn how to ask this question, and you will have further practice writing denials in the exercises of Section 1.4.

1.3 SOME IMPORTANT SPECIAL WORDS USED IN MATHEMATICAL SENTENCES

1.3.1 The Words *and, or,* and *if*. The words *and, or,* and *if* are used in mathematics to combine two given sentences so as to make a single sentence. Consider, for example, the following two sentences *p* and *q*, where we take *p* to be the sentence "Helsinki is the capital of France," and q to be the sentence "All cats scratch." Among the ways we might combine these sentences are the following:

Helsinki is the capital of France, and all cats scratch.

Either Helsinki is the capital of France, or all cats scratch.

If Helsinki is the capital of France, then all cats scratch.

Looking at things more generally, let us suppose that p and q are any two given sentences. The following examples show some of the ways p and q might be combined.

(1) The sentence "p and q" means that both p and q are true. The denial of "p and q" is the statement that either p is false or q is false, or perhaps that both of them are false.

(2) The sentence "p or q" means that at least one of the two sentences p and q must be true (which includes the possibility that both of them are true). The denial of the sentence "p or q" is the statement that both of the sentences p and q must be false.

(3) The sentence "If p then q" means that in the event that p is true, then the sentence q must also be true. Note that the only way in which the sentence "If p then q" can be false is that p should be true and q should be false. The denial of "If p then q" therefore says that "p is true and q is false." The following list gives some of the equivalent ways in which we like to write the sentence "If p then q" in mathematics:

p implies q.

$p \Rightarrow q$.

q is implied by p.

$q \Leftarrow p$.

p is true only if q is true.

If q is false, then p is false.

Either p is false, or q must be true.

p is a sufficient condition for q.

q is a necessary condition for p.

(4) The sentence "p if and only if q," which is sometimes written as "p iff q," means that the sentences p and q are equivalent. In other words, "p iff q" means that either p and q are both true, or they are both false.

(5) The pair of sentences "p. Therefore q" is very often confused with "If p then q," but "p. Therefore q" says much more. When we say "p. Therefore q," we mean the following: "I know that p is true. I also know that if p is true, then q must be true, and I therefore conclude that q must be true." To help us understand the distinction between "If p then q" and "p. Therefore q," let us look at the following example: We take p to be the sentence "It is raining" and q to be the sentence "You will get wet." The sentence "If p then q" then says that "If it is raining, then you will get wet." More elaborately, "I don't know whether or not it is raining, but if it does happen to be raining, then you will get wet." On the other hand, "p. Therefore q" says "I know that it is raining and I therefore know that you will get wet."

1.3.2 Proving a Theorem Whose Statement Contains *and, or,* or *if.*

As in the previous section, we shall suppose that p and q are sentences; and we shall suppose now that the sentences p and q have been combined into a single statement. How can we go about proving this single statement? The answer to this question depends on the way p and q have been combined, and in this section we suggest some possible approaches we might use.

First, suppose we want to prove a theorem whose conclusion is of the form "*p and q.*" We need to show that both of the statements p and q are true, and we might do so, for instance, by showing first that p is true and then showing that q is true.

Second, if we want to prove a theorem whose conclusion is of the form "*p or q,*" then we need to show that at least one of the two statements p and q is true. Among the many ways of showing this truth, the following three are quite common:

Assume that p is false, and use this assumption to show that q is true.

Assume that q is false, and use this assumption to show that p is true.

Assume that both p and q are false, and obtain a contradiction.

Third, in order to prove a theorem of the type "If p then q," we might use one of the following methods:

Assume that p is true, and use this assumption to show that q must be true.

Assume that q is false, and use this assumption to show that p is false.

Assume that p is true and q is false, and obtain a contradiction.

1.3.3 The Quantifiers *for every* and *there exists.*

The phrases *for every* and *there exists* abound in mathematics. *For every* is called the *universal quantifier;* and depending upon the context, it sometimes appears simply as *every,* sometimes as *all,* and sometimes as the symbol ∀. *There exists* is called the *existential quantifier;* and depending upon the context, it sometimes appears as *there is, we can find, it is possible to find, there must be, there is at least one, some,* or as the symbol ∃. Notice that at least one of these quantifiers appears in each of the examples of the next section and in the exercises that follow.

1.3.4 Some Examples. The examples of this section are designed to help you understand the use of the words *and, or, if, for every*, and *there exists*. Read them alongside their denials, and notice how often an awareness of the denial will help you understand the given statement.

(1) All cats scratch.
    Denial. Not all cats scratch. An alternate form of this denial is, "There exists a cat which does not scratch."

(2) There exists a cat which scratches.
    Denial. All cats do not scratch.

Warning. Do *not* confuse these two examples. Outside mathematics, it is quite common to hear people saying "All cats do not scratch" when what they really mean is "Not all cats scratch." It's wrong to confuse these sentences under any circumstances, but even if you do it outside mathematics, *don't do it here!*

(3) Some cats scratch. This sentence is example (2) again.

(4) All cats scratch and some dogs bite.
    Denial. At least one cat does not scratch or no dogs bite.

(5) If some cats scratch, then all dogs bite.
    Denial. Some cats scratch, and some dogs do not bite.

(6) Either some cats scratch, or if all dogs bite, then some birds sing.
    Denial. No cats scratch, and all dogs bite, and no birds sing.

(7) Once upon a time there was a princess.
    Denial. There has never been a princess.

(8) Once upon a time there lived a beautiful princess.
    Denial. No princess who has ever lived has been beautiful.

(9) No Irishman has ever been at a loss for words.
    Denial. At least one Irishman has on at least one occasion been at a loss for words.

(10) No one has ever seen an Englishman who is not carrying an umbrella.
    Denial. At least one person has on at least one occasion seen an Englishman who is not carrying an umbrella.

(11) For every positive number x the number $x - 1$ is also positive.
    Denial. There exists a positive number x such that $x - 1 \leq 0$.

(12) For every number x there exists a number y such that $y > x$.
    Denial. There exists a number x such that for every number y we have $y \leq x$.

(13) There is no least positive number.
   Denial.   There exists a positive number x such that for every positive number y we have $y \geq x$.

(14) For every number x either $x^2 > 1$ or $x \leq 1$.
   Denial.   There exists a number x such that $x^2 \leq 1$ and $x > 1$.

The last few examples concern two functions f and g which we assume have been given.

(15) Whenever $x > 50$, we have $f(x) = g(x)$.
   Denial.   There exists a number x such that $x > 50$ and $f(x) \neq g(x)$.

(16) There exists a number w such that $f(x) = g(x)$ for all numbers $x > w$.
   Denial.   For every number w there exists a number $x > w$ such that $f(x) \neq g(x)$.

(17) For every number x there exists a number $\delta > 0$ such that for every number t satisfying the condition $|x - t| < \delta$, we have $|f(x) - f(t)| < 1$.
   Denial.   There exists a number x such that for every positive number δ it is possible to find a number t such that $|x - t| < \delta$ and $|f(x) - f(t)| \geq 1$.

(18) There exists a number $\delta > 0$ such that for every pair of numbers x and t satisfying the condition $|x - t| < \delta$, we have $|f(x) - f(t)| < 1$.
   Denial.   For every positive number δ there exists a pair of numbers x and t such that $|x - t| < \delta$ and $|f(x) - f(t)| \geq 1$.

(19) For every number $\varepsilon > 0$ there exists a number $\delta > 0$ such that for every pair of numbers x and t satisfying the conditions $|x - 2| < \delta$ and $|t - 2| < \delta$, we have $|f(x) - f(t)| < \varepsilon$.
   Denial.   There exists a positive number ε such that for every positive number δ it is possible to find a pair of numbers x and t satisfying the conditions $|x - 2| < \delta$ and $|t - 2| < \delta$, and $|f(x) - f(t)| \geq \varepsilon$.

(20) For every number $\varepsilon > 0$ and for every number x, there exists a number $\delta > 0$ such that for every number t satisfying $t \neq x$ and $|x - t| < \delta$, we have $|f(t) - g(t)| < \varepsilon$.
   Denial.   There exists a positive number ε and there exists a number x such that for every positive number δ it is possible to find a number $t \neq x$ such that $|x - t| < \delta$ and $|f(t) - g(t)| \geq \varepsilon$.

(21) For every positive number ε there exists a positive number δ such that for every pair of numbers x and t satisfying the conditions that $x \neq t$ and $|x - t| < \delta$, we have $|f(x) - f(t)| < \varepsilon$.
   Denial.   There exists a positive number ε such that for every positive number δ it is possible to find numbers x and t such that $x \neq t$ and $|x - t| < \delta$, and $|f(x) - f(t)| \geq \varepsilon$.

1.4 EXERCISES

In the following exercises[1] f and g are given functions and A and B are given sets of real numbers. In each exercise, write the denial of the given sentence in a form that is pleasant to read and uses proper English.

- **1.** If what you said yesterday is correct, then Jim has red hair.

- **2.** Either you take me for a fool, or you must be a fool yourself.

- **3.** He walked into my office this morning, told me a pack of lies, and punched me on the nose.

- **4.** All that glitters is gold.

- **5.** You are right and I am wrong.

- **6.** You are right if and only if I am wrong.

- **7.** Either there exists a cat that does not scratch, or everyone in this room is a liar.

- **8.** I dream when I sleep. (Lewis Carroll)

- **9.** Nobody is worth listening to on military subjects, unless he can remember the battle of Waterloo. (Lewis Carroll)

- **10.** Some of us are out of breath and all of us are fat. (Lewis Carroll)

- **11.** Someone in this room is smoking.

- **12.** Fifty percent of the people in this room are smoking.

- **13.** It is with regret that I inform you that someone in this room is smoking.

- **14.** There exists a number x such that for every number $u > x$ we have $f(u) > g(u)$.

- **15.** For all numbers u and v which satisfy $u > 50$ and $v > 50$, we have $|f(u) - f(v)| < 2$.

- **16.** There exists a number p such that for all numbers u and v which satisfy $u > p$ and $v > p$, we have $|f(u) - f(v)| < 2$.

- **17.** For every number $\varepsilon > 0$ and for all numbers x and t satisfying $x > 7$ and $t > 7$, we have $|f(x) - f(t)| < \varepsilon$.

- **18.** For every number $\varepsilon > 0$ there exists a number p such that for all numbers x and t satisfying $x > p$ and $t > p$, we have $|f(x) - f(t)| < \varepsilon$.

- **19.** There exists a number p such that for every number $\varepsilon > 0$ and for all numbers x and t satisfying $x > p$ and $t > p$, we have $|f(x) - f(t)| < \varepsilon$.

---

[1] As explained in the preface, the symbol ■ is used to designate an exercise for which a solution (or partial solution) has been provided.

■ **20.** If a function h is continuous on $(0, 1)$, then there must exist a number x in $(0, 1)$ such that h is differentiable at x.

■ **21.** There exists a number w such that for every member x of A we have $x < w$.

■ **22.** Whenever x is a member of A and y is a member of B, we have $x < y$.

■ **23.** Whenever x and y are members of A, either $x = y$ or $|x - y| \geq 1$.

■ **24.** For every positive number ε it is possible to find two members x and y of A such that $x \neq y$ and $|x - y| < \varepsilon$.

■ **25.** If P and Q are any two sets of numbers, and if for every member x of P and every member y of Q, we have $x < y$, then it is possible to find a number w such that whenever x is a member of P and y is a member of Q, we have $x \leq w \leq y$.

1.5   A GUIDE TO THE PROPER INTRODUCTION OF MATHEMATICAL SYMBOLS

To help us appreciate the need for this introduction, let's ask a little question about elementary algebra:

Is it true that $(x + y)^2 = x^2 + y^2$?

Perhaps the answer ''no'' is hovering on your lips. If so, you are being a little hasty; for is it not true that $(3 + 0)^2 = 3^2 + 0^2$? As you can see, the truth or falsity of the equation $(x + y)^2 = x^2 + y^2$ depends upon precisely which numbers x and y we are talking about. Therefore, since this question says nothing about what the numbers x and y are, we must conclude that the question is meaningless as it stands.

Now let's take a look at a few of the ways this question might have been asked meaningfully:

Is it true that there exist numbers x and y such that $(x + y)^2 = x^2 + y^2$? Yes!

Is it true that for every number x there exists a number y such that $(x + y)^2 = x^2 + y^2$? Yes!

Is it true that there exists a number x such that for every number y we have $(x + y)^2 = x^2 + y^2$? Yes!

Is it true that for all numbers x and y we have $(x + y)^2 = x^2 + y^2$? No!

Notice how in each of these four meaningful variations of the question, the symbols x and y were introduced by one or other of the quantifiers *for every* and *there exists*. And in the same way, if you look back at the examples and exercises of the preceding sections, you will see that, with the exception of the few symbols that were declared as having been given at the beginning of the section, every symbol was carefully introduced when it first appeared, with one or other of the two quantifiers. Perhaps you have also noticed that as one moves from a sentence to its denial, the quantifiers *for every* and *there exists* change places. And finally, you may have noticed that it is important to know precisely where in the sentence a given symbol is introduced. For example, compare the following two sentences:

(1) For every positive number x there exists a positive number y such that $y < x$.

(2) There exists a positive number y such that for every positive number x we have $y < x$.

Although these statements may look similar, they do not say the same thing. As a matter of fact, (1) is true and (2) is false. If you look back at the sentences in Sections 1.3.4 and 1.4, you will see a number of pairs of similar-looking sentences whose meanings are quite different because of differences in the order of appearance of the symbols. So the message of this section is as follows: For a piece of mathematical writing to make sense, all symbols must be properly introduced by one or the other of the two quantifiers, and each symbol must be introduced in the right place—not too early and not too late.

1.6 HOW TO PROVE A THEOREM THAT SAYS "THERE EXISTS . . ."

We shall begin by looking at a nonmathematical example. You are standing in a room full of people, and you are asked to demonstrate that there is at least one bald man in the room. One thing you might do is go from man to man and in each case look at his head, until you have found a bald man. If your search turns up at least one bald man, you can *choose* one of the bald men you have found as an *example* of a bald man in the room; and by *giving an example,* you have proved that a bald man exists.

As a simple example of this technique in mathematics, we shall prove the following easy theorem:

There exists a prime number greater than 20.

To prove this theorem, we shall *choose* 71 as an example of a prime number greater than 20. This example proves the theorem.

Now let's look at this kind of statement in general. Suppose we wish to prove that a certain set A is not empty; in other words, we want to prove that a member of A must exist. For example, A might be the set of bald men in a given room or the set of prime numbers greater than 20. If we happen to know of an example of a member of A, then we can be sure that the set A is not empty.

We mention finally that giving an example is not the *only* method of proving existence. Suppose you were asked to verify that someone in a given room is smoking. You wouldn't have to see and identify a smoker in order to know that a smoker exists. All you would have to do is try to breathe. In mathematics, too, we

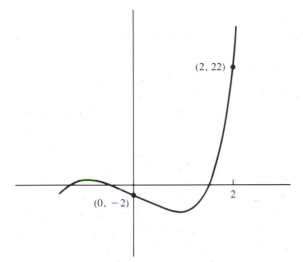

Figure 1.2

often prove existence by this sort of indirect method. As an example, we shall sketch a proof of the following statement:

There exists a number x between 0 and 2 such that $x^5 - 4x - 2 = 0$.

Some of the details of this proof will have to wait until we reach Chapter 6. But we can give a sketch of the proof here. For every real number x, define $f(x) = x^5 - 4x - 2$. Note that $f(0) < 0$ and $f(2) > 0$. But f, being a polynomial, must be *continuous*. And as you will see when you reach Section 6.8, if a function is continuous on an interval and has a negative value at one endpoint and a positive value at the other, there must be a point in the interval at which the function is zero. (See Figure 1.2.)

So even though we do not know of an example of a number x between 0 and 2 at which $f(x) = 0$, we know that such a number exists. As we have said, giving an example isn't the only way to prove existence; but it is often the easiest way, and so we give an example whenever we can.

1.7 HOW TO USE A THEOREM THAT SAYS "THERE EXISTS . . ."

If we are *given* (or if we have previously proved) that a certain set A is nonempty, and if we need a member of that set, then we may *choose* a member of the set as an example. As in the previous section, we shall illustrate this idea by looking first at a nonmathematical example: You are standing in a room full of people, and the person next to you collapses. You call, "Is there a doctor in the house?" Then having established that a doctor exists, you *choose* one of the doctors to help the person who is ill.

Now let us apply this technique to mathematics. As you may know, it is important in calculus to have a function f with the property that $f'(x) = 1/x$ for every $x > 0$. Having proved that such functions exist, we *choose* one of them [the one we choose is the one satisfying $f(0) = 1$]; and we call this function the *natural log function*. You will find the details of this procedure in Chapter 9.

When you are writing a mathematical proof, do not make the common mistake of thinking that just because a certain set A is known to be nonempty, a member of A has automatically been chosen for you. No member has been chosen until you choose it. For example, consider the following statement:

There exists a number x such that $0 < x < 1$.

Does this sentence refer to any particular number x in (0, 1)? No, it doesn't! All this sentence says is that the interval (0, 1) is nonempty, and we would be quite wrong to follow it with statements about x as if there were some x which had been introduced. For example, we would be incorrect if we wrote the following sentences:

There exists a number x such that $0 < x < 1$.
Clearly, $x^2 < x$.

On the other hand, we can legitimately say:

There exists a number x such that $0 < x < 1$.
Furthermore, for every $x \in (0, 1)$ we have $x^2 < x$.

Alternatively, we could say:

There exists a number x such that $0 < x < 1$.
Choose a number $x \in (0, 1)$. Note that $x^2 < x$.

Note that when we say that there exists a number $x \in (0, 1)$, there is nothing special about the symbol x. This statement could just as well have been written in one of the following equivalent forms:

There exists a number t such that $0 < t < 1$.
There exists a number p such that $0 < p < 1$.
There exists a number in the interval (0, 1).

1.8 HOW TO PROVE A THEOREM THAT SAYS "FOR EVERY . . ."

As in the previous two sections, we shall begin by looking at a nonmathematical example. Suppose you are standing in a room full of people and you are asked to prove that every man in the room is bald. It would surely not be good enough to produce just one bald man! You have to examine the head of every man in the room to make sure that he is bald. Should you find even one nonbald man, the proposition you are trying to prove is false. But if, after you have examined every man, you can say that they were all bald, then your proof has succeeded.

The analogue of this kind of statement in mathematics is a statement that says that all the members of a given set A must have a certain property. As an example of such a statement, we shall look again at sentence (12) of Section 1.3.4:

For every number x there exists a number y such that $y > x$.

Unfortunately, one can't prove this mathematical theorem by an exact analogue of the method we have suggested for proving that all the men in a room are bald. The trouble is that there are infinitely many numbers; and even if we were to spend the rest of our lives checking numbers one at a time to make sure that for every number x there exists a number y such that $y > x$, we would eventually die, leaving the unproved theorem as a legacy. This is where mathematicians use the word *let*. To prove the theorem, we proceed as follows:

Let x be any real number.

What this sentence really means is: "I don't know if there are any real numbers to call x and I don't care. I don't need any real numbers; but in case there are any, let x be an arbitrary one of them, coming without any restrictions, so that anything I might be able to prove about x would apply just as well to any other number. In other words, let x be an arbitrary number come to *challenge* me to prove that there exists a number y such that $y > x$."

Now that we have this challenger x in our hands, we have to prove that there exists a number y such that $y > x$. For this purpose we shall use the fact that $x + 1 > x$. We write:

Define $y = x + 1$.

Since this choice has provided us with an example of a number $y > x$, the proof is complete.

We end this section with an example of a proof that is a little different from the one just given. This time we shall prove the theorem that says that

$\sqrt{2}$ is irrational.

What this theorem really means is that for every pair of natural numbers m and n we have $m/n \neq \sqrt{2}$, or in other words, $m^2 \neq 2n^2$. The technique described above therefore suggests that in order to prove this statement, we should begin as follows:

Let m and n be natural numbers.

In the proof that follows we use a variation on this theme. First, we observe that if $\sqrt{2}$ were rational, then one could write $\sqrt{2}$ in the form m/n, where m and n are natural numbers which have no common factor (except 1). We now begin our proof of the irrationality of $\sqrt{2}$ by saying:

Let m and n be natural numbers with no common factor, and to obtain a contradiction, assume that $m^2 = 2n^2$.

The equation $m^2 = 2n^2$ implies that n^2 must be a factor of m^2. And since the numbers m^2 and n^2 have no common factor, it follows that $n = 1$. Therefore, $m^2 = 2$. Since there is clearly no natural number m such that $m^2 = 2$, we have reached the desired contradiction.

To sum up this section, a major technique used in mathematics to prove a statement that says that all members of a given set S must have a certain property, is to begin the proof with the following words:

Let x be a member of the set S.

1.9 HOW TO USE A THEOREM THAT SAYS "FOR EVERY . . ."

Let's go back to that room full of people. Suppose you have been reliably informed that all bald men are honest. And suppose you happen to be talking to a certain bald man, and this man wants to sell you oil stock. Then you may safely invest your money. Understand that the only way that the information that all bald men are honest can be of any use to you is that there should be certain bald men of special interest to you. Perhaps you are talking to them. Perhaps you are related to them. But one way or another, there are certain bald men of special interest to you, and you would like to know that *they* are honest. You may come to this conclusion because of your knowledge that *all* bald men are honest.

The mathematical analogue of this idea is almost identical. Suppose you happen to know that all the members of a given set S have a certain property, and suppose you have a special interest in some particular members of the set. Perhaps you are talking to these members, or perhaps you are related to them. One way or another, you have a special interest in certain members of S, and you would like to know that these particular members have the property mentioned. You may conclude that they do because of your knowledge that *all* members of S have the desired property.

Do not confuse the problem of *proving* that all the members of a given set S have a certain property with the question we are discussing here. As you saw in Section 1.8, when you want to *prove* that every member of a set S has a certain property, then you write, "Let x be a member of the set S." It would be a mistake to write this sentence when you know that all members of a set S have some property, and you want to *use* this fact. Once again, knowledge that all members of a set S have a certain property can only be useful to you when you have a special interest in particular members of S.

This ends our grammar lesson, and you are now ready to enter Chapter 2, where you will begin your study of mathematical analysis. As you do so, be guided by the principles you have learned here. Your success in the coming chapters will be dependent on how well you learn to implement these principles.

Chapter 2

The Completeness of the Real Number System

2.1 FUNDAMENTAL PROPERTIES OF THE REAL NUMBER SYSTEM

2.1.1 Introduction. Calculus of one variable is the theory of limits, continuity, derivatives, and integrals, in which the action takes place in the **real number system,** or ''real line'' R; and so naturally, a proper understanding of calculus must depend on a proper understanding of R. Therefore, since R, just like anything else in mathematics, needs to be given a proper definition, and since its properties need to be deduced directly from this definition, one might reasonably suppose that the starting point for a course in mathematical analysis should be the definition of the real number system. We cannot begin our course this way, however, because the process of defining the real number system and deducing its properties from the fundamental axioms of set theory is too vast a subject to be studied as the first chapter of an introductory course in mathematical analysis. Perhaps you will have the opportunity to define the real number system in your future studies. Perhaps you will go even further and study mathematical logic and then question the very axioms of set theory itself. But for now, we shall be content to begin our study of mathematical analysis with the assumption that a ready-made real number system R has somehow been placed in our hands.

Strictly speaking, we should not call the real number system R, for the real number system consists of more than just the *set* of real numbers. In addition to the set itself, the system possesses an algebraic structure which contains the special numbers 0 and 1, where $0 \neq 1$, the operations $+$ and $\times$, and the relation $<$. As you may know, $+$ and $\times$ are **binary operations** on the set R; and these operations associate to any ordered pair (x, y) of real numbers, the numbers which we write as $x + y$ and xy, respectively, and which we call the **sum** and the **product** of x and y. We shall think of the relation $<$ as associating to any ordered pair (x, y) of real numbers the statement $x < y$, which may be either true or false. Strictly speaking, we should refer to the real number system as $(R, 0, 1, +, \times, <)$, but few people actually do so. Following tradition, we shall refer to the real number system simply as R.

2.1.2 A List of the Fundamental Axioms for R.

In this section we list the fundamental axioms of R from which all its other properties may be deduced.

(1) For all numbers x, y, and z

$$x + (y + z) = (x + y) + z \qquad \text{and} \qquad x(yz) = (xy)z.$$

(2) For all numbers x and y

$$x + y = y + x \qquad \text{and} \qquad xy = yx.$$

(3) For all numbers x, y, and z, $x(y + z) = xy + xz$.

(4) For every number x, $x + 0 = x$ and $x \cdot 1 = x$.

(5) For every number x there exists a number y such that $x + y = 0$.

(6) For every number x except 0 there exists a number y such that $xy = 1$.

(7) (a) Given any real numbers x and y, one and only one of the three conditions $x < y$, $x = y$, $y < x$ is true.
(b) Given any real numbers x, y, and z, if $x < y$ and $y < z$, then $x < z$.
(c) Given any real numbers x, y, and z, if $x < y$ then $x + z < y + z$.
(d) Given any real numbers x, y, and z, if $x < y$ and $0 < z$, then $xz < yz$.

(8) The system R is *complete*. In Section 2.4 we shall say precisely what we mean by this statement.

Because the theory of limits depends heavily on axiom (7), we provide a brief review of inequalities in the next section. If you don't need this review, skip Section 2.1.3 and go on to 2.1.4.

2.1.3 A Brief Review of Inequalities.

Axiom (7) injects into the real number system a certain geometric flavor, for it is by means of this axiom that we are able to think of the real numbers as being strung out on a "number line." When this line is horizontal, our usual convention is to make the condition $x < y$ equivalent to saying that the number x lies to the *left* of the number y.

As you know, the condition $x > y$ means that $y < x$, and $x \leq y$ means that either $x < y$ or $x = y$. Note that the statement $3 \leq 3$ is true.

An important consequences of axiom (7) is the fact that for every number x we have $x^2 \geq 0$. From this result it follows that $1 = 1^2 > 0$ and $-1 < 0$, and we conclude that it is impossible to find a real number x such that $x^2 = -1$. If you are familiar with complex numbers, then you know that in the complex number system there is a number i such that $i^2 = -1$. Consequently, it is impossible to assign an order relation $<$ to the complex number system that will satisfy axiom (7). Therefore, in making the axioms of Section 2.1.2 serve as our starting point, we are restricting our attention to the system R of *real* numbers.

The inequalities that are most useful to us in analysis often involve **absolute value.** Recall that the absolute value $|x|$ of a real number x is defined to be x if $x \geq 0$, and it is defined to be $-x$ if $x < 0$. This concept is useful to us in analysis because it gives us a notion of *distance*. We can think of the absolute value of a number x as being the distance from x to 0; and given two numbers x and y, we can think of $|x - y|$ as being the distance from x to y. By thinking about absolute value in this way, we can often associate a geometric picture with an inequality that involves absolute value, which may allow us to understand the inequality more easily. For example, the inequality $|x - a| < \delta$ suggests that the distance from x to a is less than δ; and with this interpretation in mind, we can see easily that the two inequalities $|x - a| < \delta$ and $a - \delta < x < a + \delta$ are equivalent.

One of the inequalities that is most important to us in analysis is the **triangle inequality,** which says that if a, b, and c are any three numbers, then

$$|a - c| \leq |a - b| + |b - c|.$$

The triangle inequality follows at once from the inequality $|x + y| \leq |x| + |y|$ and the fact that $a - c = (a - b) + (b - c)$. The reason we call this the triangle inequality is that its two-dimensional analogue says that the sum of the lengths of any two sides of a triangle is not less than the length of the third side. (See Figure 2.1.)

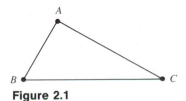

Figure 2.1

We end this section with some exercises on inequalities.

Exercises[1]

1. Prove that if x and y are any real numbers, then $|x + y| \leq |x| + |y|$. Under what circumstances is this inequality an equation?

2. Prove that if $\delta > 0$, then $|6 - \delta| < 6 + \delta$. Now prove that if $|x - 3| < \delta$, then $|x^2 - 9| < \delta(6 + \delta)$.

*3. Suppose $\delta > 0$, and that p is any number such that $p < \delta(6 + \delta)$. Prove that it is possible to find a number x such that $|x - 3| < \delta$ but $|x^2 - 9| > p$.

4. Prove that if a and b are any real numbers, then $|a| - |b| \leq |a - b|$. *Hint*: Use the fact that $a = (a - b) + b$.

5. Prove that if a and b are any real numbers, then $||a| - |b|| \leq |a - b|$. *Hint*: Look separately at the cases $|a| \leq |b|$ and $|a| > |b|$. This result will be useful when you do Exercise 4.4(19).

6. Solve each of the following inequalities for x:

   (a) $|2x - 3| < |6 - x|$.

   (b) $||x| - 5| < |x - 6|$.

   (c) $||x| - a| < |x - b|$, where a and b are given.

7. Given that a, b, and c are positive numbers and that $c < a + b$, prove that

$$\frac{c}{1 + c} < \frac{a}{1 + a} + \frac{b}{1 + b}.$$

2.1.4 The Need for a Completeness Axiom. The first seven of the eight axioms in Section 2.1.2 contain just those rules of arithmetic and inequalities that you were taught to use while you were in elementary school and during the course of your high school and college algebra. Quite apart from the system R of real numbers, there are many other mathematical systems that satisfy axioms like these, and such systems are called **totally ordered fields.** The system R is just one of these totally ordered fields. Now to help us understand what the first seven axioms are saying, let us look at a few important subsets of R and decide which of them are totally ordered fields:

> The system Z^+ of **natural numbers** (positive integers) is not a totally ordered field because Z^+ does not satisfy condition (5) or (6). As a matter of fact, since Z^+ does not contain the number 0, we should say, strictly speaking, that Z^+ does not satisfy condition (4) either.

---

[1] As explained in the preface, sections and exercises marked with a * may be omitted without loss of continuity. These digress somewhat from the main theme of the chapter.

The system Z of **integers** does a little better but is still not a totally ordered field because Z does not satisfy condition (6).

The system Q of **rational numbers** is a totally ordered field.

We see therefore that R has proper subsets like Q which are closed under the arithmetical operations of addition, subtraction, multiplication, and division. These subsets satisfy the first seven of the eight axioms. If all we ever wanted to do was add, subtract, multiply, and divide, Q would be perfectly adequate for our purposes; but the trouble with Q is that it is full of little punctures where the irrationals should be. Calculus cannot tolerate these little punctures. Calculus requires a number system in which there are no numbers "missing," as you will see very graphically as you advance through the coming chapters. Since the first seven of the eight axioms fail to distinguish between the entire set R and subsets like Q, it is axiom (8) that must make this distinction. Roughly speaking, axiom (8) is going to tell us that all the numbers that ought to be in the real number system R *are* there, and that consequently, R is a rich enough number system to support all the mathematical activity that should take place inside it.

Think back now to the early days of your schooling. The first number system you saw as a child was Z^+, and this system proved to be perfectly adequate, as long as all you ever needed to do was count, add, and multiply. As your mathematics became more sophisticated, however, you needed to be able to talk about the quotient x/y of any two numbers x and y (as long as y is not zero, of course); and for this purpose fractions (rationals) were introduced. You also needed to talk about the difference $x - y$ of any two numbers x and y, and for this purpose negative numbers were introduced. At that stage you were working in the totally ordered field Q. For a while, Q seemed to be perfectly adequate for everything you needed to do, but then came the day when you began to solve equations like $x^2 - 2 = 0$, and the radical sign $\sqrt{}$ was born.

From that moment there were irrationals in your life, but you were not necessarily aware of *all* the irrationals that "need" to exist in the real number system. Perhaps the real numbers you could imagine at that time were the numbers that we call surds. **Surds** are the numbers that can be built by using rational numbers and the operations $+$, $-$, $\times$, $\div$, and $\sqrt[n]{}$ (where n can be any natural number); for example,

$$\sqrt[13]{\frac{\left(4 - \sqrt[5]{3}\right)\left(1 + \sqrt{3 - \sqrt{2}}\right)}{\left(\sqrt[17]{7} + \sqrt{6} + \sqrt[3]{4}\right)\left(6 - \sqrt[6]{10}\right)}} + \sqrt[7]{\frac{\sqrt[4]{5}}{4 - \sqrt[3]{2}} + 1}.$$

For the rest of this section we shall denote the set of surds by S.

We see that S is a strictly wider number system than Q; and clearly, S is a totally ordered field. From the point of view of high school algebra there is no doubt that S is a better number system than Q. For example, if a, b, and c are numbers in S and $b^2 - 4ac$ is not negative, then the quadratic equation $ax^2 + bx + c = 0$ can always be solved in S because the solutions of this equation are, of course,

$$x = \frac{-b \pm \sqrt{b^2 - 4ac}}{2a}.$$

On the other hand, the inadequacy of the number system S becomes apparent the moment we turn our attention to *cubic* equations. Strange as it may seem, the cubic equation $8x^3 - 6x - 1 = 0$ has three real solutions, but none of the solutions of this innocent-looking equation lie in S. It can also be shown that the numbers $\cos \pi/9$, $\log_3 4$, e, and π do not belong to S. We therefore conclude that S is not a rich enough number system to support all the activity that goes on in our mathematics.

As a matter of fact, the completeness condition (8) will ensure that the real number system R is *immensely* bigger than S. What we mean precisely is explained in Appendix A, which deals with the important notion of *countability*. If you choose to follow your reading of this chapter with a reading of Appendix A, you will see that the set S is **countable**, while the set R of all real numbers is not; and you will see that countable sets can be thought of as being much smaller than uncountable sets. However, even if you do not read Appendix A, you might still enjoy the following statement which Cantor made about the set Q of rational numbers:

> The rationals are spotted in the line like stars in a black sky while the dense blackness is the firmament of the irrationals.

He could just as well have been speaking of the set S. Appendix A provides a way of making Cantor's statement precise.

If, after you have finished Chapter 8 (the chapter on integration), you decide to read Appendix B, then you will have an even deeper insight about the way sets like Q and S are so "small" that they are like a scattering of stars in a black sky. In the meantime, keep in mind that in some sense, if the real number system R is laid out on a "number line," then only a very "few" of the points on that line will correspond to points of S. On the other hand, the axiom of completeness will guarantee that there are enough real numbers in R to correspond to *all* the points on the number line. In other words, the axiom will tell us that there are enough real numbers to make the number line "solid," to make it "connected," to make it *complete*.

2.2 UPPER AND LOWER BOUNDS OF A SET OF NUMBERS

In order to make ourselves ready for a precise statement of the completeness axiom, we shall now introduce the important concepts of upper and lower bound of a set of numbers.

2.2.1 Definitions of Upper Bound and Lower Bound. Suppose that $A \subseteq \boldsymbol{R}$ and that $\alpha \in \boldsymbol{R}$. We say that α is an **upper bound** of A when no member of A is greater than α. In other words, α is an upper bound of A when for every member x of A we have $x \leq \alpha$. The concept of a lower bound is introduced similarly: We say that α is a **lower bound** of A when no member of A is less than α, or equivalently, when for every member x of A we have $x \geq \alpha$.

Notice that if α is an upper bound of a set A, then every number larger than α must also be an upper bound of A. Similarly, if α is a lower bound of a set A, then so is every number smaller than α.

2.2.2 Some Examples. The following four examples illustrate the definition of an upper bound.

(1) 6 is an upper bound of (0, 1) and -2 is not.

(2) 1 is an upper bound of [0, 1) and 0 is not.

(3) 1 is an upper bound of [0, 1] and $\frac{3}{4}$ is not.

(4) 11 is an upper bound of $\{-3, 2, 5\}$ and 4 is not.

We can learn a lesson from these examples: An upper bound of a set does not have to be *strictly* greater than all the members of the set. As we saw in example (3), 1 is an upper bound of the set [0, 1] even though $1 \in [0, 1]$.

With these examples in mind, we make a simple but important observation in the next subsection.

2.2.3 When a Bound of a Set Lies in the Set. An upper bound of a set need not belong to that set; but if it does, then it is the **largest member** of the set, and no smaller number can be an upper bound of the set. A lower bound of a set need not belong to the set; but if it does, then it is the **smallest member** of the set, and no larger number can be a lower bound of the set.

2.2.4 The Maximum and Minimum of a Set. If a set of numbers has a largest member, then we call that largest member the **maximum** of the set. If a set has a smallest member, then we call that smallest member the **minimum** of the set. But note that there is no guarantee that a given set of numbers will have a maximum or a minimum. Look again at the examples of Section 2.2.2. Obviously, in example (3), 1 is the largest member of [0, 1] and 0 is the smallest. The number 0 is also the smallest member of the set [0, 1). But we can see that (0, 1) has *no* smallest member: For given any member x of (0, 1), we see at once that $x/2$ is a smaller member.

In the same sort of way, we can see that neither of the sets (0, 1) and [0, 1)

can have a largest member. For given any member x, we see that $(x + 1)/2$ is a larger member.

2.2.5 Another Look at the Examples of Section 2.2.2.

Suppose A is one of the intervals $(0, 1)$, $[0, 1)$, or $[0, 1]$. As we know, any number α satisfying $\alpha \geq 1$ must be an upper bound of A. Now what about a number α satisfying $\alpha < 1$? Although it may seem clear that such a number cannot be an upper bound of A, we must prove this precisely from the definition.

We assume, then, that $\alpha < 1$, and we want to show that there exists a member x of A such that $\alpha < x$. To prove this assertion, we shall consider two cases.

Case 1: $\alpha < 0$.

In this case we take $x = \frac{1}{2}$, and we have certainly found an example of a member x of A such that $\alpha < x$.

Case 2: $\alpha \geq 0$.

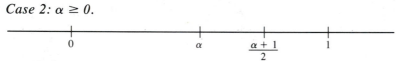

In this case we take $x = (\alpha + 1)/2$, and once again, we have found an example of a member x of A such that $\alpha < x$.

We therefore conclude that among all the upper bounds of A, the number 1 is least. In other words, the number 1 is the *least upper bound* of A.

2.2.6 Least Upper Bounds and Greatest Lower Bounds.

If a number α is an upper bound of a set A, and if no number smaller than α is an upper bound of A, then α is, of course, the **least upper bound of** A. In the same way, if a number α is a lower bound of a set A, and if no number larger than α is a lower bound of A, then α is the **greatest lower bound of** A.

The notion of least upper bound is very important; and in order to understand it properly, we shall look at it a little more closely. When we say that a number α is the least upper bound of a set A, what we are really saying is that

α is an upper bound of A; but for every number $\beta < \alpha$, β is not an upper bound of A.

We can rewrite this condition more fully:

For every member x of A we have $x \leq \alpha$; but if β is any number less than α, then there exists a member x of A such that $x > \beta$.

A good exercise at this stage is to write down the corresponding conditions for a number to be the greatest lower bound of a set.

If a set A happens to have a least upper bound, then the least upper bound of A is called the **supremum** of A and is written as sup A. If a set A happens to have a greatest lower bound, then the greatest lower bound of A is called the **infimum** of A and is written as inf A. From the previous examples we see that sup $(0, 1) = 1$ and inf $(0, 1) = 0$.

2.2.7 Bounded Sets. Suppose A is a set of real numbers.

    (1) If there exists a number which is an upper bound of A, then we say that A is **bounded above**.

    (2) If there exists a number which is a lower bound of A, then we say that A is **bounded below**.

    (3) If A is bounded both above and below, then we say that A is **bounded**.

2.3 EXERCISES

    1. Show that if a set A has a largest member α, then α is the least upper bound of A.

    2. Show that if a set A has an upper bound α and α is a member of A, then α is the largest member of A.

    3. Show that if a set A has a least upper bound α, then the set of all upper bounds of A is $[\alpha, \infty)$.

■ **4.** Suppose that a set A has a least upper bound α and that $x < \alpha$. What conclusions can you draw about x?

■ **5.** Suppose that a set A has a least upper bound α and that $x > \alpha$. What conclusions can you draw about x?

    6. Given that $A \subseteq B$ and α is an upper bound of B, explain why α is an upper bound of A.

    7. Prove that if α is an upper bound of A and β is an upper bound of B, then the larger of the two numbers α and β is an upper bound of $A \cup B$.

■ **8.** Show that if $\emptyset$ is the empty set, then every number is both an upper bound and a lower bound of $\emptyset$.

■ **9.** Set A is a bounded set of real numbers and $B = \{|x| \mid x \in A\}$. Prove that B is bounded.

   10. If A and B are sets of real numbers, then the set $A + B$ is defined to be the set $\{x \mid \exists\, a \in A \text{ and } b \in B \text{ such that } x = a + b\}$. Work out the set $A + B$ in each of the following cases:

    (a) $A = [0, 1]$ and $B = [-1, 0]$.
    (b) $A = [0, 1]$ and $B = \{1, 2, 3\}$.
    (c) $A = (0, 1)$ and $B = \{1, 2, 3\}$.

11. Prove that if two sets A and B are bounded above, then so is $A + B$.

12. Complete the following sentence: A number α is the greatest lower bound of a set A if and only if . . . and . . . .

2.4 PROPERTY (8) OF *R*: THE AXIOM OF COMPLETENESS

As we have said, the axiom of completeness must tell us that the number line is "solid," in other words, that none of its points are "missing." Roughly speaking, the axiom gives the line its solidity by saying that whenever a set A is nonempty and bounded above, then there is a point on the number line that sits at the "top" of A. Of course, if a set A has a largest member, then this largest member must be the "top" of A. But what if A has no largest member? Look again at the set $[0, 1)$. This set has no largest member, but who would doubt that the "top" of this set is the number 1? Quite simply, 1 is the "top" of $[0, 1)$ because 1 is the *least upper bound* of $[0, 1)$. Finally, let us look at the set $A = \{x > 0 \mid x^2 < 2\}$, whose supremum is more interesting. As you will see in Exercise 2.8(5), A has no largest member; but if $s = \sup A$, then $s = \sqrt{2}$. The existence of sup A follows from the axiom of completeness, and in this way the axiom guarantees that the number $\sqrt{2}$ is not "missing."

2.4.1 Statement of the Axiom of Completeness.
Every set of real numbers which is nonempty and bounded above has a least upper bound.

2.4.2 Theorem: Infimum Version of the Completeness Axiom.
Suppose A is a set of numbers which is nonempty and bounded below. Then A has a greatest lower bound.

■ **Proof.** Define B to be the set of all lower bounds of A. Because A is bounded below, we know that B is nonempty. We shall now show that every member of A is an upper bound of B. To see this, let $y \in A$. We need to see that $x \leq y$ for every point $x \in B$. But whenever $x \in B$, x is a lower bound of A and, as such, cannot exceed the member y of A.

    So every member of A is an upper bound of B; and because A is not empty, we conclude that B is bounded above. We therefore know from the completeness axiom that B has a supremum. Define $\alpha = \sup B$. What we would now like to show is that this number α is the greatest lower bound of A, in other words, that α is the largest member of B. To do this, all we need to show is that $\alpha \in B$. But since every member x of A is *an* upper bound of B, and α is the *least* upper bound of B, we have $\alpha \leq x$ for every $x \in A$. This conclusion tells us that α is a lower bound of A, in other words, that $\alpha \in B$ as required. ■

2.4.3 An Exercise with Its Solution. Suppose that A is a nonempty set of real numbers which is bounded above but has no largest member, and that $a \in A$. Prove that sup $A = \sup(A \backslash \{a\})$.

Define $B = A \backslash \{a\}$. In order to guarantee that sup B is meaningful, we need to show that B is nonempty and bounded above. Using the fact that a is not the largest member of A, we choose a member b of A such that $a < b$. Clearly, $b \in B$. Therefore, B is nonempty. Since for every member x of B we have $x \leq \sup A$, we see that sup A is an upper bound of B. Therefore, B is bounded above.

Now since sup A is *an* upper bound of B and sup B is the *least* upper bound of B, it is clear that sup $B \leq \sup A$. It remains to show that sup $A \leq \sup B$, and we do so by showing that sup B is an upper bound of A. Let $x \in A$. We need to show that $x \leq \sup B$. But in the event that $x > a$, we have $x \in B$; and so $x \leq \sup B$. And in the event that $x \leq a$, we have $x \leq a < b \leq \sup B$.

2.5 EXERCISES

1. Given that A and B are sets of real numbers, that A is nonempty, that B is bounded above, and that $A \subseteq B$, prove that sup $A \leq \sup B$.

2. Given that A is a nonempty bounded set of real numbers, prove that inf $A \leq \sup A$.

3. Is statement (25) in Section 1.4 true or false?

4. Given that a set A of real numbers is nonempty and bounded above, prove that A has a largest member if and only if sup $A \in A$.

■ 5. It is given that A and B are nonempty bounded sets of real numbers and that for every point $x \in A$ there is a point $y \in B$ such that $x < y$; and that for every point $y \in B$ there exists a point $x \in A$ such that $x > y$. Prove that sup $A = \sup B$.

■ 6. It is given that A and B are nonempty sets of real numbers and that for all points $x \in A$ and $y \in B$, we have $x < y$. Prove that sup $A \leq \inf B$. Give an example to show that it is possible to have sup $A = \inf B$.

■ 7. Given sets A and B as in exercise 6, and given that sup $A = \inf B$, prove that for every $\delta > 0$ it is possible to find a point $x \in A$ and a point $y \in B$ such that $x + \delta > y$.

8. Given sets A and B as in exercise 6, and given that for every $\delta > 0$ there exist $x \in A$ and $y \in B$ such that $x + \delta > y$, prove that sup $A = \inf B$.

9. It is given that A is a nonempty set of real numbers, that $A \neq \mathbf{R}$, and that for every $x \in A$ and every real number $y < x$ we have $y \in A$. Prove, first, that A must be bounded above. Now prove that there exists a number α such that for all numbers $x < \alpha$ we have $x \in A$, and for all numbers $x > \alpha$ we have $x \notin A$.

10. It is given that A is a nonempty set of real numbers, that A is bounded above, that $q \in \boldsymbol{R}$, and that $B = \{x + q \mid x \in A\}$. Prove that sup $B = q +$ sup A.

11. It is given that A is a nonempty set of real numbers, that A is bounded above, and that $B = \{-x \mid x \in A\}$. Prove that B is nonempty and bounded below and that inf $B = -$sup A.

▪ **12.** It is given that the sets A and B are nonempty and bounded above and that $C = A + B$. [See Exercise 2.3(10) for the definition of $A + B$.] Prove that sup $C =$ sup $A +$ sup B.

▪ **13.** Given that A is a nonempty bounded set and that $B = \{x - t \mid x \in A$ and $t \in A\}$ and that $C = \{|x - t| \mid x \in A$ and $t \in A\}$, prove that sup $A -$ inf $A =$ sup $B =$ sup C.

14. It is given that A is nonempty and bounded above, that $\alpha =$ sup A, and that $\alpha \notin A$. Prove that for every number $\beta < \alpha$ there are at least two different points of A lying between β and α.

15. State and prove a result that says more than exercise (14).

16. It is given that A is nonempty and bounded above, that $\delta > 0$, and that whenever x and y are two different points of A, we have $|x - y| \geq \delta$. Prove that A has a largest member. *Hint:* Use exercises (4) and (14).

▪ **17.** A set A of real numbers is said to be **convex** if, whenever $x < y < z$ and both x and z belong to A, then the number y must also belong to A. (Strictly speaking, we should call this kind of convex set **order convex**. The idea of convexity in higher-dimensional spaces is more complicated than it is in $\boldsymbol{R}$.)
   (a) Explain why the set [2, 3) is convex, and the set [0, 1] $\cup$ [2, 3) is not.
   (b) Prove that if A is convex and unbounded both above and below, then $A = \boldsymbol{R}$.
   (c) Prove that if A is convex, bounded below, and unbounded above, and if inf $A = \alpha$, then A is one of the sets $[\alpha, \infty)$, (α, ∞).
   (d) Prove that if a set A is convex, then A must have one of the following forms: $\emptyset$, $\boldsymbol{R}$, $(-\infty, \beta)$, $(-\infty, \beta]$, (α, ∞), $[\alpha, \infty)$, (α, β), $[\alpha, \beta)$, $(\alpha, \beta]$, $[\alpha, \beta]$. Note that a set which has one of these ten forms is called an **interval**. Therefore, in this exercise you are showing that every convex set of real numbers has to be an interval. A solution to this exercise can be found in Section 6.8.

2.6 INTEGERS, NATURAL NUMBERS, AND MATHEMATICAL INDUCTION

Although we have mentioned integers and natural numbers informally in the course of our discussion of the fundamental axioms for $\boldsymbol{R}$, we should note that the only integers that are actually mentioned in these axioms are the numbers 0 and 1, and that the word *integer* doesn't appear there at all. Therefore, if we are to work strictly from the axioms, we must say precisely what we mean by a natural number and what we mean by an integer. Roughly speaking, a **natural number** is a real number that can be obtained by adding the number 1 repeatedly to itself, for instance, $2 = 1 + 1$, $3 = 1 + 1 + 1, \ldots$. Looking at the set $\mathbf{Z}^+$ of natural numbers in this way, we see that $\mathbf{Z}^+$ is closed under addition and that 1 is its least member. We can now define an **integer** to be a real number x such that either $x = 0$ or x is a natural number or $-x$ is a natural number. We note that the set $\mathbf{Z}$ of integers is closed under addition and subtraction, and that 1 is the least positive integer.

We must admit that this definition of a natural number is not really precise. However, in order to give a precise definition and then to use this definition to prove that the set of integers is closed under addition and subtraction and that 1 is its least positive member, we would have to make use of certain concepts from set theory that lie beyond our scope. We shall therefore make do with the above definitions of *natural number* and *integer,* and we shall leave the task of writing the definitions more precisely for a more advanced course.

We begin this section, then, with the observation that the set $\mathbf{Z}$ of integers is closed under addition and subtraction, and that the least positive integer is 1.

2.6.1   Theorem.   Every nonempty set of integers which is bounded above must have a greatest member.

■ *Proof.*   Suppose that A is a nonempty set of integers and that A is bounded above. Define $\alpha = \sup A$. We would like to show that α is the greatest member of A; and to show this, all we have to see is that α is a member of A. To obtain a contradiction, let us suppose that α does not belong to A. Now since $\alpha - 1 < \alpha$ and α is the *least* upper bound of A, we see that $\alpha - 1$ is not an upper bound of A. Choose $x \in A$ such that $\alpha - 1 < x$.

Since α is an upper bound of A but not a member of A, we must have $x < \alpha$. Therefore, x, being less than the least upper bound of A, cannot be an upper bound of A. Choose a member y of A such that $y > x$. But now we have found two integers x and y such that $0 < y - x < 1$, which contradicts the fact that 1 is the least positive integer.   ■

2.6.2   Theorem.   Every nonempty set of integers which is bounded below must have a smallest member.

■ *Proof.*   This proof is left as an exercise. Try to do it two ways:

(1) Copy the method of proof of Theorem 2.6.1 to show that if A is a nonempty set of integers and A is bounded below, then inf A is the smallest member of A.

(2) Given a nonempty set A of integers which is bounded below, look at the set $\{x \mid -x \in A\}$. What can you say about the largest member of this set?

∎

2.6.3 Theorem. The statement of the following theorem is known as the **Archimedean property** of the totally ordered field R.

(1) The set Z^+ of natural numbers is not bounded above.

(2) The set Z of integers is unbounded both above and below.

∎ **Proof.** If Z^+ were bounded above, it would have a largest member; but for every $n \in Z^+$ the number $n + 1$ is a still larger member. This proves (1), and the proof of (2) is similar. ∎

2.6.4 Theorem: The Principle of Mathematical Induction. Suppose that to each natural number n there is associated a certain statement p_n. Suppose that the statement p_1 is true; and that for every natural number n for which the statement p_n happens to be true, the statement p_{n+1} is also true. Then the statement p_n must be true for every $n \in Z^+$.

∎ **Proof.** To obtain a contradiction, let us assume that there are natural numbers n for which the statement p_n is false. Define

$$E = \{n \in Z^+ \mid \text{the statement } p_n \text{ is false}\}.$$

Using Theorem 2.6.2, we define n to be the least member of E. Since the statement p_1 is true, we know that $n > 1$. Define $m = n - 1$. Then $m \in Z^+$; and since $m < n$, we see that p_m is true. From our assumptions it follows that p_{m+1} is true, in other words, that p_n is true, contradicting the choice of n. ∎

2.6.5 Some Examples of Proofs by Induction. In our first example we shall use the principle of mathematical induction to prove that the product of two natural numbers is a natural number.

(1) For each natural number n we let p_n be the statement that for every natural number k the number kn is natural. The statement p_1 is obviously true. Now suppose n is any natural number for which the statement p_n happens to be true. Then given any natural number k, we see that the number $k(n + 1)$, being the sum of the two natural numbers kn and k, must be a natural number. This shows that the statement p_{n+1} is true. The truth of p_n for every natural number n now follows by mathematical induction, and we conclude that the product of any two natural numbers is a natural number. One may deduce easily from this result that the product of any two integers is an integer.

(2) In this example we shall use the principle of induction to prove **Bernoulli's inequality,** which states that if x is any positive number and n is any natural number, then $(1 + x)^n \geq 1 + nx$. Let $x > 0$, and for each natural number n let p_n be the statement that $(1 + x)^n \geq 1 + nx$. The statement p_1 is obviously true. Now suppose n is any natural number for which the statement p_n happens to be true. Then we see that

$$(1 + x)^{n+1} = (1 + x)^n(1 + x) \geq (1 + nx)(1 + x)$$
$$= 1 + (n + 1)x + nx^2 \geq 1 + (n + 1)x,$$

which establishes the truth of the statement p_{n+1}.

(3) In this example we shall illustrate the principle of induction by using it to prove a familiar theorem about differentiation. We shall assume that you are familiar with the idea of a derivative and the product rule for differentiation, ideas which will be developed very carefully in Chapter 7. The result we shall prove here is as follows: For every natural number n, if f is the polynomial defined by $f(x) = x^n$ for all real numbers x, then for every x we have $f'(x) = nx^{n-1}$.

To prove this theorem, we proceed as follows: For every natural number n, let p_n be the following statement: If f is the polynomial defined by $f(x) = x^n$ for all real numbers x, then for every x we have $f'(x) = nx^{n-1}$. The statement p_1 is obviously true. Now let us suppose that n is any natural number for which the statement p_n happens to be true, and define $f(x) = x^{n+1}$ for every real number x. Then since $f(x) = x^n x$ for every x, it follows from the product rule for differentiation that for every x we have

$$f'(x) = nx^{n-1}x + x^n 1 = (n + 1)x^{n+1-1},$$

which establishes the truth of p_{n+1}.

We shall use the principle of mathematical induction to prove a number of theorems in this book, including Theorem 3.3.2 and Theorem 11.11.4.

2.7 THE DENSENESS OF THE SET Q OF RATIONAL NUMBERS

The **denseness** of Q refers to the fact that between any two distinct real numbers there will always be some rationals. We begin with a special case of this statement which is of interest in its own right.

2.7.1 Theorem. For every positive number x it is possible to find a natural number n such that $1/n < x$.

■ **Proof.** Let $x > 0$. Since $\mathbf{Z}^+$ is not bounded above, $1/x$ is not an upper bound of $\mathbf{Z}^+$. Therefore, we can choose a member n of $\mathbf{Z}^+$ such that $n > 1/x$, and we see at once that $1/n < x$. ■

2.7.2 Theorem. Suppose that a and b are any real numbers such that $a < b$. Then $Q \cap (a, b)$ is not empty; in other words, there exists a rational number r such that $a < r < b$.

■ **Proof.** Choose a natural number n such that $n > 1/(b - a)$. Note that $nb - na > 1$. Define $A = \{x \in Z \mid x > na\}$. Since Z is not bounded above, A is nonempty; therefore, since A is clearly bounded below, it follows from Theorem 2.6.2 that A must have a least member. Define m to be the least member of A.

Since $m - 1 \le na$, we have $m \le 1 + na < nb$. Therefore, $na < m < nb$, and it follows that $a < m/n < b$. ■

2.8 EXERCISES

1. Use mathematical induction to prove that if a set has exactly n members, then it has exactly 2^n subsets.

2. Use mathematical induction to prove that if a set A has exactly n members, then given an integer k such that $0 \le k \le n$, the set A has exactly $n!/(n - k)!k!$ subsets with k members.

3. Given that $A = Q \cap (0, 1)$, prove that sup $A = 1$.

4. Given two numbers a and b such that $a < b$, prove that there exists an irrational number α such that $a < \alpha < b$. *Hint:* Use Theorem 2.7.2 to find a rational number between $a\sqrt{2}$ and $b\sqrt{2}$. Strictly speaking, the method you are using here is illegal, because we have not yet established the existence of the irrational number $\sqrt{2}$. You can remedy this fault straight away by doing the next exercise, or if you prefer, you can wait until Chapter 6 where the existence of nth roots will drop out of the properties of continuous functions.

5. (a) Given that $x > 0$, $x^2 < \alpha$, $0 < h < x$, and $h < (\alpha - x^2)/3x$, show that $(x + h)^2 < \alpha$.
   (b) Given that $x > 0$, $x^2 > \alpha$, and $0 < h < (x^2 - \alpha)/2x$, show that $(x - h)^2 > \alpha$.
   (c) Suppose α is a positive number and that $A = \{x > 0 \mid x^2 < \alpha\}$. Show that A is nonempty and bounded above. Now show that if sup $A = s$, then $s^2 = \alpha$.
   (d) Deduce that every positive real number has a square root.

6. Go to the library and browse through some old calculus books until you find the **Leibniz formula** for the nth derivative of the product of two functions. Prove the formula by mathematical induction. Don't read the proof unless you have to.

2.9 THE EXTENDED REAL NUMBER SYSTEM: THE SYMBOLS ∞ AND −∞

Although the complete number system R, which we have described in the preceding sections, is adequate for the needs of mathematical analysis, we can simplify the statements of many theorems and reduce the number of cases that have to be considered in some proofs by introducing the two infinity symbols, ∞ and $-\infty$. This extends our number system slightly, giving us what we call the **extended real number system** $[-\infty, \infty] = \boldsymbol{R} \cup \{\infty, -\infty\}$.

2.9.1   Arithmetic in [−∞, ∞].   First, we define

$$\infty + \infty = \infty \cdot \infty = (-\infty)(-\infty) = \infty \quad \text{and} \quad -\infty - \infty = (-\infty)\infty = \infty(-\infty) = -\infty.$$

Next, given any real number x, we define

$$\infty + x = x + \infty = \infty \quad \text{and} \quad -\infty + x = x - \infty = -\infty,$$

and $x/\infty = 0$. Finally, if $x > 0$, we define $\infty \cdot x = x \cdot \infty = \infty/x = \infty$; and if $x < 0$, we define $\infty \cdot x = x \cdot \infty = \infty/x = -\infty$.

Our purpose in giving these definitions is to simplify the statements of the arithmetical rules for limits of sequences (see Chapter 4) and functions (see Chapter 5). For example, the definition $\infty + \infty = \infty$ reflects the theorem that says that if $\lim_{x \to a} f(x) = \infty$ and $\lim_{x \to a} g(x) = \infty$, then $\lim_{x \to a} [f(x) + g(x)] = \infty$. On the other hand, if $\lim_{x \to a} f(x) = \infty$ and $\lim_{x \to a} g(x) = \infty$, then $\lim_{x \to a} [f(x) - g(x)]$ is quite unpredictable, which is why we have left the expression $\infty - \infty$ undefined. For similar reasons, we have also left the expressions $-\infty + \infty, 0 \cdot \infty, \infty \cdot 0$, and ∞/∞ undefined.

2.9.2   The Order Relation < in [−∞, ∞].   We extend the order relation $<$ in $\boldsymbol{R}$ to $[-\infty, \infty]$ by agreeing that $-\infty < \infty$, and that for all real numbers x we have $-\infty < x$ and $x < \infty$.

2.9.3   Suprema and Infima in [−∞, ∞].   Given any subset A of $[-\infty, \infty]$, it is clear that ∞ is an upper bound of A and that $-\infty$ is a lower bound. Consequently, if A is a set of real numbers, we shall stick to our earlier convention and call A bounded above when A is bounded above in the system $\boldsymbol{R}$. In other words, a set A is understood to be bounded above when there is a *real number* α such that α is an upper bound of A. Similarly, a set A is understood to be bounded below when there is a real number α such that α is a lower bound of A.

When a set A of numbers is not bounded above, then ∞ is the only upper bound of A; and therefore, sup $A = \infty$. Similarly, if A is not bounded below, then inf $A = -\infty$.

This concludes the mainstream of Chapter 2. At this stage you can choose to read some or all of Appendix A on sets, functions, and countability, which follows, or you can omit this appendix and proceed with Chapter 3. See the Preface for more details on the role of Appendix A.

Appendix *A* *(Optional Appendix to Chapter 2)*

Sets, Functions, and Countability

Our notion of set in this book is the one Cantor used in the theory of sets that he evolved during the nineteenth century and in which one views a *set* as being simply a collection of objects. In this book the words *set*, *class*, *family*, and *collection* all mean the same thing. It is important to understand that the members of a given set might themselves be sets; for example, the members of the set {3, 8, {2, 3}} are the numbers 3 and 8 and the set {2, 3}. As a matter of fact, it is precisely because members of a set might be sets that the theory which was conceived by Cantor has some logical flaws. As we said in the Prelude, these flaws first came to light toward the end of the nineteenth century in the form of a number of para-doxes. What is a paradox? In this context, a **paradox** is an argument which deduces a contradiction from the fundamental axioms upon which mathematics is founded. A paradox therefore demonstrates that mathematics as we know it does not work. It is not the purpose of this appendix to dwell on the subject of these paradoxes nor to show how modern set theory has been developed in such a way that it is (we hope) paradox-free. We leave such matters for an advanced course on the theory of sets. On the other hand, we feel that this appendix would be incomplete if it did not provide at least some inkling of what these paradoxes are and how they come about. Therefore, before launching into our main subject matter, we shall explore just one of them, the paradox of Bertrand Russell which was quoted in the Prelude; and we shall see how this paradox casts doubt on Cantor's set theory.

A.1   RUSSELL'S PARADOX

As we have said, an important requirement of the mathematical theory of sets is that we allow sets to have members which are themselves sets. This requirement leads us to question whether a given set E might possibly be one of its own members. As Russell himself put it, the set of all cows is not a cow and is therefore not one of its own members. On the other hand, the set of all those objects which are *not* cows is also not a cow, and therefore this set is one of its own members. So given a set E, we may reasonably ask whether or not E is a member of itself. If it is, then we shall say that the set E is *self-possessed*; in other words, a set E is self-possessed when $E \in E$. Russell's paradox asks the following question:

> If A is the set of all those sets that are not members of themselves, is A a member of itself?

However one answers this question, the answer leads to a contradiction. To see why, suppose that A is the set of all those sets which are not self-possessed. Now if we had $A \in A$, then A would be a self-possessed member of A, which is impossible, because none of the members of A are self-possessed. We therefore conclude that $A \notin A$. But then A is not self-possessed; and since every set which is not self-possessed belongs to A, we have $A \in A$ after all. So the very existence of this set forces a contradiction upon us, and we have to conclude (rather sadly) that the concept of a set as a collection of objects is not sound.

It would be impossible for us to repair this terrifying chasm in our understanding at the level at which we are working in this book. In a more advanced course in the theory of sets you will learn how one may avoid this kind of paradox by working with a collection of sets which are all subsets of one particular set U which is known not to cause any trouble. Furthermore, you will see that all the sets one might expect to encounter in the mainstream of mathematics may be fitted into a set U of this type, and we shall make the assumption that this has been done for all the sets you will encounter in this book.

A.2   UNION, INTERSECTION, AND DIFFERENCE OF SETS

If A and B are two given sets, then the **union** of the two sets A and B, which is written as $A \cup B$, is defined to be the set of all those objects which belong to *at least one* of the sets A and B. Another way of stating this definition is to say that $A \cup B = \{x \mid x \in A \text{ or } x \in B\}$. The **intersection** of two sets A and B is defined to be the set of all those objects that belong to *both* A and B; in other words, $A \cap B =$

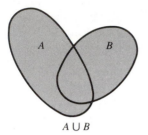
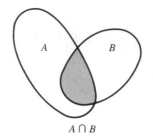
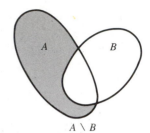

$A \cup B$ $A \cap B$ $A \setminus B$

Figure A.1

$\{x \mid x \in A \text{ and } x \in B\}$. Finally, if A and B are two given sets, then the difference $A \setminus B$ between A and B is defined to be the set of all those objects that belong to A but not to B. In other words, $A \setminus B = \{x \mid x \in A \text{ and } x \notin B\}$. See Figure A.1.

The notions of union and intersection can be extended to more general families of sets: If $\mathscr{F}$ is a family of sets, then the *union* $\cup \mathscr{F}$ of all the members of $\mathscr{F}$ is defined by saying that an object x belongs to $\cup \mathscr{F}$ when there is *at least one* member A of $\mathscr{F}$ such that $x \in A$. In other words, $\cup \mathscr{F} = \{x \mid \text{there exists } A \in \mathscr{F} \text{ such that } x \in A\}$. If a family $\mathscr{F}$ of sets is nonempty, then the *intersection* $\cap \mathscr{F}$ of all the members of $\mathscr{F}$ is defined to be the set of all those objects which belong to *every* member of $\mathscr{F}$; in other words, $\cap \mathscr{F} = \{x \mid x \in A \text{ for every } A \in \mathscr{F}\}$. In the event that the family $\mathscr{F}$ is finite, say $\mathscr{F} = \{A_1, A_2, \ldots, A_n\}$, the union and intersection of all the members of $\mathscr{F}$ are written as $\cup_{i=1}^{n} A_i$ and $\cap_{i=1}^{n} A_i$, respectively. If for each natural number n, A_n is a given set [in other words, if (A_n) is a sequence of sets], then we define

$$\bigcup_{n=1}^{\infty} A_n = \{x \mid x \in A_n \text{ for at least one } n \in \mathbf{Z}^+\}$$

and

$$\bigcap_{n=1}^{\infty} A_n = \{x \mid x \in A_n \text{ for every } n \in \mathbf{Z}^+\}.$$

(For a discussion of the set $\mathbf{Z}$ of integers and the set $\mathbf{Z}^+$ of natural numbers, see Section 2.6.)

Some examples of unions and intersections of this type are the following:

(1) $\cup_{n=1}^{\infty} (n - 1, n)$ is the set of all those positive numbers which are not natural numbers.

(2) $\cap_{n=1}^{\infty} [0, 1/n] = \{0\}$.

A.3 FUNCTIONS

A.3.1 Definition of a Function. Given two sets A and B, the **Cartesian product** $A \times B$ of A and B is defined to be the set of all ordered pairs (x, y) for which $x \in A$ and $y \in B$. In other words, $A \times B = \{(x, y) \mid x \in A \text{ and } y \in B\}$. If f is a subset of

$A \times B$, and if for every point $x \in A$ there is precisely one point $y \in B$ such that $(x, y) \in f$, then we say that f is a **function** *from A into B*; and we call A the **domain** of the function f. If f is a function from A into B, then we write $f: A \to B$, and we say f *maps A into B*.

A.3.2 Some Examples

(1) Let $A = B = \mathbf{R}$, and define $f = \{(x, x^2) \mid x \in A\}$.

(2) Let $A = B = \mathbf{R}$, and define $f = \{(x, x^3) \mid x \in A\}$.

(3) Let $A = [0, \infty)$ and $B = \mathbf{R}$, and define $f = \{(x, x^2) \mid x \in A\}$.

(4) Let $A = [-1, 1]$ and $B = [0, 1]$, and define $f = \{(x, x^2) \mid x \in A\}$.

(5) Let $A = B = \mathbf{Z}^+$ and $f = \{(2n, 2^n) \mid n \in \mathbf{Z}^+\} \cup \{(2n - 1, 3^n) \mid n \in \mathbf{Z}^+\}$.

A.3.3 Function Notation. Suppose that A and B are sets and that f is a function from A into B. For each point $x \in A$ we know that there is one and only one point $y \in B$ such that $(x, y) \in f$; and the usual notation is to write this point y as $f(x)$. With this notation we have $f = \{(x, f(x)) \mid x \in A\}$; and if we want to describe the function f, then all we have to do is say what $f(x)$ is for each point $x \in A$. For practical purposes, we may think of f as being a process of associating an object called $f(x)$ to each point $x \in A$. If E is any subset of A, then the *restriction* of f to E means the function g from E to B defined by $g(x) = f(x)$ for every point $x \in E$; and we define $f[E] = \{f(x) \mid x \in E\}$. We define the **range** of the function f to be the set $f[A] = \{f(x) \mid x \in A\}$.

We now take another look at the examples of Section A.3.2.

(1) For every $x \in \mathbf{R}$ we have $f(x) = x^2$. The domain of f is $\mathbf{R}$, and the range of f is $[0, \infty)$.

(2) For every $x \in \mathbf{R}$ we have $f(x) = x^3$. The domain of f is $\mathbf{R}$, and the range of f is also $\mathbf{R}$.

(3) For every $x \in [0, \infty)$ we have $f(x) = x^2$. The interval $[0, \infty)$ is both the domain and the range of f.

(4) For every $x \in [-1, 1]$ we have $f(x) = x^2$. The domain of f is $[-1, 1]$, and the range of f is $[0, 1]$.

(5) In this example, if n is any natural number, then we have $f(2n) = 2^n$ and $f(2n - 1) = 3^n$. The domain of f is $\mathbf{Z}^+$, and the range of f is the set $\{2^n \mid n \in \mathbf{Z}^+\} \cup \{3^n \mid n \in \mathbf{Z}^+\}$.

A.3.4 The Concepts One–One and Onto. Suppose that A and B are sets and that $f: A \to B$. We say that the function f is **one–one**—or alternately that f is an *injection*—if for every point $y \in B$ there is at most one point $x \in A$ such that $y = f(x)$. Another way of saying that f is one–one is to say that whenever x_1 and x_2 are two different points of the set A, we have $f(x_1) \neq f(x_2)$. The function f is said to be **onto** the set B if for every point $y \in B$ there is at least one point $x \in A$ such that $y = f(x)$. Another way of saying that f is onto the set B is to say that B is the range of f.

Note that the functions in examples (2), (3), and (5) of Section A.3.2 are one–one while the functions in examples (1) and (4) are not. In example (1), f is a function from $\mathbf{R}$ into $\mathbf{R}$ but not onto $\mathbf{R}$. On the other hand, this function maps $\mathbf{R}$ onto $[0, \infty)$.

A.4 EXERCISES

■ **1.** Given $f(x) = (3x - 2)/(x + 1)$ for all $x \in \mathbf{R}\backslash\{-1\}$, determine whether or not f is one–one, and find the range of f.

■ **2.** Given $f(x) = (3x - 2)/(x + 1)$ for all $x \in (0, 1)$, find the range of f.

3. Given that f is a function from a set A to a set B, and that g is a function from B to a set C, the **composition** $g \circ f$ of f and g is defined by $g \circ f(x) = g(f(x))$ for every $x \in A$. (See also Section 5.7.) Prove that if both f and g are one–one, then so is their composition $g \circ f$.

4. Given that f is a function from A onto B and that g is a function from B onto C, prove that $g \circ f$ maps A onto C.

■ **5.** Given that $f : A \to B$, that $g : B \to C$, and that the function $g \circ f$ is one–one, is it possible to prove that f and g are one–one? What if f maps A onto B?

6. Given that $f : A \to B$, that g is a one–one function from B into C, and that the function $g \circ f$ maps A onto C, prove that f maps A onto B and that g maps B onto C.

■ **7.** Suppose that A and B are nonempty sets and that f is a one–one function from A into B. Prove that there exists a function g from B onto A.

■ **8.** Suppose that A and B are nonempty sets and that g is a function from B onto A. Prove that there exists a one–one function f from A into B.

A.5 EQUIVALENCE OF SETS

A.5.1 Definition of Equivalence. Given two sets A and B, we say that A is **equivalent** to B and write $A \sim B$ when there exists a one–one function from A onto B.

A.5.2 Some Simple Properties of Equivalence

(1) Suppose A and B are sets and that f is a one–one function from A onto B. The **inverse function** f^{-1} of f is defined to be the function that sends each point $y \in B$ to the unique point $x \in A$ for which $y = f(x)$. Clearly, f^{-1} is a

one–one function from B onto A. Thus if $A \sim B$, then we also have $B \sim A$. (See Section 6.11 for a discussion of inverse functions of one–one continuous functions.)

(2) Given sets A, B, and C, a one–one function f from A onto B, and a one–one function g from B onto C, the composition $g \circ f$ is one–one from A onto C. It therefore follows that if $A \sim B$ and $B \sim C$, then $A \sim C$.

(3) Given sets A and B, the condition that A is equivalent to a subset of B is the condition that there should exist a one–one function from A into B. When A is equivalent to a subset of B, we write $A \sim\subseteq B$. It is easy to see that if $A \sim\subseteq B$ and $B \sim\subseteq C$, then $A \sim\subseteq C$.

(4) In view of Exercises A.4(7) and A.4(8), we see that if A and B are nonempty sets, then the condition $A \sim\subseteq B$ is equivalent to the condition that there should exist a function g from B onto A.

A.5.3 Some Examples.   These examples illustrate the notion of equivalence in the familiar context of sets of real numbers.

(1) Let P be the set of all nonnegative integers. For every $n \in P$, define $f(n) = n + 1$. Then f is a one–one function from P onto $\mathbf{Z}^+$, and we deduce that $P \sim \mathbf{Z}^+$.

(2) Let A be the set of all even natural numbers, and for every $n \in \mathbf{Z}^+$, define $f(n) = 2n$. Then f is a one–one function from $\mathbf{Z}^+$ onto A, and we deduce that $\mathbf{Z}^+ \sim A$.

(3) Whenever $x \geq 0$, define $f(x) = x^2/(1 + x^2)$; and whenever $x < 0$, define $f(x) = -x^2/(1 + x^2)$. See Figure A.2. Since f is a one–one function from $\mathbf{R}$ onto $(-1, 1)$, we deduce that $\mathbf{R} \sim (-1, 1)$.

(4) Given any real numbers a and b satisfying $a < b$, the interval (a, b) is equivalent to $(0, 1)$. To see why, note that the function f defined by $f(x) = a + (b - a)x$ for all $x \in (0, 1)$ is a one–one function from $(0, 1)$ onto (a, b). Thus we see that any two open, bounded intervals of positive

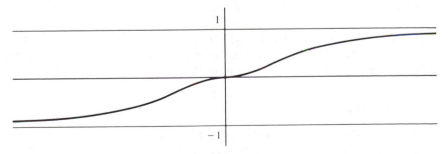

Figure A.2

length are equivalent; and using example (3), we conclude that every open, bounded interval of positive length is equivalent to R. An unbounded, open interval is also equivalent to R. We leave the proof of this assertion as an exercise. In Section A.6.4 we shall go further and show that any subset of R which includes an interval of positive length must be equivalent to R.

A.6 THE SCHROEDER–BERNSTEIN EQUIVALENCE THEOREM

A.6.1 Introduction. Given two sets A and B, the condition $A \sim B$ tell us that the members of A and B can be paired like partners in some giant dance hall and that therefore the sets A and B are of "equal size." The condition $A \sim\subseteq B$ suggests that A should be a "smaller" set than B, but the previous examples show us that even if A is equivalent to what seems to be quite a small subset of B, we might still have $A \sim B$. This phenomenon suggests a number of interesting questions:

(1) If a set A seems to be smaller than a set B, how can we tell when it is *really* smaller? For example, it is clear that $Z^+ \sim\subseteq R$, but it is not immediately obvious (is it?) whether or not we have $Z^+ \sim R$.

(2) Given two sets A and B, can we be sure that at least one of the conditions $A \sim\subseteq B$ and $B \sim\subseteq A$ will hold? In other words, if A isn't smaller than B, can we be sure that B is smaller than A?

(3) Suppose that A and B are two given sets, that $A \sim\subseteq B$, and that $B \sim\subseteq A$. Do these conditions guarantee that $A \sim B$?

Well? What do you think? We shall deal with question (3) in Section A.6.3, where we shall study an important theorem known as the **Bernstein equivalence theorem.** This theorem tells us that the answer to question (3) is *yes*. The lemma given in the next subsection is a special case of the equivalence theorem and contains the technicalities that we need.

A.6.2 Lemma. Suppose that $C \subseteq B \subseteq A$ and that $A \sim C$. Then $A \sim B$.

■ *Proof.* Using the fact that $A \sim C$, choose a one–one function f from A onto C. (See Figure A.3.) We shall prove the theorem by giving an example of a one–one function g from A onto B. We begin by defining $P_1 = B \backslash C$. Having done this, we define

$$P_2 = (B \backslash C) \cup f[P_1] = (B \backslash C) \cup \{f(x) \mid x \in P_1\},$$

$$P_3 = (B \backslash C) \cup f[P_2] = (B \backslash C) \cup \{f(x) \mid x \in P_2\},$$

$$P_4 = (B \backslash C) \cup f[P_3] = (B \backslash C) \cup \{f(x) \mid x \in P_3\}.$$

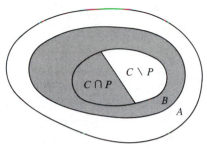

Figure A.3

In general, for each natural number n, having defined P_n, we define

$$P_{n+1} = (B\backslash C) \cup f[P_n] = (B\backslash C) \cup \{f(x) \mid x \in P_n\};$$

and having defined P_n for each natural number n, we define

$$P = \bigcup_{n=1}^{\infty} P_n.$$

This set P has three very important properties: First, $B\backslash C \subseteq P$. Second, given any natural number n and any point $x \in P_n$, we have $f(x) \in P_{n+1}$; and from this fact it follows that whenever $x \in P$, we have $f(x) \in P$. Finally, whenever $f(x) \in P$, we must have $x \in P$. To see why, suppose that $f(x) \in P$, and choose a natural number n such that $f(x) \in P_n$. Since it is impossible to have $f(x) \in B\backslash C$, we must have $f(x) \in f[P_{n-1}]$; and this means that for some point $t \in P_{n-1}$ we have $f(x) = f(t)$. From the fact that f is one–one we see that $x = t$, and we conclude that $x \in P$.

We now define a function g on the set A as follows:

$$g(x) = \begin{cases} x, & \text{whenever } x \in P, \\ f(x), & \text{whenever } x \in A\backslash P. \end{cases}$$

We need to show that this function g is one–one and that g maps A onto B. To see that g is one–one, let x and t belong to A, and suppose that $x \neq t$. If both x and t belong to P, then $g(x) = x \neq t = g(t)$; and if both x and t belong to $A\backslash P$, we have $g(x) = f(x) \neq f(t) = g(t)$. The final possibility is that one of the points x and t lies in P and the other does not; and in this case, clearly, the same is true of $g(x)$ and $g(t)$; and so once again, we have $g(x) \neq g(t)$. Therefore, g is one–one. To see that g maps A onto B, let $y \in B$. In the event that $y \in P$, we have $y = g(y)$. In the event that y is not a point of P, then $y \in C$ and we can write $y = f(x)$ for some point x which also does not lie in P. In this case we have $y = f(x) = g(x)$, and we conclude that g maps A onto B.   ■

A.6.3   The Schroeder-Bernstein Equivalence Theorem.   Given any two sets A and B, if $A \sim \subseteq B$ and $B \sim \subseteq A$, then we have $A \sim B$.

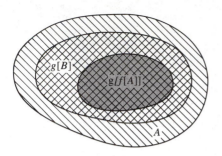
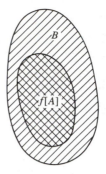

Figure A.4

■ **Proof.** Choose a one–one function f from A into B and a one–one function g from B into A. See Figure A.4. Note that $A \sim f[A]$ and that the restriction of g to $f[A]$ is a one–one function from $f[A]$ onto $g[f[A]]$. Therefore, $A \sim g[f[A]]$, and since $g[f[A]] \subseteq g[B] \subseteq A$, it follows from the previous lemma that $A \sim g[B]$. But $g[B] \sim B$, and we may therefore conclude that $A \sim B$. ■

A.6.4 Some Applications of the Equivalence Theorem. Now that we are in possession of the equivalence theorem, we can give simple proofs of a number of interesting results. In the first proof we keep the promise which we made at the end of Example A.5.3(4).

(1) From what was said in Example A.5.3(4), it follows that if A is any set of real numbers which includes an interval of positive length, then $\boldsymbol{R} \sim\subseteq A$. And since it is automatically true that $A \sim\subseteq \boldsymbol{R}$, it follows from the equivalence theorem that $A \sim \boldsymbol{R}$.

(2) If we define $f(m, n) = 2^m 3^n$ for every point $(m, n) \in \boldsymbol{Z}^+ \times \boldsymbol{Z}^+$, then f is a one–one function from $\boldsymbol{Z}^+ \times \boldsymbol{Z}^+$ into $\boldsymbol{Z}^+$, and so $\boldsymbol{Z}^+ \times \boldsymbol{Z}^+ \sim\subseteq \boldsymbol{Z}^+$. On the other hand, it is easy to find a one–one function from $\boldsymbol{Z}^+$ into $\boldsymbol{Z}^+ \times \boldsymbol{Z}^+$; for example, we might define $g(n) = (1, n)$ for every point $n \in \boldsymbol{Z}^+$, and so $\boldsymbol{Z}^+ \sim\subseteq \boldsymbol{Z}^+ \times \boldsymbol{Z}^+$. It therefore follows from the equivalence theorem that $\boldsymbol{Z}^+ \times \boldsymbol{Z}^+ \sim \boldsymbol{Z}^+$.

(3) In this example we shall improve upon example (2) and find a one–one function from $\boldsymbol{Z} \times \boldsymbol{Z}$ into $\boldsymbol{Z}^+$. Given a point $(m, n) \in \boldsymbol{Z} \times \boldsymbol{Z}$, we define $f(m, n)$ as follows:

> If $m \geq 0$ and $n \geq 0$, we define $f(m, n) = 2^m 3^n$.
>
> If $m \geq 0$ and $n < 0$, we define $f(m, n) = 2^m 7^{-n}$.
>
> If $m < 0$ and $n \geq 0$, we define $f(m, n) = 5^{-m} 3^n$.
>
> If $m < 0$ and $n < 0$, we define $f(m, n) = 5^{-m} 7^{-n}$.

It therefore follows from the equivalence theorem that $\boldsymbol{Z} \times \boldsymbol{Z} \sim \boldsymbol{Z}^+$.

(4) We look now at the set Q of rational numbers. Because there are rational numbers between any two different real numbers, Q seems to be a much larger set than Z^+, but as we shall see, $Q \sim Z^+$. To prove this assertion, we shall make use of the reduced form of a rational number. We say that m/n is the *reduced form* of a rational number x when $x = m/n$, m and n are integers with no common factor, and $n > 0$. The reduced form of 0 is understood to be $0/1$. For every rational x with reduced form m/n, define $f(x) = (m, n)$. The function f defined this way is a one–one function from Q into $Z \times Z$, and it follows from example (3) that $Q \sim\subseteq Z^+$. Since we obviously have $Z^+ \sim\subseteq Q$, we deduce from the equivalence theorem that $Q \sim Z^+$.

A.7 FINITE SETS, INFINITE SETS, AND COUNTABLE SETS

A.7.1 Introduction. Suppose A is a nonempty set and that $x_1 \in A$. There are two possibilities: Either $A = \{x_1\}$, or we can choose a point $x_2 \in A\backslash\{x_1\}$. If x_2 has been chosen, then there are again two possibilities: Either $A = \{x_1, x_2\}$, or we can choose a point $x_3 \in A\backslash\{x_1, x_2\}$. In general, for each natural number n, if we have already chosen points $x_1, x_2, \ldots, x_n$ in the set A, then unless $A = \{x_1, x_2, \ldots, x_n\}$, we must be able to choose a point x_{n+1} in the set $A\backslash\{x_1, x_2, \ldots, x_n\}$.

We see therefore that if A is a given nonempty set, we may be able to find a natural number n and points $x_1, x_2, \ldots, x_n$ in A such that $A = \{x_1, x_2, \ldots, x_n\}$. On the other hand, the process of choosing the points $x_1, x_2, \ldots$ might not terminate, and in this case we can define a one–one function f from Z^+ into A by defining $f(n) = x_n$ for each $n \in Z^+$.

A.7.2 Finite Sets and Infinite Sets. A set A is said to be **finite** if A is empty, or if for some natural number n it is possible to write A in the form $A = \{x_1, x_2, \ldots, x_n\}$. A set which isn't finite is said to be **infinite,** and in view of Section A.7.1, we see that a set A is infinite if and only if $Z^+ \sim\subseteq A$.

A.7.3 Countable Sets. A set A is said to be **countable** if A is a finite set, or if the members of A can be listed in an infinite sequence

$x_1, x_2, x_3, \ldots$.

More precisely, a set A is said to be countable if A is a finite set or $A \sim Z^+$.

A.7.4 Another Way of Looking at the Definition of a Countable Set. We have seen that if A is an infinite set, then $Z^+ \sim\subseteq A$. Therefore, in view of the equivalence theorem, if we want to show that a given infinite set A is countable, then all we need to show is that $A \sim\subseteq Z^+$. Now the condition $A \sim\subseteq Z^+$ obviously holds when A is finite; and we therefore conclude that if A is any given set, then A

is countable if and only if $A \sim \subseteq \mathbf{Z}^+$. Still another way of looking at countable sets can be obtained from Property A.5.2(4). A set A is countable if and only if there exists a function from $\mathbf{Z}^+$ onto A.

A.7.5 Some Examples of Countable Sets. In view of the examples in Section A.6.4, we see that all the sets $\mathbf{Z}$, $\mathbf{Z}^+ \times \mathbf{Z}^+$, $\mathbf{Z} \times \mathbf{Z}$, and $\mathbf{Q}$ are countable, because each of these sets is equivalent to $\mathbf{Z}^+$.

Our first theorem on countable sets, given in the next subsection, can be interpreted as saying that among all infinite sets the countable sets are the smallest.

A.7.6 Theorem. A subset of a countable set is countable.

■ *Proof.* Suppose that $B \subseteq A$ and that A is countable. Choose a one–one function f from A into $\mathbf{Z}^+$. Then the restriction of f to B is a one–one function from B into $\mathbf{Z}^+$.   ■

We shall see in the next subsection that in spite of the fact that the set $\mathbf{Q}$ of all rational numbers is countable, the set $\mathbf{R}$ of all real numbers is not. From the standpoint of countability, the set $\mathbf{Q}$ can therefore be thought of as being much smaller than $\mathbf{R}$, and this is what Cantor had in mind when he spoke of the rationals being like stars in a black sky. (See Sections 2.1.4 and 8.22.3.) Not surprisingly, the fact that $\mathbf{R}$ is uncountable depends upon the completeness of the real number system, and the proof of the theorem that follows therefore makes use of some of the ideas developed in Chapter 2. A second proof of the uncountability of $\mathbf{R}$ can be found in Section A.9.5. That proof makes use of the so-called Cantor diagonalization method and the concept of decimal described in Section 4.10.2.

A.7.7 Theorem. The set $\mathbf{R}$ of all real numbers is uncountable.

■ *Proof.* To obtain a contradiction, let us assume that $\mathbf{R}$ is countable, and write $\mathbf{R}$ in the form $\{x_1, x_2, x_3, \ldots\}$. We begin our argument by choosing two real numbers a_1 and b_1 such that $a_1 < b_1$ and $x_1 \notin [a_1, b_1]$. Having done this, we choose numbers a_2 and b_2 such that $a_1 \leq a_2 < b_2 \leq b_1$ and $x_2 \notin [a_2, b_2]$. We continue in this way. For each n, having chosen a_n and b_n, we choose a_{n+1} and b_{n+1} such that $a_n \leq a_{n+1} < b_{n+1} \leq b_n$ and $x_{n+1} \notin [a_{n+1}, b_{n+1}]$:

Notice that for each n none of the numbers $x_1, x_2, \ldots, x_n$ can lie in the interval $[a_n, b_n]$. Now we define $A = \{a_n \mid n \in \mathbf{Z}^+\}$ and $\alpha = \sup A$. Of course, α is a real number, and so for some natural number N we must have $\alpha = x_N$. Since α is an upper bound of A, we see that $a_N \leq \alpha$; and since b_N is an upper bound of A and

α is the *least* upper bound of A, we see that $\alpha \leq b_N$. We have therefore deduced that $x_N \in [a_N, b_N]$, contradicting the choice of a_N and b_N, and this contradiction shows that the set R is uncountable. ∎

A.7.8 Theorem. If (A_n) is a sequence of countable sets, then $\cup_{n=1}^{\infty} A_n$ is countable.

■ **Proof.** We define $A = \cup_{n=1}^{\infty} A_n$. We shall prove that A is countable by showing that $A \sim \subseteq \mathbf{Z}^+ \times \mathbf{Z}^+$. For each $n \in \mathbf{Z}^+$, using the fact that the set A_n is countable, choose a one–one function f_n from A_n into $\mathbf{Z}^+$. Now given any point $x \in A$, there is a least natural number n for which $x \in A_n$. For each $x \in A$ we define $g(x) = (n, f_n(x))$, where n is the least natural number for which $x \in A_n$. It is easy to see that this function g is a one–one function from A into $\mathbf{Z}^+ \times \mathbf{Z}^+$. ∎

A.8 EXERCISES

■ 1. Prove that if A and B are countable sets, then the set $A \times B$ is countable.

2. Prove that if A is countable and f is a function from A onto B, then B is countable.

3. Given a set A and a natural number n, the set A^n is defined to be the set of all finite sequences of the form $(x_1, x_2, \ldots, x_n)$, where for each $i = 1, 2, \ldots, n$ we have $x_i \in A$. Prove that if A is a countable set and n is any natural number, then the set A^n is countable. *Hint:* Use exercise 1 and prove the result by induction.

*■ 4. Prove that if $A = (x_1, y_1)$ and $B = (x_2, y_2)$ are points in the coordinate plane $\mathbf{R}^2$, then there exists a circle that passes through A and B but does not pass through any point (x, y) in $\mathbf{R}^2 \backslash \{A, B\}$ for which both x and y are rational. *Hint:* Find an uncountable set of circles that intersect at A and B, and use the fact that $\mathbf{Q} \times \mathbf{Q}$ is countable.

5. Prove that if n is a natural number and $\mathcal{P}_n$ is the set of all polynomials with rational coefficients whose degrees do not exceed n, then the set $\mathcal{P}_n$ is countable. *Hint:* Use exercise 3.

6. Prove that if $\mathcal{P}$ is the set of all polynomials with rational coefficients, then $\mathcal{P}$ is countable. *Hint:* Use exercise 5 and Theorem A.7.8.

7. A real number x is said to be **algebraic** if there exists a polynomial f with rational coefficients such that $f(x) = 0$. Prove that the set of all algebraic numbers is countable. *Hint:* Use exercise 6 and Theorem A.7.8, together with the fact that a nonzero polynomial can have only finitely many roots.

■ 8. Prove that there exist real numbers that are not algebraic. Such numbers are said to be **transcendental**.

■ **9.** Given $f: S \to (0, \infty)$ and that S is uncountable, prove that there exists a number $\delta > 0$ such that the set $\{x \in S \mid f(x) > \delta\}$ is infinite.

■ **10.** Prove that if A and B are countable sets of numbers, then the set $A + B = \{x + y \mid x \in A \text{ and } y \in B\}$ is countable. Make similar definitions for the sets $A - B$, AB, and $A \div B$, and show that these sets are also countable.

11. For the sake of these exercises, if A is a set of real numbers, we define $\text{rad}(A)$ (the set of *radicals* of the numbers $|x|$ for $x \in A$) to be $\{|x|^{1/n} \mid n \in \mathbf{Z}^+ \text{ and } x \in A\}$. Prove that if A is a countable set of real numbers, then so is $\text{rad}(A)$.

12. We define $S_1 = \text{rad}(\mathbf{Q})$. Now given any natural number n for which the set S_n is defined, we define $A_n = S_n + S_n$, $B_n = A_n - A_n$, $C_n = B_n B_n$, and $D_n = C_n \div C_n$; and we define $S_{n+1} = \text{rad}(D_n)$. The set $S = \bigcup_{n=1}^{\infty} S_n$ is the set of *surds* which was mentioned in the introduction to Chapter 2. Prove that S is countable.

■ **13.** (This exercise makes use of the material of Chapter 3.) Prove that every uncountable subset of $\mathbf{R}$ has a limit point in $\mathbf{R}$.

■ **14.** (This exercise makes use of the material of Chapter 3.) Prove that every nonempty open set is uncountable.

■ **15.** (This exercise makes use of the material of Chapter 3.) Prove that if H is a nonempty, closed subset of $\mathbf{R}$ and if every point of H is a limit point of H, then H is uncountable. *Hint:* Use the method of proof of Theorem A.7.7. Choose your intervals $[a_n, b_n]$ in such a way that $H \cap [a_n, b_n]$ is infinite for each n.

*16. (This exercise makes use of the material of Chapters 3 and 4.) Give an example of a countable, closed, bounded set H such that if $H_1 = \mathcal{L}(H)$, and $H_{n+1} = \mathcal{L}(H_n)$ for every $n \in \mathbf{Z}^+$, then $H_n \neq \emptyset$ for every n.

17. (This exercise makes use of the material of Chapter 6.) Given that $f: I \to \mathbf{R}$, where I is an interval, and given that the function f is increasing, prove that the set $\{x \in I \mid f \text{ is discontinuous at } x\}$ is countable. *Hint:* Use Theorem 6.10.6. For each point x at which f is discontinuous, choose a rational number r_x such that $\lim_{t \to x-} f(t) < r_x < \lim_{t \to x+} f(t)$.

18. (Like exercise 13, this exercise makes use of the material of Chapter 3.) Prove that if A is an uncountable set of real numbers, then $\mathcal{L}(A)$ is uncountable.

A.9 THE POWER SET OF A SET

A.9.1 Definition of a Power Set. Given any set A, the **power set** $\pi(A)$ of A is defined to be the set of all subsets of A.

For example, the power set of the empty set $\emptyset$ is the set $\{\emptyset\}$ whose sole member is $\emptyset$. If $A = \{a, b, c\}$, then $\pi(A)$ has eight members. In fact, $\pi(A) = \{\emptyset, \{a\}, \{b\}, \{c\}, \{a, b\}, \{a, c\}, \{b, c\}, \{a, b, c\}\}$. As an exercise, you might prove by induction that if a set A has n members, then $\pi(A)$ has 2^n members.

A.9.2   Theorem.   Given two sets A and B, if $A \sim B$, then we have $\pi(A) \sim \pi(B)$.

■ *Proof.*   Choose a one–one function f from A onto B. We now define a function g from $\pi(A)$ into $\pi(B)$ by defining $g(E) = f[E]$ for every subset E of A. It is easy to see that g is one–one and onto $\pi(B)$.   ■

A.9.3   Theorem: Cantor's Inequality.   If A is any set, and f is any function from A into $\pi(A)$, then the function f is not onto $\pi(A)$.

The technique of proof of this theorem is sometimes called the **Cantor diagonalization method**.

■ *Proof.*   Suppose f is a function from A into $\pi(A)$. We need to find a member B of $\pi(A)$ (in other words, a subset B of A) such that B does not lie in the range of f. What we require of B is that for every point $x \in A$ we have $f(x) \neq B$. We define B as follows:

$$B = \{x \in A \mid x \notin f(x)\}.$$

(This definition of B may remind you of Russell's paradox.) Given any point $x \in A$, $f(x)$ is a subset of A; and we can reasonably ask whether or not the point x belongs to the set $f(x)$. The set B is the set of all those points $x \in A$ for which $x \notin f(x)$. We must now show that for every point $x \in A$ we have $f(x) \neq B$. Let $x \in A$. There are two possibilities: Either $x \in B$ or $x \notin B$. If $x \in B$, then by the definition of B we have $x \notin f(x)$; and it follows that $B \neq f(x)$. If $x \notin B$, then by the definition of B we have $x \in f(x)$; and once again, we see that $B \neq f(x)$.   ■

A.9.4   Corollary.   Given any set A, we have $A \sim\subseteq \pi(A)$ but not $A \sim \pi(A)$.

■ *Proof.*   In view of Cantor's inequality, all that remains to be proved is that $A \sim\subseteq \pi(A)$. For every point $x \in A$, we define $f(x) = \{x\}$. Clearly, f is an example of a one–one function from A into $\pi(A)$.   ■

The next theorem provides a second way of showing that the set $\boldsymbol{R}$ is uncountable.

A.9.5   Theorem

(1) $\boldsymbol{R} \sim\subseteq \pi(\boldsymbol{Q})$.

(2) $\pi(\boldsymbol{Z}^+) \sim\subseteq \boldsymbol{R}$.

(3) The three sets R, $\pi(Q)$, and $\pi(Z^+)$ are equivalent to one another and are all uncountable.

■ **Proof of (1).** For each point x of R, we define $f(x) = \{r \in Q \mid r < x\}$. Using the fact that between any two distinct real numbers there are rationals, it is easy to show that this function f is one–one. ■

■ **Proof of (2).** This part of the theorem is harder to prove than part (1) because it depends on the completeness of R. We need to make use of decimals in this proof, and we shall assume that if (a_n) is any sequence in the set $\{0, 1, 2, 3, 4, 5, 6, 7, 8, 9\}$, then the decimal

$$\frac{a_1}{10^1} + \frac{a_2}{10^2} + \frac{a_3}{10^3} + \cdots + \frac{a_n}{10^n} + \cdots = \sum_{n=1}^{\infty} \frac{a_n}{10^n}$$

converges to a real number in $[0, 1]$. Details of this proof can be found in example (4) of Section 4.10.2 and example (8) of Section 10.2. We shall also be using the idea of a *characteristic function*, which is defined in Section 8.5.1.

To prove that $\pi(Z^+) \sim \subseteq R$, we define a function f from $\pi(Z^+)$ into R by letting $f(A) = \sum_{n=1}^{\infty} \frac{\chi_A(n)}{10^n}$ for every subset A of Z^+. Note that if $A \subseteq Z^+$, then $f(A)$ is a decimal whose digits are all zeros and ones, and since no two such decimals with different digits can be equal, the function f is obviously one–one. ■

■ **Proof of (3).** We saw in Example A.6.4(4) that $Q \sim Z^+$, and therefore it follows from Theorem A.9.2 that $\pi(Q) \sim \pi(Z^+)$. The fact that each of these sets is equivalent to R therefore follows at once from parts (1) and (2) and the equivalence theorem, and the fact that these sets are uncountable follows from the fact that there is no function from Z^+ onto $\pi(Z^+)$. ■

A.10 EXERCISES

■ **1.** Given any two sets A and B, the set A^B is defined to be the set of all functions from B into A. Prove that for any set A we have $\{0, 1\}^A \sim \pi(A)$.

■ **2.** Prove that for any sets A, B, and C we have $(A^B)^C \sim A^{(B \times C)}$.

■ **3.** Prove that $R^{Z^+} \sim R$. *Hint:* By exercise 1 and Theorem A.9.5, $R \sim \{0, 1\}^{Z^+}$. Now use exercise 2.

■ **4.** Prove that if n is any natural number, then $R^n \sim R$. Prove that there are as many numbers between 0 and 0.0001 as there are points in the universe. (First, give a meaning to this statement.)

5. Imagine that you are giving a talk to a junior high school class. You are speaking about an infinite flock of sheep that can be arranged into an infinite sequence (S_n).

(a) Explain how the sheep may be shared between two people in such a way that each of them appears to get a complete flock. Do this again with three people. Now, explain how the sheep may be shared among an infinite sequence of people (P_n) in such a way that each of them gets a complete flock.

(b) We now introduce the Greedian robbers. These nasty people come in gangs to steal sheep, and each gang consists of an infinite sequence of robbers. Each robber has a big nose which is either pink or blue; and when one looks at a gang of Greedian robbers, one immediately sees the *color scheme* of their noses. Prove that if to every possible color scheme there is a gang of robbers matching that scheme, then it is impossible to portion out the sheep to the gangs in such a way that every gang gets at least one sheep. Write your proof well enough to allow you to give a successful talk, one that will be properly understood by those who are listening to you.

(c) Make a careful comparison between the proof you gave in part (b) and the proof of Cantor's inequality.

Chapter 3

Elementary Topology of the Real Line

3.1 INTRODUCTION

At the heart of calculus lies the notion of a limit, and in a sense, everything we shall be doing from now on will be concerned with limits of one kind or another. In Chapter 4 we shall study limits of sequences, and we shall then apply what we have learned to the study of limits and continuity of functions in Chapters 5 and 6. This in turn will prepare us for the study of derivatives and integrals, which is our main objective.

The title of the present chapter refers to the **topology** of the real line; and what this means, roughly speaking, is that part of the structure of R upon which the idea of a limit depends. Let us look, for the moment, at the important idea of limit of a function and try to see what we need of our number system for the idea of limit to make sense. The definition as one might see it in elementary calculus is as follows:

If f is a function defined on an interval $[a, b]$ and $c \in (a, b)$, and α is some number, then the condition $f(x) \to \alpha$ as $x \to c$ means that one can make the number $f(x)$ lie as close as we like to α provided that we take $x \neq c$ and x sufficiently close to c.

At this stage you may even remember that there is an ε–δ form in which this definition may be written more carefully, but don't even think about it. We shall leave such things for the later chapters where they belong. Notice, however, that the operative word in this rough definition of a limit is *close*. What we really need in order to be able to speak about limits is a notion that certain numbers are close to each other and that other numbers are not. We have such a notion, for as you saw in Section 2.1.3, the absolute value function gives us a definition of **distance** between numbers. Given two numbers x and y, the distance from x to y is defined to be the number $|x - y|$; and therefore, two numbers x and y can be thought of as being close to each other when $|x - y|$ is small. It is this notion that leads us to the important idea of neighborhood with which we begin this chapter. Note that if a is any number and $\delta > 0$, the set of numbers x whose distance from a is less than δ is the set of those numbers x satisfying the inequality $|x - a| < \delta$, or equivalently, $a - \delta < x < a + \delta$.

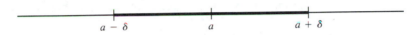

3.2 NEIGHBORHOODS

3.2.1 Definition of a Neighborhood. Given a number x and a set A of real numbers, we say that A is a **neighborhood** of x when there exists a number $\delta > 0$ such that $(x - \delta, x + \delta) \subseteq A$.

Roughly speaking, A is a neighborhood of x when x is a member of A which lies "well inside" A, away from the "boundary" of A. Notice that if A is a neighborhood of x and $A \subseteq B$, then obviously, B is also a neighborhood of x.

3.2.2 Some Examples. The following examples illustrate the definition of a neighborhood.

(1) Given $a < x < b$, the interval (a, b) is a neighborhood of x. To see this, suppose that $a < x < b$, and define δ to be the smaller of the two numbers $x - a$ and $b - x$.

Then $\delta > 0$, and since $x - \delta \geq x - (x - a) = a$ and $x + \delta \leq x + (b - x) = b$, we have $(x - \delta, x + \delta) \subseteq (a, b)$.

(2) Given $a < b$, the interval $[a, b]$ is not a neighborhood of either of the numbers a and b.

(3) The set Q is not a neighborhood of any number because, as we saw in Exercise 2.8(4), no interval of positive length can be included in Q. In a similar way, we can see that the set $R\backslash Q$ of irrational numbers is not a neighborhood of any number.

3.3 INTERSECTION OF NEIGHBORHOODS

We begin by looking at the intersection of two neighborhoods of a given point.

3.3.1 Theorem. Given two neighborhoods U and V of a number x, the set $U \cap V$ must also be a neighborhood of x.

■ **Proof.** Choose $\delta_1 > 0$ such that $(x - \delta_1, x + \delta_1) \subseteq U$, and choose $\delta_2 > 0$ such that $(x - \delta_2, x + \delta_2) \subseteq V$. Define δ to be the smaller of the two numbers δ_1 and δ_2. Then $\delta > 0$.

But $(x - \delta, x + \delta) \subseteq (x - \delta_1, x + \delta_1) \subseteq U$ and $(x - \delta, x + \delta) \subseteq (x - \delta_2, x + \delta_2) \subseteq V$; so $(x - \delta, x + \delta) \subseteq U \cap V$, and it follows that $U \cap V$ is a neighborhood of x. ■

Now using the principle of mathematical induction (Theorem 2.6.4), we shall obtain an extension of the above theorem which deals with the intersection of finitely many neighborhoods of a point.

3.3.2 Theorem. The intersection of finitely many neighborhoods of a number is a neighborhood of that number.

■ **Proof.** Let x be any real number; and for each natural number n, define p_n to be the following statement:

Whenever $U_1, U_2, U_3, \ldots, U_n$ are neighborhoods of x, then $\bigcap_{j=1}^{n} U_j$ is a neighborhood of x.

The statement p_1 is obviously true. Now suppose that n is a natural number for which the statement p_n happens to be true, and in order to demonstrate the truth of p_{n+1}, suppose that $U_1, U_2, U_3, \ldots, U_{n+1}$ are all neighborhoods of x. Then since

$$\bigcap_{j=1}^{n+1} U_j = \left[\bigcap_{j=1}^{n} U_j \right] \cap U_{n+1},$$

the set $\bigcap_{j=1}^{n+1} U_j$ is the intersection of two neighborhoods of x and must therefore be a neighborhood of x. So p_{n+1} is true whenever p_n is true, and the result therefore follows by induction.

■

We mention, finally, that the intersection of infinitely many neighborhoods of a number does not have to be a neighborhood of that number. Suppose, for example, that $U_n = (-1/n, 1/n)$ for each natural number n. The intersection of all these sets U_n is the set of all those numbers which belong to U_n for every natural number n, and we write this set as $\cap_{n=1}^{\infty} U_n$. Now although every set U_n is a neighborhood of 0, it follows simply from Theorem 2.7.1 that $\cap_{n=1}^{\infty} U_n = \{0\}$, which is not a neighborhood of 0.

3.4 EXERCISES

1. Given $x \in R$ and $U \subseteq R$, prove that U is a neighborhood of x if and only if it is possible to find two numbers a and b such that $a < x < b$ and $(a, b) \subseteq U$.

2. Prove that if x and y are two distinct real numbers, then it is possible to find a neighborhood U of x and a neighborhood V of y such that $U \cap V = \emptyset$.

3. Given that A is nonempty and bounded above, prove that neither A nor $R \backslash A$ can be a neighborhood of sup A.

■ 4. Prove that if $x \in A$ and U is a neighborhood of y, then $A + U$ is a neighborhood of $x + y$.

5. Referring to the example given at the end of the previous section, explain carefully why $\cap_{n=1}^{\infty}(-1/n, 1/n) = \{0\}$.

6. Complete the following sentence: "A set U fails to be a neighborhood of a number x when for every number $\delta > 0, \ldots$"

3.5 OPEN SETS AND CLOSED SETS

3.5.1 Definition. A set A of numbers is said to be **open** if for every member x of A the set A is a neighborhood of x. A set A of numbers is said to be **closed** if its complement $R \backslash A$ is open. To see this definition another way, note that if $U = R \backslash A$, then $A = R \backslash U$. We can therefore say that a set A is closed when A is the complement of an open set.

3.5.2 Some Examples. We illustrate the notion of open and closed sets with the following examples.

(1) The set R is open and therefore $\emptyset = R \backslash R$ is closed.

(2) The set $\emptyset$ is open. To see this, notice that the only way a set A can fail to be open is that there should exist a member x of A such that A is not a neighborhood of x; and this requirement demands, in particular, that there must exist a member x of A. So $\emptyset$ is open, and we conclude that $R = R \backslash \emptyset$ is closed.

(3) Given $a < b$, the interval (a, b) is open. We saw this result in Example 3.2.2(1).

(4) Given any number a, the intervals $(-\infty, a)$ and (a, ∞) are open. We leave the proof of this fact as a simple exercise.

(5) Given any number a, since $(-\infty, a] = R \backslash (a, \infty)$, we see that $(-\infty, a]$ is closed. One can show similarly that $[a, \infty)$ is closed.

(6) The set $R \backslash Z$ is open and therefore Z is closed. To see this, let $x \in R \backslash Z$. Using the results of Section 2.6, define n to be the largest integer that does not exceed x. Then clearly, $n < x < n + 1$. Since $(n, n + 1)$ is a neighborhood of x, it follows at once that $R \backslash Z$ is a neighborhood of x. In a similar way, we can show that Z^+ is closed.

(7) The set Q is neither open nor closed because neither Q nor $R \backslash Q$ can be a neighborhood of any number. (Prove the latter assertion.)

(8) The interval $[0, 1)$ is neither open nor closed. (Prove this assertion.)

(9) Suppose $A = \{1/n \mid n \in Z^+\}$. Then since $A \subseteq Q$, A is not a neighborhood of any number; and so A is not open. On the other hand, $0 \in R \backslash A$, and it follows from Theorem 2.7.1 that $R \backslash A$ is not a neighborhood of 0. Therefore, $R \backslash A$ is not open, and we deduce that A is neither open nor closed.

(10) Suppose A is the set given in example (9). Then $A \cup \{0\}$ is closed. We leave the proof of this assertion as an exercise.

Do not make the common mistake of thinking that whenever a set isn't open, it must be closed. In fact, most sets are neither open nor closed, and we have seen two sets that are both open and closed. We shall make this statement more precise in Section 3.7.

3.5.3 Theorem

(1) The intersection of finitely many open sets is open.

(2) The union of finitely many closed sets is closed.

(3) The union of any family (finite or infinite) of open sets is open.

(4) The intersection of any family (finite or infinite) of closed sets is closed.

■ *Proof*

(1) Suppose that $\{U_1, U_2, \ldots, U_n\}$ is a finite family of open sets and that x lies in their intersection. We need to see that $\cap_{i=1}^{n} U_i$ is a neighborhood of x, but this assertion follows at once from Theorem 3.3.2.

(2) Suppose that $\{H_1, H_2, \ldots, H_n\}$ is a finite family of closed sets. Then for each $i = 1, 2, \ldots, n$ the set $R \backslash H_i$ is open, and since

$$\textbf{\textit{R}} \setminus \bigcup_{i=1}^{n} H_i = \bigcap_{i=1}^{n} (\textbf{\textit{R}} \setminus H_i),$$

which is open by part (1) of the theorem, it follows that $\cup_{i=1}^{n} H_i$ is closed.

(3) For the definition of the union of an arbitrary family of sets, refer to Section A.1 in Appendix A. Suppose $\mathcal{F}$ is any family of open sets, and write the union of all these open sets as U. We need to show that the set U is open. Let $x \in U$. This means that x must belong to at least one member V of the family $\mathcal{F}$. Choose a member V of $\mathcal{F}$ such that $x \in V$. Since V is a neighborhood of x and $V \subseteq U$, we deduce that U must be a neighborhood of x; and this is all we needed to show that U is open.

(4) Suppose that $\mathcal{F}$ is a family of closed sets, and write the intersection of all these sets as H. We want to show that H is closed and to do so, we shall show that $\textbf{\textit{R}} \setminus H$ is open. Let $x \in \textbf{\textit{R}} \setminus H$. We need to show that $\textbf{\textit{R}} \setminus H$ is a neighborhood of x. Choose a member F of the family $\mathcal{F}$ such that $x \notin F$. Since F is closed, $\textbf{\textit{R}} \setminus F$ is a neighborhood of x. But since $H \subseteq F$, we have $\textbf{\textit{R}} \setminus F \subseteq \textbf{\textit{R}} \setminus H$; and it follows that $\textbf{\textit{R}} \setminus H$ is a neighborhood of x. ∎

3.6 EXERCISES

1. Say briefly why a set A is open iff for every number $x \in A$, there exists a number $\delta > 0$ such that $(x - \delta, x + \delta) \subseteq A$.

2. Prove that a set A is closed iff for every number $x \in \textbf{\textit{R}} \setminus A$, there exists a number $\delta > 0$ such that $(x - \delta, x + \delta) \cap A = \emptyset$.

3. Prove that for every number x, the singleton $\{x\}$ is closed.

4. Prove that every finite set is closed. *Hint:* Use exercise 3 and Theorem 3.5.3(2).

5. Given that $\textbf{\textit{Q}} \subseteq A$ and that A is closed, prove that $A = \textbf{\textit{R}}$.

▪ 6. Prove that if a set A is nonempty, closed, and bounded above, then A has a largest member.

7. Prove that if U is open and H is closed, then the set $U \setminus H$ is open.

8. Prove that if U is open and H is closed, then the set $H \setminus U$ is closed.

▪ 9. Given $A \subseteq \textbf{\textit{R}}$ and U is an open subset of $\textbf{\textit{R}}$, prove that the set $A + U$ is open.

*▪ 10. Suppose $\{H_n \mid n \in \textbf{\textit{Z}}^+\}$ is a family of nonempty closed bounded sets, and that for each n we have $H_{n+1} \subseteq H_n$ and $\alpha_n = \inf H_n$. Prove that if $A = \{\alpha_n \mid n \in \textbf{\textit{Z}}^+\}$ and $\alpha = \sup A$, then $\alpha \in \cap_{n=1}^{\infty} H_n$.

3.7 THE SETS THAT ARE BOTH OPEN AND CLOSED

We have already seen that the sets $\emptyset$ and R are both open and closed. Now we shall see that apart from these two sets, there are no sets which are both open and closed.

3.7.1 Theorem. Suppose that a set A of numbers is both open and closed. Then either $A = R$ or $A = \emptyset$.

■ **Proof.** To obtain a contradiction, assume that A is a nonempty, open, closed subset of R and assume that the complement B of A is nonempty. Note that B, like A, is both open and closed. Choose a number a in A and a number b in B. Now either $a < b$ or $b < a$. We shall assume without loss of generality that $a < b$, and we leave the almost identical proof for the case $b < a$ as an exercise. Define $E = A \cap (-\infty, b)$. Note that E is nonempty (because $a \in E$) and is bounded above (because b is an upper bound). Now because both A and $(-\infty, b)$ are open, it follows that E is open. But we also have $E = A \cap (-\infty, b]$ (because $b \notin A$); and therefore, since both A and $(-\infty, b]$ are closed, we deduce that E is closed. Define $\alpha = \sup E$. From Exercise 3.4(3) it follows that neither E nor $R \backslash E$ can be a neighborhood of α. But both of these sets are open, and α has to lie in one of them. This is the desired contradiction. ■

3.8 THE CLOSURE OF A SET

Given a number x and a set A of real numbers, we say that x belongs to the **closure** of the set A when for every neighborhood U of x we have $U \cap A \neq \emptyset$. The closure of A is denoted by $\bar{A}$.

Note that $A \subseteq \bar{A}$, for if $x \in A$ and U is a neighborhood of x, then the point x itself belongs to $U \cap A$. In general, however, there might be numbers that belong to $\bar{A}$ even though they do not belong to A.

3.8.1 Theorem. Suppose A is a set of real numbers and $x \in R$. Then the following two conditions are equivalent:

(1) $x \in \bar{A}$.

(2) For every number $\delta > 0$ we have $(x - \delta, x + \delta) \cap A \neq \emptyset$.

■ **Proof.** First we shall prove that (1) $\Rightarrow$ (2): Suppose that (1) holds, and let $\delta > 0$. Since $(x - \delta, x + \delta)$ is a neighborhood of x, it follows at once that $(x - \delta, x + \delta) \cap A \neq \emptyset$.

Now to prove that (2) $\Rightarrow$ (1), assume that (2) holds and let U be a neighborhood of x. Using the definition of a neighborhood, we now choose a number $\delta > 0$ such that $(x - \delta, x + \delta) \subseteq U$. Because of our assumption that (2) holds, we know that $(x - \delta, x + \delta) \cap A \neq \emptyset$, and it follows at once that $U \cap A \neq \emptyset$. ■

3.8.2 Examples. The following examples illustrate the notion of closure.

(1) Define $A = (0, 1)$. If $x < 0$, then $(-\infty, 0)$ is a neighborhood of x which does not intersect with A. Therefore, no negative number can lie in $\bar{A}$, and it follows similarly that no number greater than 1 can lie in $\bar{A}$. Now even though the number 0 does not belong to A, we can see that $0 \in \bar{A}$, because if $\delta > 0$, then the set $A \cap (-\delta, \delta)$ contains all numbers which lie between 0 and the smaller of the two numbers 1 and δ.

Therefore, $\bar{A} = [0, 1]$.

(2) Using Theorem 2.7.2, one may see that $\bar{Q} = R$. Similarly, one may use Exercise 2.8(4) to show that $\overline{R\backslash Q} = R$.

(3) Define $A = \{1/n \mid n \in Z^+\}$. In this case $\bar{A} = \{1/n \mid n \in Z^+\} \cup \{0\}$.

3.8.3 Theorem. Whenever $A \subseteq B$, we have $\bar{A} \subseteq \bar{B}$.

■ *Proof.* Suppose that A and B are sets of numbers and that $A \subseteq B$. Let $x \in \bar{A}$ and let U be a neighborhood of x. Since $U \cap A \neq \emptyset$, it follows at once that $U \cap B \neq \emptyset$. ■

3.8.4 Theorem. For every set A of numbers the set $\bar{A}$ must be closed.

■ *Proof.* Let $A \subseteq R$. We need to show that $R\backslash\bar{A}$ is open. Let $x \in R\backslash\bar{A}$. We would like to know that there exists a number $\delta > 0$ such that $(x - \delta, x + \delta) \subseteq R\backslash\bar{A}$. Using Theorem 3.8.1 and the fact that $x \notin \bar{A}$, choose a number $\delta > 0$ such that $(x - \delta, x + \delta) \cap A = \emptyset$. If the interval $(x - \delta, x + \delta)$ contained any point y of $\bar{A}$, then being a neighborhood of y, this interval would have to intersect with A, which it does not. Therefore, the interval $(x - \delta, x + \delta)$ contains no points of $\bar{A}$, and it follows that $(x - \delta, x + \delta) \subseteq R\backslash\bar{A}$. ■

3.8.5 Theorem. A set A of real numbers is closed if and only if $A = \bar{A}$.

■ *Proof.* If $A = \bar{A}$, then the fact that A is closed follows from Theorem 3.8.4. Suppose now that A is closed. To show that $A = \bar{A}$, we need to show that every point of $\bar{A}$ belongs to A. Let $x \in \bar{A}$; and to obtain a contradiction, assume that $x \notin A$. Then because A is closed, we see that $R\backslash A$ is a neighborhood of x; and therefore, $(R\backslash A) \cap A \neq \emptyset$, which is a contradiction. ■

3.8.6 Theorem. Given any set A of real numbers, we have $\bar{\bar{A}} = \bar{A}$.

■ *Proof.* The result follows at once from Theorems 3.8.4 and 3.8.5. ■

3.8.7 Theorem. Suppose A and B are any two sets of numbers. Then $\overline{A \cup B} = \bar{A} \cup \bar{B}$.

■ **Proof.** Since $A \subseteq A \cup B$, we have $\bar{A} \subseteq \overline{A \cup B}$; and similarly, $\bar{B} \subseteq \overline{A \cup B}$. It follows that $\bar{A} \cup \bar{B} \subseteq \overline{A \cup B}$. Now on the other hand, $A \cup B \subseteq \bar{A} \cup \bar{B}$, and so $\overline{A \cup B} \subseteq \overline{\bar{A} \cup \bar{B}}$. But $\bar{A} \cup \bar{B}$ is closed, and so from Theorem 3.8.6 we deduce that $\overline{A \cup B} = \bar{A} \cup \bar{B}$. ■

3.9 EXERCISES

■ 1. Given a number x and a set A of real numbers, prove that $x \in \bar{A}$ if and only if $R \backslash A$ is not a neighborhood of x.

2. Given two sets A and B, show that $\overline{A \cap B} \subseteq \bar{A} \cap \bar{B}$. By considering the case $A = (0, 1)$ and $B = (1, 2)$, show that the two sides of this inequality need not be the same.

3. Prove that if A is nonempty and bounded above, then $\sup A \in \bar{A}$. Prove that if A is nonempty and bounded below, then $\inf A \in \bar{A}$.

■ 4. Given that $A \subseteq H$ and that H is closed, prove that $\bar{A} \subseteq H$.

5. Given a subset A of R, prove that $\bar{A} = R$ if and only if for every nonempty open set U we have $A \cap U \neq \emptyset$.

■ 6. Given two open sets U and V satisfying $\bar{U} = \bar{V} = R$, prove that $\overline{U \cap V} = R$.

7. Two sets A and B are said to be **separated** when $A \cap \bar{B} = \bar{A} \cap B = \emptyset$. Which of the following pairs of sets are separated?
   (a) $[0, 1]$ and $[2, 3]$.   **(b)** $(0, 1)$ and $(1, 2)$.
   (c) $(0, 1]$ and $(1, 2)$.   **(d)** Q and $R \backslash Q$.

8. Prove that if A and B are two mutually disjoint, closed sets, then A and B are separated.

9. Prove that if A and B are two mutually disjoint, open sets, then A and B are separated.

10. Given $A \subseteq R$, prove that if A and $R \backslash A$ are separated, then A is both open and closed. What, therefore, can you deduce about a set A if A and $R \backslash A$ are separated?

*11. A nonempty set G of real numbers is said to be a **group**[1] if whenever x and y belong to G, so do $x + y$ and $x - y$.
   (a) Determine which of the following sets are groups: R, $\emptyset$, $\{1, 0, -1\}$, Q, Z, Z^+, Q^+, the set of even integers, the set of positive even integers, $\{0\}$, $\{m + n\sqrt{2} \mid m$ and n are integers$\}$.

[1] The definition of a group given here is a special case of the more general concept of a group that you might see in your algebra courses.

(b) Show that every group must contain the number 0. Show that if G is any group other than $\{0\}$, then G must contain infinitely many positive numbers.

(c) Suppose that G is a group other than $\{0\}$, and that $\alpha = \inf\{x \in G \mid x > 0\}$. Show that if $\alpha > 0$, then α is the smallest positive member of G, that $G = \{n\alpha \mid n \in \mathbf{Z}\}$, and that G is closed.

(d) Show that if G is a group other than $\{0\}$ and G has no smallest positive member, then $\bar{G} = \mathbf{R}$. *Hint:* Use the method of proof of Theorem 2.7.2.

(e) Suppose that α is irrational, that $H = \{n\alpha \mid n \in \mathbf{Z}\}$, and that $G = \mathbf{Z} + H$. Prove that although $\mathbf{Z}$ and H are closed groups and G is a group, G is not closed.

3.10 LIMIT POINTS (SOMETIMES CALLED ACCUMULATION POINTS)

Given a number x and a set A of real numbers, we say that x is a **limit point** of A if for every neighborhood U of x we have $U \cap A\backslash\{x\} \neq \emptyset$. The set of all real numbers which are limit points of A is denoted by $\mathscr{L}(A)$. You should take careful note of the distinction between the notions of limit point of a set and a point that lies in the closure of the set. Look at the definitions again: If $x \in \mathbf{R}$ and $A \subseteq \mathbf{R}$, then $x \in \bar{A}$ means that for every neighborhood U of x we have $U \cap A \neq \emptyset$; and $x \in \mathscr{L}(A)$ means that for every neighborhood U of x, $U \cap A\backslash\{x\} \neq \emptyset$. Notice that it is a little harder for x to belong to $\mathscr{L}(A)$ than it is for x to belong to $\bar{A}$; for in order for us to have $x \in \mathscr{L}(A)$, not only must every neighborhood of x intersect with A, but in fact, every neighborhood of x must intersect with the slightly smaller set $A\backslash\{x\}$. In other words, $x \in \mathscr{L}(A)$ iff $x \in \overline{A\backslash\{x\}}$.

3.10.1 Example. We take $A = (0, 1) \cup \{2\}$.

We see that $2 \in A$ and so, of course, $2 \in \bar{A}$; but $2 \notin \mathscr{L}(A)$.

3.10.2 Remark. Given a number x and a set A of real numbers, if it so happens that $x \notin A$, then $A = A\backslash\{x\}$; and therefore, the conditions $x \in \mathscr{L}(A)$ and $x \in \bar{A}$ are equivalent. In other words, the sets $\bar{A}$ and $\mathscr{L}(A)$ can only be different at points of the given set A. Refer to Example 3.10.1 and note the following:

If $x < 0$, then x belongs to none of the three sets A, $\bar{A}$, or $\mathscr{L}(A)$. We could say the same thing if $1 < x < 2$ or $x > 2$.

If x is either 0 or 1, then x belongs to the sets $\mathscr{L}(A)$ and $\bar{A}$ but not to A.

If $x = 2$, then x belongs to the sets A and $\bar{A}$ but not to $\mathscr{L}(A)$.

If $0 < x < 1$, then x belongs to all three of the sets A, $\bar{A}$, and $\mathscr{L}(A)$.

3.10.3 Theorem. Given any set A of real numbers, we have $\bar{A} = A \cup \mathcal{L}(A)$.

■ **Proof.** From the remarks we have made above, it is clear that both A and $\mathcal{L}(A)$ are subsets of $\bar{A}$. Now suppose that $x \in \bar{A}$. Unless $x \in A$, as we have explained above, the condition $x \in \bar{A}$ is equivalent to the condition $x \in \mathcal{L}(A)$. So either $x \in A$ or $x \in \mathcal{L}(A)$.
■

3.10.4 Theorem. A set A of real numbers is closed if and only if $\mathcal{L}(A) \subseteq A$.

■ **Proof.** This result follows at once from Theorems 3.8.5 and 3.10.3.
■

3.10.5 Theorem. Suppose x is a limit point of A. Then every neighborhood of x contains infinitely many points of A.

■ **Proof.** Let U be a neighborhood of x; and to obtain a contradiction, assume that the set $U \cap A$ is finite. Define $E = U \cap A\backslash\{x\}$; and noting that E is nonempty, write E in the form $E = \{y_1, y_2, \ldots, y_n\}$. Define δ to be the smallest member of the set $\{|x - y_j| \mid 1 \leq j \leq n\}$. Then $U \cap (x - \delta, x + \delta)$ is a neighborhood of x which does not intersect with $A\backslash\{x\}$, contradicting the fact that x is a limit point of A.■

3.10.6 Corollary. No finite set can have a limit point.

3.11 EXERCISES

1. Prove that $\mathcal{L}(\mathbf{Z}) = \phi$.

2. Prove that $\mathcal{L}\{1/n \mid n \in \mathbf{Z}^+\} = \{0\}$.

3. Prove that if $B \subseteq A$, then $\mathcal{L}(B) \subseteq \mathcal{L}(A)$.

4. Given sets A and B, prove that $\mathcal{L}(A \cup B) = \mathcal{L}(A) \cup \mathcal{L}(B)$.

5. Prove that $\mathcal{L}(\mathbf{Q}) = \mathbf{R}$.

■ **6.** Prove that if a set U is open, then $\mathcal{L}(U) = \bar{U}$.

7. Given that a set A is nonempty and bounded above, but that A has no largest member, prove that $\sup A \in \mathcal{L}(A)$.

■ **8.** Given that A is bounded and that $\mathcal{L}(A) = \phi$, prove that every nonempty subset of A must have both a greatest member and a smallest member.

3.12* THE SMALL–DISTANCE PROPERTY: A NECESSARY CONDITION FOR A SET TO HAVE A LIMIT POINT

This section can be omitted without loss of continuity.

3.12.1 Definition. If $A \subseteq R$, then we say that A has the **small-distance property** if for every $\varepsilon > 0$ there exist two different points x and y of A such that $|x - y| < \varepsilon$.

3.12.2 Theorem. If a set A has a limit point, then A must have the small-distance property.

■ *Proof.* In order to show that A has the small-distance property, let $\varepsilon > 0$. Choose $\alpha \in \mathcal{L}(A)$. Then since $(\alpha - \varepsilon/2, \alpha + \varepsilon/2)$ is a neighborhood of α, the set $(\alpha - \varepsilon/2, \alpha + \varepsilon/2) \cap A$ must be infinite. Choose two different points x and y in $(\alpha - \varepsilon/2, \alpha + \varepsilon/2) \cap A$.

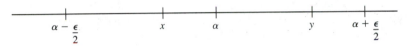

Then clearly, $|x - y| < \varepsilon$. ■

3.12.3 Example. A set with the small-distance property does not always have a limit point. Look, for instance, at the set $Z^+ \cup \{n + 1/n \mid n \in Z^+\}$:

3.13 THE BOLZANO–WEIERSTRASS LIMIT POINT THEOREM

Although the statement of the Bolzano-Weierstrass theorem which follows may look quite simple, this theorem is really a sophisticated restatement of the fact that the system R of real numbers is complete. We shall use the Bolzano-Weierstrass theorem in Chapter 4 to help us prove some important theorems about sequences which depend on completeness, and these theorems in turn will be used when we come to the results about continuous functions, derivatives, and integrals which depend on the completeness of R. We mention in passing that many of these theorems can be used as alternative forms of the axiom of completeness. See Olmsted [12] for details.

The statement of the **Bolzano-Weierstrass theorem** is given in Theorem 3.13.1.

3.13.1 Theorem. Every bounded infinite subset of R must have a limit point in R.

■ *Proof.* Suppose A is a bounded infinite subset of R. In order to prove that A has a limit point, we shall consider two cases.

Case 1. There exists a nonempty subset B of A such that B has no least member.

In this case, choose such a subset B of A. Define $\alpha = \inf B$. We shall show that α is a limit point of A. Let $\delta > 0$. We need to show that the set $(\alpha - \delta, \alpha + \delta) \cap A\backslash\{\alpha\}$ is not empty. Since $\alpha + \delta > \alpha$, it follows that $\alpha + \delta$ is not a lower bound of B. Choose a member x of B such that $x < \alpha + \delta$. Then since $\alpha < x < \alpha + \delta$, it follows that $x \in (\alpha - \delta, \alpha + \delta) \cap A\backslash\{\alpha\}$.

Case 2. Every nonempty subset of A has a least member.

Define x_1 to be the least member of A. Since A is infinite, the set $A\backslash\{x_1\}$ is nonempty and therefore has a least member. Define x_2 to be the least member of the set $A\backslash\{x_1\}$. Of course, $x_1 < x_2$. Now since A is infinite, the set $A\backslash\{x_1, x_2\}$ is nonempty; and we define x_3 to be the least member of this set. Notice that $x_1 < x_2 < x_3$. We continue this process: For each natural number n, having chosen x_1, $x_2, \ldots, x_n$, we define x_{n+1} to be the least member of the set $A\backslash\{x_1, x_2, x_3, \ldots, x_n\}$; and we see that $x_n < x_{n+1}$.

Having defined x_n for each natural number n, we define $B = \{x_n \mid n \in \mathbf{Z}^+\}$; and we notice that B is a nonempty subset of A and that B does not have a greatest member. We can now use a method similar to that of Case 1 to show that if $\alpha = \sup B$, then $\alpha \in \mathscr{L}(A)$. We leave the details as an exercise. ∎

3.14 EXERCISES

1. Given that a set A is both closed and bounded, prove that every infinite subset of A must have a limit point that belongs to A.

■2. Prove that if every nonempty subset of a given set A has both a greatest and a least member, then the set A must be finite.

3. Rewrite the proof of the Bolzano-Weierstrass theorem, taking as your Case 1 the condition that there exists a nonempty subset B of A such that B has no greatest member.

*4. Write a different proof of the Bolzano-Weierstrass theorem by completing the details of the following:
   (a) Define $B = \{x \in \mathbf{R} \mid A \cap (x, \infty) \text{ is infinite}\}$, and show that B is nonempty and bounded above.
   (b) Define $\alpha = \sup B$, and show that $\alpha \in \mathscr{L}(A)$.

The following exercises on the small-distance property can be omitted without loss of continuity.

*5. Prove that no finite set can have the small-distance property.

*6. Give an example of an infinite set which does not have the small-distance property.

*7. Given a set A with the small-distance property and $x \in A$, prove that $A \backslash \{x\}$ has the small-distance property.

*8. Prove that every infinite bounded set must have the small-distance property.

3.15 NEIGHBORHOODS OF INFINITY

In Section 2.9 we introduced the two infinity symbols, and we extended the order relation $<$ and the operations of arithmetic to the extended real number system $[-\infty, \infty]$. At this stage we shall show how the important idea of neighborhood may be extended to the system $[-\infty, \infty]$.

Given $U \subseteq \mathbf{R}$, we say that U is a **neighborhood of** ∞ if there exists a number w such that $(w, \infty) \subseteq U$. Similarly, we say that U is a neighborhood of $-\infty$ if there exists a number w such that $(-\infty, w) \subseteq U$.

3.15.1 Some Simple Facts. The following results describe some important properties of neighborhoods of ∞.

(1) If U is a neighborhood of ∞ and $U \subseteq V$, then V is a neighborhood of ∞.

(2) If U and V are both neighborhoods of ∞, then so is $U \cap V$.

(3) If U is a neighborhood of ∞ and B is unbounded above, then $U \cap B \neq \emptyset$.

We leave the proofs of these three statements as exercises. While you are thinking about them, you should also think about their analogues for neighborhoods of $-\infty$.

3.15.2 ∞ and $-\infty$ as Limit Points. We now extend the concept of limit point to include the symbols ∞ and $-\infty$. Given $A \subseteq \mathbf{R}$, we say that ∞ is a *limit point of A* if for every neighborhood U of ∞ we have $U \cap A \neq \emptyset$. Similarly, if $A \subseteq \mathbf{R}$, then we say that $-\infty$ is a *limit point of A* if for every neighborhood U of $-\infty$ we have $A \cap U \neq \emptyset$. Note, however, that the symbol $\mathcal{L}(A)$, which we defined in Section 3.10, stands for the set of *real* numbers which are limit points of A. This set never contains ∞ or $-\infty$. So, for example, even though both ∞ and $-\infty$ are limit points of $\mathbf{Z}$, since there are no real numbers that are limit points of $\mathbf{Z}$, we have $\mathcal{L}(\mathbf{Z}) = \emptyset$.

3.15.3 Some Simple Exercises

1. Prove that if a set A is bounded above, then ∞ is not a limit point of A. State and prove an analogue for $-\infty$.

2. Prove that if a set A is unbounded above, then ∞ is a limit point of A. State and prove an analogue for $-\infty$.

3. Given that a set A is infinite, prove that A has a limit point which is either in $\mathcal{L}(A)$ or in $\{\infty, -\infty\}$.

Chapter *4*

Sequences

4.1 THE BASIC DEFINITIONS

4.1.1 Definition of a Sequence. The usual definition of a *sequence of real numbers*—or, alternatively, a *sequence in **R***—is that it is a function f whose domain is the set $\mathbf{Z}^+$ of natural numbers such that for every natural number n, $f(n)$ is a real number.

This definition is not really satisfactory because it prevents us from including examples like $f(n) = \sqrt{n - 5}$ for all naturals $n \geq 5$. This function "should" be a sequence, but it can't be if we demand that the domain of a sequence be the entire set $\mathbf{Z}^+$ of natural numbers. We shall therefore take a definition of sequence which is just a little wider than the one given above.

We define a **sequence of real numbers** to be a real-valued function f defined on a set of the form $\{n \in \mathbf{Z}^+ \mid n \geq N\}$, where N is some natural number. In other words, if f is a sequence, then for some natural number N the domain of f is $\{n \in \mathbf{Z}^+ \mid n \geq N\}$; and $f(n)$ is a real number for every natural number $n \geq N$.

Traditionally, when speaking about sequences, we do not use the usual function notation; and instead of speaking about a sequence f, we shall typically denote the sequence which takes the value x_n at each natural number n (in its domain) by (x_n). There is, of course, nothing special about the letter n of the alphabet. We could just as well call the above sequence (x_m) or (x_i). In following this tradition, one should be careful not to confuse the symbols x_n, which stands for the value of the sequence at the given natural number n, and (x_n), which stands for the sequence itself. Given a sequence (x_n), the symbol x_n is meaningful only when we have a particular natural number n in mind; and when we speak of x_n, we are speaking of a *real number*. We often call x_n the *nth term of the sequence* (x_n).

Here are some examples of sequences.

(1) $x_n = n^2 + n - 6$ for every natural number n.

(2) $x_n = (-1)^n$ for every natural number n.

(3) $x_n = 1/(n - 1)(n - 7)$ for all natural numbers $n \geq 8$.

(4) $x_n = 3$ if n is even and $x_n = 1/n$ if n is odd.

(5) $x_n = k/m$ if there exist natural numbers k and m such that $n = 2^k 3^m$, and $x_n = 0$ otherwise.

4.1.2 The Range of a Sequence. Considering that a sequence is just a function with domain $\{n \in \mathbf{Z}^+ \mid n \geq N\}$, where N is some natural number, the **range** of a given sequence (x_n) is defined to be the set $\{x_n \mid n \geq N\}$. Thus if (x_n) is the sequence in Example 4.1.1(2), then the range of (x_n) is the finite set $\{-1, 1\}$; and in Example 4.1.1(5) the range of (x_n) is $\mathbf{Q} \cap [0, \infty)$. Can you give an example of a sequence whose range is $\mathbf{Q}$?

If you are familiar with the notion of countability (discussed in Appendix A), then you might like to observe that a set can be the range of a sequence if and only if the set is countable.

4.1.3 Upper and Lower Bounds of a Sequence. Given a sequence (x_n) and a number α, we say that α is an *upper bound of the sequence* (x_n) if α is an upper bound of its range. In other words, α is an upper bound of (x_n) when $x_n \leq \alpha$ for every n in the domain of (x_n). Similarly, we say that a number α is a *lower bound of the sequence* (x_n) if $\alpha \leq x_n$ for every n in the domain of (x_n). It should now be clear what we mean when we say that a given sequence is *bounded above* or *below* or that a given sequence is *bounded*. When a sequence is bounded above, then the *supremum* of the sequence is defined to be the supremum of the range of the sequence. The *infimum* of a sequence is defined similarly.

4.1.4 Increasing and Decreasing Sequences. We say that a sequence (x_n) is **increasing** if $x_n \leq x_{n+1}$ for every n in its domain; and if $x_n < x_{n+1}$ for every n, then we say that the sequence (x_n) is **strictly increasing**. *Decreasing sequences* and *strictly decreasing sequences* are defined similarly. A sequence which is either increasing or decreasing is said to be **monotone,** and a sequence which is either strictly increasing or strictly decreasing is said to be **strictly monotone**.

4.1.5* Subsequences. This section on subsequences can be omitted without loss of continuity.

Roughly speaking, a *subsequence* of a given sequence (x_n) is a sequence that is made by leaving out some of the terms. For example, the first term of a subsequence might be x_3, the second term x_7, the third term x_8, the fourth term x_{104}, and so on. What we are looking at here is a strictly increasing sequence (n_i) of natural numbers, and we are saying that the subsequence is $x_{n_1}, x_{n_2}, x_{n_3}, \ldots$. More precisely, we are saying that for each natural number i the ith term of the subsequence is x_{n_i}, which suggests the following precise definition: A sequence (y_n) is said to be a **subsequence** of a sequence (x_n) if there exists a strictly increasing sequence (n_i) of natural numbers such that for every natural number i we have $y_i = x_{n_i}$.

We illustrate the definition of a subsequence with the following examples.

(1) Suppose (x_n) is a given sequence. For each natural number i define $n_i = 2i$, and define $y_i = x_{n_i}$. Then for each i we have $y_i = x_{2i}$, and so we commonly refer to the sequence (y_n) as (x_{2n}).

(2) Suppose $x_n = (-1)^n$ for each n. Then the subsequence (x_{2n}) is the sequence which takes the constant value 1, and the subsequence (x_{2n-1}) is the sequence which takes the constant value -1.

(3) Suppose $x_n = n(-1)^{n+1}$ for each n. Then the subsequence (x_{6n}) is the sequence whose value at each n is $-6n$.

(4) Suppose (y_n) is a subsequence of a sequence (x_n) and that (z_n) is a subsequence of (y_n). Then (z_n) is a subsequence of (x_n). We leave the proof of this assertion as an exercise.

4.2 THE CONCEPTS "FREQUENTLY" AND "EVENTUALLY"

4.2.1 Definition. Suppose that (x_n) is a sequence of real numbers and that $A \subseteq R$. We say that the sequence (x_n) is **in the set** A if $x_n \in A$ for every natural number n. We say that the sequence (x_n) is **eventually in the set** A if there exists a natural number N such that for all natural numbers $n \geq N$ we have $x_n \in A$. We say that the sequence (x_n) is **frequently in the set** A if we have $x_n \in A$ for infinitely many natural numbers n. In other words, (x_n) is frequently in A when the set $\{n \in Z^+ \mid x_n \in A\}$ is infinite. Notice that if a sequence (x_n) is eventually in a set A, then (x_n) must be frequently in A. It is also worth noting that when a sequence (x_n) is not eventually in a set A, it must be frequently in $R \backslash A$. Similarly, one may see that when a sequence (x_n) is eventually in a set A, it cannot be frequently in $R \backslash A$.

4.2.2 Some Simple Examples. The examples that follow help illustrate the notions of *frequently* and *eventually*.

(1) Suppose $x_n = 1$ for $n \leq 20$, and $x_n = 0$ for $n \geq 21$. Then whenever a set A contains the number 0, the sequence (x_n) will be eventually in A.

(2) Suppose that $x_n = 1 + (-1)^n$ for each natural number n. Then (x_n) is frequently in $\{2\}$ and frequently in $\{0\}$ and eventually in any set that includes $\{0, 2\}$.

(3) Suppose $x_n = 1/n$ for every natural number n and that $\varepsilon > 0$. Then (x_n) is eventually in $(0, \varepsilon)$. To see why, choose a natural number N such that $N > 1/\varepsilon$. Then $x_n \in (0, \varepsilon)$ whenever $n \geq N$.

4.2.3 Note on the Use of Language. Suppose (x_n) is a given sequence and $A \subseteq \mathbf{R}$. The condition that (x_n) be frequently in the set A requires that there should be infinitely many natural numbers n such that $x_n \in A$. Do not make the common mistake of phrasing the condition that (x_n) be frequently in A by saying that "infinitely many x_n's belong to A." The reason this phrasing is wrong is that it suggests that there should be infinitely many *different* numbers of the form x_n that belong to A. Look at the sequence (x_n), where $x_n = (-1)^n$ for all natural numbers n. There are, of course, infinitely many natural numbers n such that $x_n \in \{1\}$, and so (x_n) is frequently in $\{1\}$; but there are certainly *not* infinitely many numbers of the form x_n in the finite set $\{1\}$.

We end this section with a theorem that relates the concept "frequently" with the idea of a subsequence. Unless you are including subsequences in your reading of this chapter, skip the theorem and proceed to Section 4.3.

4.2.4* Theorem. A sequence (x_n) is frequently in a set A if and only if (x_n) has a subsequence in A.

■ *Proof.* Suppose (x_n) has a subsequence (y_n) in A, and choose a strictly increasing sequence (n_i) of natural numbers such that $y_i = x_{n_i}$ for each i. Since there are infinitely many natural numbers of the form n_i, and since $x_{n_i} \in A$ for each i, it is clear that (x_n) is frequently in A.

Assume now that (x_n) is frequently in A, and define E to be the set of all those natural numbers n for which $x_n \in A$. Now using the fact that the set E is infinite, we shall construct a strictly increasing sequence (n_i) in E. We define n_1 to be the least member of E. Having done this we define n_2 to be the least member of $E \backslash \{n_1\}$. Note that $n_2 > n_1$. In general, if i is any natural number for which the numbers $n_1, n_2, \ldots, n_i$ have been defined, we define n_{i+1} to be the least member of the infinite set $E \backslash \{n_1, n_2, \ldots, n_i\}$. It is clear that if $y_i = x_{n_i}$ for each i, then (y_n) is in A and is a subsequence of (x_n). ■

4.3 LIMITS AND PARTIAL LIMITS OF SEQUENCES

4.3.1 The Definitions.

(1) Given a sequence (x_n) and given any $x \in [-\infty, \infty]$, we say that x is a **limit** of the sequence (x_n) if for every neighborhood U of x the sequence (x_n) is

eventually in U. When x is a limit of (x_n), we write $x_n \to x$ as $n \to \infty$. If the symbol n is understood, we sometimes write this more simply as $x_n \to x$.

(2) Given a sequence (x_n) and given any $x \in [-\infty, \infty]$, we say that x is a **partial limit**[1] of the sequence (x_n) if for every neighborhood U of x the sequence (x_n) is frequently in U.

Notice that it is harder for a number to be a limit of a given sequence than it is to be a partial limit. Any limit of a sequence will automatically be a partial limit. We begin our discussion of limits and partial limits with some simple theorems which provide useful alternative forms of the definitions.

4.3.2 Theorem. Suppose that (x_n) is a given sequence and that x is a given real number. Then the following two conditions are equivalent:

(1) $x_n \to x$ as $n \to \infty$.

(2) For every number $\varepsilon > 0$ the sequence (x_n) is eventually in the set $(x - \varepsilon, x + \varepsilon)$.

■ **Proof.** To prove that (1) $\Rightarrow$ (2), assume that (1) holds. Let $\varepsilon > 0$. Then since $(x - \varepsilon, x + \varepsilon)$ is a neighborhood of x, we may conclude that (x_n) is eventually in $(x - \varepsilon, x + \varepsilon)$. Now to prove that (2) $\Rightarrow$ (1), assume that (2) holds. Let U be a neighborhood of x. Choose a number $\varepsilon > 0$ such that $(x - \varepsilon, x + \varepsilon) \subseteq U$. Then since (x_n) is eventually in $(x - \varepsilon, x + \varepsilon)$, it follows at once that (x_n) is eventually in U. ■

4.3.3 Theorem. Suppose (x_n) is a given sequence and that x is a given real number. Then the following two conditions are equivalent:

(1) x is a partial limit of (x_n).

(2) For every number $\varepsilon > 0$ the sequence (x_n) is frequently in the set $(x - \varepsilon, x + \varepsilon)$.

The proof of this theorem is almost identical to that of Theorem 4.3.2 and will be left as an exercise.

4.3.4 Theorem. Suppose (x_n) is a given sequence. Then the following two conditions are equivalent:

(1) $x_n \to \infty$ as $n \to \infty$.

(2) For every real number w the sequence (x_n) is eventually in (w, ∞).

---

[1] The concept of a partial limit of a sequence that is introduced in this section is sometimes known as a *cluster point* (Kelley [7]), sometimes as a *limit point* (Gelbaum and Olmsted [6]), and sometimes as a *subsequential limit* (Rudin [13]). We have chosen the term *partial limit* because the terms *cluster point* and *limit point* might be confused with the idea of limit point of a set, and the term *subsequential limit* is not appropriate when the theory of sequences is applied to more general spaces.

■ *Proof.* To prove that (1) ⇒ (2), assume that (1) holds. Let w be any real number. Since (w, ∞) is a neighborhood of ∞, it follows at once that (x_n) is eventually in (w, ∞). Now to prove that (2) ⇒ (1), assume that (2) holds. Let U be a neighborhood of ∞. Choose a number w such that $(w, \infty) \subseteq U$. Then since (x_n) is eventually in (w, ∞), it follows at once that (x_n) is eventually in U.   ■

4.3.5 Theorem. Suppose (x_n) is a given sequence. Then the following two conditions are equivalent:

(1) ∞ is a partial limit of (x_n).

(2) For every real number w the sequence (x_n) is frequently in (w, ∞).

We leave the proof of this theorem as an exercise.

4.3.6 Some Examples. The examples which follow illustrate the notions of limit and partial limit. They will also acquaint you with some important techniques which are needed for a proper understanding of the central theorems of the chapter.

(1) Given that $x \in \mathbf{R}$ and that $x_n = x$ for every $n \in \mathbf{Z}^+$, we have $x_n \to x$ as $n \to \infty$.

(2) Suppose that $x \in \mathbf{R}$, that (x_n) is a given sequence, and that $x_n = x$ for all $n \geq 100$. Then $x_n \to x$ as $n \to \infty$.

(3) Suppose that $x_n = 1/n$ for every natural number n. Then $x_n \to 0$ as $n \to \infty$.

■ *Proof.* We shall use the form of convergence given in Theorem 4.3.2(2). Let $\varepsilon > 0$:

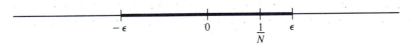

Choose a natural number $N > 1/\varepsilon$ and note that $1/N < \varepsilon$. Therefore, whenever $n \geq N$, we have $0 < x_n = 1/n \leq 1/N < \varepsilon$; and it follows that whenever $n \geq N$, we have $x_n \in (-\varepsilon, \varepsilon)$. We have therefore shown that (x_n) is eventually in $(-\varepsilon, \varepsilon)$. ■

(4) Suppose $x_n = n$ for every natural number n. Then $x_n \to \infty$ as $n \to \infty$.

■ *Proof.* Let U be a neighborhood of ∞. Choose a number α such that $(\alpha, \infty) \subseteq U$:

Now choose a natural number $N > \alpha$. Then whenever $n \geq N$, we have $x_n > \alpha$; and so for all $n \geq N$, we have $x_n \in U$. This shows that (x_n) is eventually in U.   ■

(5) Suppose that $x_n = n/(n + 1)$ for every natural number n. Then $x_n \to 1$ as $n \to \infty$.

■ **Proof.** First, we observe that for each n we have $|x_n - 1| = 1/(n + 1)$. Let $\varepsilon > 0$:

Now for any given n the condition $x_n \in (1 - \varepsilon, 1 + \varepsilon)$ is equivalent to the condition $|x_n - 1| < \varepsilon$; in other words, $1/(n + 1) < \varepsilon$. But the latter inequality is equivalent to the condition that $n > 1/\varepsilon - 1$, and this tells us how to finish the proof. We choose a natural number N such that $N > 1/\varepsilon - 1$ and we observe that whenever $n \geq N$, we have $|x_n - 1| < \varepsilon$; so that $x_n \in (1 - \varepsilon, 1 + \varepsilon)$. Thus (x_n) is eventually in $(1 - \varepsilon, 1 + \varepsilon)$. ■

(6) Suppose $x_n = (3 + 5n)/(2 - 8n)$ for every natural number n. Then $x_n \to -\frac{5}{8}$ as $n \to \infty$.

■ **Proof.** First, we observe that for each n we have

$$\left| x_n - \left(-\frac{5}{8} \right) \right| = \left| \frac{3 + 5n}{2 - 8n} - \left(-\frac{5}{8} \right) \right| = \left| \frac{34}{8(2 - 8n)} \right| = \frac{34}{8(8n - 2)} < \frac{5}{8n - 2}.$$

Now let $\varepsilon > 0$. We observe that for any given n the condition $|x_n - (-\frac{5}{8})| < \varepsilon$ will be satisfied whenever $5/(8n - 2) < \varepsilon$. But the latter inequality is equivalent to the condition that $n > \frac{1}{8}(5/\varepsilon + 2)$, and this tells us how to finish the proof. We choose a natural number N such that $N > \frac{1}{8}(5/\varepsilon + 2)$, and we observe that whenever $n \geq N$, we have $|x_n - (-\frac{5}{8})| < \varepsilon$. ■

(7) Suppose that $x_n = 2$ for n even and $x_n = 1$ for n odd. Then the set of partial limits of (x_n) is $\{1, 2\}$.

■ **Proof.** It is clear that both 1 and 2 are partial limits of (x_n). Now suppose that x is any other real number. Then $R\setminus\{1, 2\}$ is a neighborhood of x. Since (x_n) is not frequently in this neighborhood, we conclude that x is not a partial limit of (x_n). So (x_n) has no partial limits other than the numbers 1 and 2. ■

(8) Suppose $x_n = 1$ for n even and $x_n = n$ for n odd. Then the set of partial limits of (x_n) is $\{1, \infty\}$.

(9) Suppose that $x_n = 1$ when n is a multiple of 3, that $x_n = 0$ when n is one more than a multiple of 3, and that $x_n = 6$ when n is two more than a multiple of 3. Then the set of partial limits of (x_n) is $\{0, 1, 6\}$.

(10) Suppose that (x_n) is the sequence of Example 4.1.1(5). As we have observed, the range of (x_n) is $Q \cap [0, \infty)$. It is therefore clear that given $\alpha < 0$, the sequence (x_n) is not frequently in the neighborhood $(-\infty, 0)$ of α. Consequently, whenever $\alpha < 0$, α is not a partial limit of (x_n). On the other hand, if $\alpha \geq 0$, then every neighborhood of α must contain infinitely many positive rational numbers; and therefore, every neighborhood of α must contain x_n for infinitely many natural numbers n. It follows that the set of partial limits of (x_n) is $[0, \infty]$.

(11) This example is just like example (10) except that we shall keep the sequence between 0 and 1. If n has the form $n = 2^k 3^m$, where k and m are natural numbers and $k \leq m$, then we define $x_n = k/m$. For all other values of n we define $x_n = 0$. We see that the range of (x_n) is $\mathbf{Q} \cap [0, 1]$, and it follows easily that the set of partial limits of (x_n) is $[0, 1]$.

(12) Just one more variation on the theme: If n has the form $n = 2^k 3^m$, where k and m are natural numbers and either $k \leq m$ or $2m \leq k \leq 3m$, then we define $x_n = k/m$; and for all other values of n we define $x_n = 0$. This time the range of (x_n) turns out to be the set of all rational numbers in $[0, 1] \cup [2, 3]$, and so the set of partial limits of (x_n) is $[0, 1] \cup [2, 3]$.

4.3.7 The Theorem of Uniqueness of Limits.

Suppose that a sequence (x_n) has a limit x and that $x \neq y$. Then y cannot be a partial limit of (x_n).

■ *Proof.* We shall suppose that $x < y$. The case $y < x$ is similar.

Choose a number α such that $x < \alpha < y$. Using the facts that $(-\infty, \alpha)$ is a neighborhood of x and that $x_n \to x$, choose a natural number N such that for all natural numbers $n \geq N$ we have $x_n < \alpha$. It follows that (x_n) is not frequently in the neighborhood (α, ∞) of y, and so y cannot be a partial limit of the sequence. ■

4.3.8 Corollary.

No sequence can have more than one limit.

4.3.9 Limit Notation.

If (x_n) has a limit x, then since x is the only limit of the sequence, we can give this limit a name. We call it $\lim_{n \to \infty} x_n$, or more simply, $\lim x_n$.

4.3.10 Convergent and Divergent Sequences.

A sequence (x_n) is said to be **convergent** if (x_n) has a limit and $\lim x_n$ is neither ∞ nor $-\infty$. In other words, (x_n) is convergent when $\lim x_n$ exists and is a real number. Any sequence which is not convergent is said to be **divergent**. Note that there are two ways for a sequence to be divergent: Either the limit is $\pm\infty$, or the sequence has no limit at all. Roughly speaking, what makes a sequence fail to have a limit is that the sequence "bounces around" or "wobbles" or "oscillates." In Sections 4.9, 4.10, and 4.11 we shall see some theorems that provide some precise interpretations of this idea.

4.3.11 The Sandwich Theorem. Suppose (x_n), (y_n), and (z_n) are three given sequences and $x \in \mathbf{R}$. Suppose that $x_n \to x$ and $z_n \to x$, and suppose that for every natural number n we have $x_n \leq y_n \leq z_n$. Then $y_n \to x$.

■ **Proof.** Let $\varepsilon > 0$:

Now we use the fact that both (x_n) and (z_n) must eventually be in $(x - \varepsilon, x + \varepsilon)$. Choose a natural number N_1 such that for all natural numbers $n \geq N_1$ we have $x_n \in (x - \varepsilon, x + \varepsilon)$, and choose a natural number N_2 such that for all $n \geq N_2$ we have $z_n \in (x - \varepsilon, x + \varepsilon)$. We define N to be the larger of the two numbers N_1 and N_2. Then for all $n \geq N$ we have $x - \varepsilon < x_n \leq y_n \leq z_n < x + \varepsilon$; and therefore, for all natural numbers $n \geq N$ we have $y_n \in (x - \varepsilon, x + \varepsilon)$. ■

4.3.12 Theorem. Given any sequence (x_n) of real numbers, the following conditions are equivalent:

(1) (x_n) is bounded above.

(2) ∞ is not a partial limit of (x_n).

■ **Proof.** First, we shall prove that (1) $\Rightarrow$ (2). Assume that (1) holds. Choose a real number α such that α is an upper bound of (x_n). We observe that (α, ∞) is a neighborhood of ∞ and that (x_n) is not frequently in this neighborhood. Therefore, ∞ is not a partial limit of (x_n).

Now we shall prove that (2) $\Rightarrow$ (1). Assume that (2) holds. Choose a neighborhood U of ∞ such that (x_n) is not frequently in U, and then choose a real number α such that $(\alpha, \infty) \subseteq U$. We notice that the set F defined by $F = \{n \in \mathbf{Z}^+ \mid x_n > \alpha\}$ is finite. Now the range of (x_n) is the union of the two sets $\{x_n \mid n \in F\}$ and $\{x_n \mid n \in \mathbf{Z}^+ \backslash F\}$. Therefore, since the first of these two sets, being finite, is bounded above, and since the second is bounded above by α, we conclude that (x_n) is bounded above. ■

4.3.13 Theorem. Given any sequence (x_n) of real numbers, the following conditions are equivalent:

(1) (x_n) is bounded below.

(2) $-\infty$ is not a partial limit of (x_n).

We leave the proof of this theorem as an exercise.

4.3.14 Corollary. A sequence (x_n) of real numbers is bounded if and only if neither ∞ nor $-\infty$ is a partial limit of (x_n).

We end our discussion of limits and partial limits with a theorem that relates the concept of partial limit with limits of subsequences. Unless you are including subsequences in your reading of this chapter, skip the theorem and proceed to the exercises in Section 4.4.

4.3.15* Theorem. If (x_n) is a sequence of real numbers and $x \in [-\infty, \infty]$, then the following two conditions are equivalent:

(1) x is a partial limit of (x_n).

(2) There exists a subsequence of (x_n) whose limit is x.

■ **Proof.** To prove that $(2) \Rightarrow (1)$, assume that (2) holds. Choose a subsequence (y_n) of (x_n) such that $y_n \to x$, and choose a strictly increasing sequence (n_i) such that $y_i = x_{n_i}$ for each i. Now given any neighborhood U of x, since $x_{n_i} \in U$ for all sufficiently large i, it is clear that (x_n) is frequently in U.

The proof that $(1) \Rightarrow (2)$ is a little harder. We shall consider first the case in which $x \in \mathbf{R}$. Assume then that x is a real number which is a partial limit of a sequence (x_n). We need to find a subsequence of (x_n) which converges to x. For each natural number i we define $U_i = (x - 1/i, x + 1/i)$. Then for each i, U_i is a neighborhood of x, and so (x_n) is frequently in U_i. Choose a natural number n_1 such that $x_{n_1} \in U_1$. Now choose $n_2 > n_1$ such that $x_{n_2} \in U_2$; and in general, having chosen n_i, choose a natural number $n_{i+1} > n_i$ such that $x_{n_{i+1}} \in U_{i+1}$. Since $|x_{n_i} - x| < 1/i$ for each i, it follows easily that $x_{n_i} \to x$ as $i \to \infty$.

Now suppose that ∞ is a partial limit of (x_n), and for each natural number i we define $U_i = (i, \infty)$. Each U_i is a neighborhood of ∞; and by using a simple adaptation of the previous argument, we can find a subsequence of (x_n) which tends to ∞. We leave the case $x = -\infty$ as an exercise. ■

4.4 EXERCISES

1. Give an example of a sequence with exactly five partial limits.

2. Give an example of a sequence whose set of partial limits is $\{-\infty, 0, \infty\}$.

■ **3.** Give an example of a sequence whose set of partial limits is $\{1\} \cup [2, 3]$.

■ **4.** Suppose that $x_n \to x > 0$. Prove that there exists a natural number N such that for all $n \geq N$ we have $x_n > 0$. You should be able to write one simple proof that handles the cases $x = \infty$ and $x \in \mathbf{R}$ together.

■ **5.** Given that $x_n \geq 0$ for every natural number n and that x is a partial limit of (x_n), prove that $x \geq 0$.

6. Given a sequence (x_n) and $x \in \mathbf{R}$, prove that the condition $x_n \to x$ is equivalent to the condition that for every number $\varepsilon > 0$ there exists a natural number N such that for all $n \geq N$ we have $|x_n - x| < \varepsilon$. *Hint:* The condition $|x_n - x| < \varepsilon$ is equivalent to saying that $x_n \in (x - \varepsilon, x + \varepsilon)$.

7. Given that $x_n = (3 + 2n)/(5 + n)$ for each n, prove that $x_n \to 2$.

8. Given that $x_n = 1/2^n$ when n is even and $x_n = 1/(n^2 + 1)$ when n is odd, prove that (x_n) is convergent, and find its limit.

■ 9. Prove that $2^n/n! \to 0$ as $n \to \infty$.

10. Prove that $n!/n^n \to 0$ as $n \to \infty$.

■ 11. For every natural number k we define $x_n = 1/k$ whenever $2^{k-1} \le n \le 2^k - 1$. Prove that $x_n \to 0$.

■ 12. Given a sequence (x_n) and $x \in \mathbf{R}$, prove that the condition $x_n \to x$ is equivalent to the condition that for every number $\varepsilon > 0$ there exists a natural number N such that for all $n \ge N$ we have $|x_n - x| < 5\varepsilon$.

13. Given a sequence (x_n) and $x \in \mathbf{R}$, prove that the condition $x_n \to x$ is equivalent to the condition that for every number $\varepsilon > 0$ there exists a natural number N such that for all $n \ge N$ we have $|x_n - x| < \varepsilon/2$.

14. Given a sequence (x_n) and a number x, prove that $x_n \to x$ if and only if $|x_n - x| \to 0$.

15. Suppose that (x_n) and (y_n) are given sequences, that $x \in \mathbf{R}$, that $y_n \to 0$, and that for every n we have $|x_n - x| \le y_n$. Prove that $x_n \to x$.

16. Given a sequence (x_n), prove that the condition $x_n \to \infty$ is equivalent to the condition that for every number w there exists a natural number N such that for all $n \ge N$ we have $x_n > w$.

■ 17. Prove that $(n^3 - n^2 + 1)/(n^2 + 5) \to \infty$ as $n \to \infty$.

18. Given that $x_n \ge y_n$ for every n and that $y_n \to \infty$, prove that $x_n \to \infty$.

19. Given that $x_n \to x$, prove that $|x_n| \to |x|$. *Hint:* Use Exercise 2.1.3(5).

■ 20. We say that two sequences (x_n) and (y_n) are *eventually close* if for every number $\varepsilon > 0$ there exists a natural number N such that for all $n \ge N$ we have $|x_n - y_n| < \varepsilon$.
   (a) Prove that if (x_n) and (y_n) are eventually close and $x_n \to x$, then $y_n \to x$. (Consider separately the cases $x \in \mathbf{R}$ and $x = \pm\infty$.)
   (b) Prove that if (x_n) and (y_n) are eventually close and x is a partial limit of (x_n), then x must be a partial limit of (y_n).
   (c) Prove that if (x_n) and (y_n) are eventually close, then they have the same partial limits.
   (d) Give an example of two sequences with the same partial limits which are not eventually close.

■ 21. This exercise introduces the notion of a Cauchy sequence (which will be discussed in depth in Section 4.11; you will find a solution to this exercise at the beginning of the proof of Theorem 4.11.3). A sequence (x_n) is said to be a **Cauchy sequence** if for every number $\varepsilon > 0$ there exists a natural number N

such that for all natural numbers m and n such that $m \geq N$ and $n \geq N$ we have $|x_m - x_n| < \varepsilon$. Given that (x_n) is convergent to a number x, prove that (x_n) must be a Cauchy sequence.

22. Given that $x_n \to x$, that p is some integer, and that $y_n = x_{n+p}$ for all sufficiently large natural numbers n, prove that $y_n \to x$.

*23. In this question you are invited to explore an alternative (and superior) definition of a subsequence: Given two sequences (x_n) and (y_n), we say that (y_n) is a *subsequence* of (x_n) if for every natural number N there exists a natural number M such that $\{y_n \mid n \geq M\} \subseteq \{x_n \mid n \geq N\}$. Using this definition, develop the theory of subsequences and obtain some of their properties. State and prove analogues of Theorems 4.2.4 and 4.3.15.

4.5 THE ARITHMETICAL RULES FOR LIMITS

The arithmetical rules for limits are the theorems which display the relationship between the behavior of limits and the arithmetical operations (addition, subtraction, multiplication, and division). The central theorem of this section is Theorem 4.5.2.

4.5.1 Theorem. Suppose that (x_n) is a bounded sequence and that $y_n \to 0$. Then $x_n y_n \to 0$.

■ *Proof.* Choose a positive number b such that for every natural number n we have $-b \leq x_n \leq b$. Let $\varepsilon > 0$. We shall show that for n sufficiently large we have $|x_n y_n| < \varepsilon$. Now for each n we see that $|x_n y_n| \leq b\,|y_n|$, and so all we need to know is that $|y_n| < \varepsilon/b$ for n sufficiently large. Using the facts that $y_n \to 0$ and that $(0 - \varepsilon/b, 0 + \varepsilon/b)$ is a neighborhood of 0, we now choose a natural number N such that whenever $n \geq N$, we have $y_n \in (0 - \varepsilon/b, 0 + \varepsilon/b)$; in other words, $|y_n| < \varepsilon/b$. Then for all naturals $n \geq N$ we have $|x_n y_n| < b\varepsilon/b = \varepsilon$. ■

4.5.2 Theorem. Suppose that (x_n) and (y_n) are sequences, that x and y are points in $[-\infty, \infty]$, and that $x_n \to x$ and $y_n \to y$. Then each of the following conditions holds provided that its right-hand side is defined:

(1) $x_n + y_n \to x + y$.

(2) $x_n - y_n \to x - y$.

(3) $x_n y_n \to xy$.

(4) $x_n / y_n \to x/y$.

This theorem is actually an efficient way of stating a host of different results. In part (1), for example, x and y might both be real numbers; or we might have $x = \infty$, in which case y can be anything except $-\infty$. In part (3), x and y might be real numbers; or we might have $x = \infty$, in which case y can be anything except 0, and there are two main cases to consider: $y < 0$ or $y > 0$. We have hardly begun to list the many possibilities that are included in this theorem. If we were to state them all in separate theorems, the list of these theorems would seem endless; and one of the main reasons for our introduction of the extended real number system in Section 2.9.1 was that it gives us the ability to state Theorem 4.5.2 briefly and simply. As you have probably guessed, the proof of Theorem 4.5.2 is going to be quite long, but it isn't hard. It is long because one needs to write a separate approach for each of the many cases that have to be considered. We shall give a detailed proof in only some of these cases. It is your responsibility to study the proofs that are given and then to write out sufficiently many others to guarantee that you can deal with them all.

■ *Proof of Part (1) When x and y Are Real Numbers.* Let $\varepsilon > 0$. We need to show that for n sufficiently large we have $|(x_n + y_n) - (x + y)| < \varepsilon$. The key to the proof is the inequality

$$|(x_n + y_n) - (x + y)| = |(x_n - x) + (y_n - y)| \leq |x_n - x| + |y_n - y|.$$

Now using the facts that $(x - \varepsilon/2, x + \varepsilon/2)$ is a neighborhood of x and that $x_n \to x$, choose a natural number N_1 such that whenever $n \geq N_1$, we have $x_n \in (x - \varepsilon/2, x + \varepsilon/2)$; in other words, $|x_n - x| < \varepsilon/2$. In a similar fashion, choose a natural number N_2 such that whenever $n \geq N_2$, we have $|y_n - y| < \varepsilon/2$. Define N to be the larger of the two numbers N_1 and N_2. Then for all natural numbers $n \geq N$ we have

$$|(x_n + y_n) - (x + y)| = |(x_n - x) + (y_n - y)|$$

$$\leq |x_n - x| + |y_n - y| < \frac{\varepsilon}{2} + \frac{\varepsilon}{2} = \varepsilon. \qquad \blacksquare$$

■ *Proof of Part (1) When x = ∞.* Since $y \neq -\infty$, the sequence (y_n) must be bounded below. Choose a lower bound α of (y_n). Since $y + \infty = \infty$, we have to show that $x_n + y_n \to \infty$. Let U be a neighborhood of ∞, and choose a number w such that $(w, \infty) \subseteq U$. Using the fact that $(w - \alpha, \infty)$ is a neighborhood of ∞, choose a natural number N such that whenever $n \geq N$, we have $x_n \in (w - \alpha, \infty)$; in other words, $x_n > w - \alpha$. Then whenever $n \geq N$, we have $x_n + y_n > w - \alpha + \alpha = w$; and so for $n \geq N$ we have $x_n + y_n \in U$. $\qquad \blacksquare$

■ *Proof of Part (3) When x and y Are Real Numbers.* Since $x_n \to x$ and $y_n \to y$, we know that $x_n - x \to 0$ and $y_n - y \to 0$. We need to show that $x_n y_n - xy \to 0$. Now for each n we have

$$x_n y_n - xy = x_n y_n - x_n y + x_n y - xy = x_n(y_n - y) + y(x_n - x),$$

and therefore, by part (1) of the theorem, all we need to show is that each of the two terms on the right side of this identity tends to zero. Now since the convergent sequence (x_n) is bounded and $y_n - y \to 0$, we deduce from Theorem 4.5.1 that

$x_n(y_n - y) \to 0$. The fact that $y(x_n - x) \to 0$ follows in the same way, because the constant sequence (y) is bounded and $x_n - x \to 0$. ∎

■ *Proof of Part (3) When $x > 0$ and $y = \infty$.* Since $x \cdot \infty = \infty$, we need to show that $x_n y_n \to \infty$. Choose a number δ such that $0 < \delta < x$. (Alternatively, one might want to define $\delta = x/2$.)

Let U be a neighborhood of ∞, and choose a number w such that $(w, \infty) \subseteq U$. We need to show that for n sufficiently large we have $x_n y_n > w$. Using the fact that (δ, ∞) is a neighborhood of x, choose a natural number N_1 such that whenever $n \geq N_1$, we have $x_n > \delta$; and using the fact that $(w/\delta, \infty)$ is a neighborhood of ∞, choose a natural number N_2 such that whenever $n \geq N_2$, we have $y_n > w/\delta$. Define N to be the larger of the two numbers N_1 and N_2. Then for all natural numbers $n \geq N$ we have $x_n y_n > \delta w/\delta = w$; therefore, for all $n \geq N$ we have $x_n y_n \in U$.

■ *Proof of Part (4) When y Is a Real Number.* Since $x_n/y_n = x_n(1/y_n)$ for each n, the result will follow at once from part (3) when we have shown that $1/y_n \to 1/y$. Now in order for x/y to be defined, we must have $y \neq 0$. Choose a number δ such that $0 < \delta < |y|$:

Using the fact that $R \setminus [-\delta, \delta]$ is a neighborhood of y, choose a natural number N_1 such that whenever $n \geq N_1$, we have $|y_n| > \delta$. Then, of course, for all $n \geq N_1$, $1/y_n$ is defined and $|1/y_n| < 1/\delta$. Therefore, ignoring the finitely many numbers y_n for $n < N_1$, we can say that the sequence $(1/y_n)$ is bounded. The fact that $1/y_n - 1/y \to 0$ now follows from Theorem 4.5.1 in view of the identity $1/y_n - 1/y = (y - y_n)/y_n y$, which holds for all $n \geq N_1$, the boundedness of the sequence $(1/y_n y)$, and the fact that $y - y_n \to 0$. ∎

■ *Proof of Part (4) When $y = \infty$.* Because of the identity $x_n/y_n = x_n(1/y_n)$ and part (3) of the theorem, all we have to show is that $1/y_n \to 0$. Let $\varepsilon > 0$. Using the fact that $(1/\varepsilon, \infty)$ is a neighborhood of ∞, choose a natural number N such that whenever $n \geq N$, we have $y_n > 1/\varepsilon$. Then for all $n \geq N$ we see that $-\varepsilon < 0 < 1/y_n < \varepsilon$. ∎

4.6 EXERCISES

1. Find several cases of Theorem 4.5.2 that were not proved, and prove them yourself.

2. Given that $x_n \to x$ and y is a partial limit of (y_n), prove that $x + y$ is a partial limit of $(x_n + y_n)$.

3. State and prove some results like exercise 2 which refer to subtraction, multiplication, and division.

4. Give an example of a pair of sequences (x_n) and (y_n) and partial limits x and y of (x_n) and (y_n), respectively, such that $x + y$ is not a partial limit of $(x_n + y_n)$.

5. Give an example of two divergent sequences (x_n) and (y_n) such that $x_n + y_n \to 3$.

6. Give examples of pairs of sequences (x_n) and (y_n) such that $x_n \to 0$ and $y_n \to \infty$ and which satisfy the following conditions:
   (a) $x_n y_n \to 0$.
   (b) $x_n y_n \to 6$.
   (c) $x_n y_n \to \infty$.
   (d) $(x_n y_n)$ is bounded but has no limit.

7. Is it true that if both the sequences (x_n) and $(x_n + y_n)$ are convergent, then (y_n) must be convergent?

▪ 8. Given that $x_n \to 0$, prove that $(x_1 + x_2 + x_3 + \cdots + x_n)/n \to 0$.

▪ 9. Repeat exercise 8 with 0 replaced by a real number x.

10. Given that $x_n - y_n \to 0$ and that at least one of the sequences (x_n) and (y_n) does not have 0 as a partial limit, prove that neither of the sequences (x_n) and (y_n) can have 0 as a partial limit. *Hint:* Use exercise 2.

11. Given that $x_n - y_n \to 0$ and that at least one of the sequences (x_n) and (y_n) does not have 0 as a partial limit, prove that $x_n/y_n \to 1$.

12. Give an example to show that in exercise 11 the given condition about 0 not being a partial limit is really needed.

13. Given that $x_n/y_n \to 1$ and that at least one of the sequences (x_n) and (y_n) is bounded, prove that $x_n - y_n \to 0$. Give an example to show that this result can fail if the sequences (x_n) and (y_n) are unbounded.

14. Suppose that (x_n) and (y_n) are given sequences, that $y_n \to 0$, and that for each n we have $z_n = x_n - y_n$. Prove that the sequences (x_n) and (z_n) have the same partial limits.

▪ 15. Suppose that (x_n) and (y_n) are given sequences, that $y_n \to 1$, and that for each n we have $z_n = x_n y_n$. Prove that the sequences (x_n) and (z_n) have the same partial limits.

4.7 SEQUENCES IN A SET; USING SEQUENCES TO DESCRIBE CLOSURE

4.7.1 Theorem. Suppose that $A \subseteq \mathbf{R}$, that (x_n) is a sequence in the set A, and that x is a partial limit of (x_n). Then $x \in \bar{A}$.

■ *Proof.* Let U be a neighborhood of x. Since (x_n) is frequently in U, we can certainly choose one natural number n such that $x_n \in U$. Obviously, $x_n \in U \cap A$, and therefore, $U \cap A \neq \varnothing$. ■

4.7.2 Theorem. Suppose that $A \subseteq \boldsymbol{R}$, that (x_n) is a sequence which is frequently in the set A, and that $x_n \to x$. Then $x \in \bar{A}$.

■ *Proof.* Let U be a neighborhood of x. Choose a natural number N such that whenever $n \geq N$, we have $x_n \in U$. Now using the fact that (x_n) is frequently in A, choose a natural number $n \geq N$ such that $x_n \in A$. Obviously, $x_n \in U \cap A$, and therefore, $U \cap A \neq \varnothing$. ■

4.7.3 Theorem. Suppose $A \subseteq \boldsymbol{R}$ and that $x \in \bar{A}$. Then there exists a sequence (x_n) in A such that $x_n \to x$.

■ *Proof.* We start by observing that for every natural number n, since $(x - 1/n, x + 1/n)$ is a neighborhood of x, we must have $(x - 1/n, x + 1/n) \cap A \neq \varnothing$. For every natural number n, choose a member (which we shall call x_n) of the set $(x - 1/n, x + 1/n) \cap A$. This choice defines a sequence (x_n) in A, and for every $n \in \boldsymbol{Z}^+$ we have $|x_n - x| < 1/n$. It now follows that $x_n \to x$. [See Exercise 4.4(15) and Example 4.3.6(3).] ■

4.7.4 Theorem. Suppose $A \subseteq \boldsymbol{R}$. Then the following two conditions are equivalent:

(1) A is unbounded above.

(2) There exists a sequence (x_n) in A such that $x_n \to \infty$.

■ *Proof.* To prove that $(2) \Rightarrow (1)$, assume that (2) holds. Choose a sequence (x_n) in A such that $x_n \to \infty$. It follows from Theorem 4.3.12 that (x_n) is unbounded above, and therefore, A is unbounded above.

Now to prove that $(1) \Rightarrow (2)$, assume that (1) holds. For every natural number n we observe that n is not an upper bound of A; and we choose a member of A, which we call x_n, such that $x_n > n$. This choice defines a sequence (x_n) in A, and it is easy to show [see Exercise 4.4(18)] that $x_n \to \infty$. ■

4.8 EXERCISES

■ **1.** Given $A \subseteq \boldsymbol{R}$, prove that A is unbounded below iff there exists a sequence (x_n) in A such that $x_n \to -\infty$.

■ **2.** Given a set A of real numbers and an upper bound α of A, prove that $\alpha = \sup A$ iff there exists a sequence (x_n) in A such that $x_n \to \alpha$.

3. Given a sequence (x_n) which is frequently in a set A and given a partial limit x of (x_n), is it necessarily true that $x \in \bar{A}$?

4. Given $A \subseteq \boldsymbol{R}$, prove that A is closed iff no sequence in the set A can have a limit in $\boldsymbol{R} \backslash A$. *Hint:* The condition that A be closed is equivalent to saying that $\bar{A} = A$.

5. Given $A \subseteq \boldsymbol{R}$, prove that A is closed iff no sequence which is frequently in the set A can have a limit in $\boldsymbol{R} \backslash A$.

6. Given $A \subseteq \boldsymbol{R}$, prove that A is closed iff no sequence in the set A can have a partial limit in $\boldsymbol{R} \backslash A$.

7. Given $A \subseteq \boldsymbol{R}$, prove that A is both closed and bounded iff no sequence in the set A can have a limit in $[-\infty, \infty] \backslash A$.

8. Can you state and prove any more results of this type?

9. Given $A \subseteq \boldsymbol{R}$ and $x \in \boldsymbol{R}$, prove that $x \in \mathcal{L}(A)$ iff there exists a sequence (x_n) in A such that x does not belong to the range of (x_n) and $x_n \to x$. *Hint:* $x \in \mathcal{L}(A)$ iff $x \in \overline{A \backslash \{x\}}$.

PROPERTIES OF SEQUENCES WHICH DEPEND UPON COMPLETENESS

In the next five sections we come face to face with some of the most powerful results about sequences, these being the results which depend on the *completeness* of the real number system $\boldsymbol{R}$. In later chapters you will see a number of important properties of continuous functions, derivatives, and integrals, which also depend on the completeness of the real number system, and you will see how very useful the present results on sequences will be when we come to prove those theorems. The central theorem of Section 4.9 is Theorem 4.9.3. Notice how we make use of the Bolzano-Weierstrass theorem in the proof of this result.

4.9 THE EXISTENCE OF PARTIAL LIMITS

4.9.1 Theorem. Suppose that (x_n) is a sequence of numbers and that x is a real number which is a limit point of the range of (x_n). Then x must be a partial limit of (x_n).

■ *Proof.* Let U be a neighborhood of x. Using Theorem 3.10.5, we see that U contains infinitely many different points of the set $\{x_n \mid n \in \boldsymbol{Z}^+\}$ and therefore that U must contain points of the form x_n for infinitely many natural numbers n. It follows at once that x must be a partial limit of (x_n). ■

4.9.2   Remark.   It is by no means true that a partial limit of a sequence (x_n) has to be a limit point of the range of (x_n). To see an extreme case where this result fails, look at a constant sequence: Suppose that $x \in \mathbf{R}$, and define $x_n = x$ for every $n \in \mathbf{Z}^+$. Then x is a partial limit (in fact, the *limit*) of the sequence (x_n). But the range of (x_n) is the finite set $\{x\}$, which has no limit points at all.

4.9.3   Theorem.   Suppose that H is a closed, bounded set of real numbers and that (x_n) is a sequence which is frequently in the set H. Then (x_n) must have at least one partial limit that belongs to the set H.

■ *Proof.*   Define $A = H \cap \{x_n \mid n \in \mathbf{Z}^+\}$. In order to prove the theorem, we shall consider two cases.

Case 1: The set A is finite.   In this case we use a form of what is known as the **pigeonhole principle**. This principle says that if infinitely many pigeons have to share a finite number of nests, then at least one nest will have to hold infinitely many pigeons. There are infinitely many natural numbers n for which x_n lies in the finite set A, and there must therefore exist a point x of A such that $x_n = x$ for infinitely many natural numbers n. Such a number x will certainly be a partial limit of the sequence (x_n).

Case 2: The set A is infinite.   Since H is bounded, A must be bounded; and it therefore follows from the Bolzano-Weierstrass limit point theorem (Theorem 3.13) that the set A must have at least one real limit point. Choose $x \in \mathscr{L}(A)$. Since $A \subseteq H$, we see that $x \in \mathscr{L}(H)$; and because H is closed, it follows that $x \in H$. Finally, since $A \subseteq \{x_n \mid n \in \mathbf{Z}^+\}$, we see that x is a limit point of the range of (x_n); and it follows from Theorem 4.9.1 that x is a partial limit of (x_n).   ■

4.9.4   Theorem.

(1) Every sequence has at least one partial limit.

(2) Every bounded sequence has at least one real partial limit.

■ *Proof.*   From Corollary 4.3.14 it follows that if (x_n) is an unbounded sequence, then either $-\infty$ or ∞ is a partial limit of (x_n). Suppose, then, that a sequence (x_n) is bounded. Choose real numbers α and β such that $\alpha \le x_n \le \beta$ for all n. It follows at once from Theorem 4.9.3 that (x_n) must have a partial limit in the closed, bounded set $[\alpha, \beta]$.   ■

4.9.5   Theorem.   A sequence has a limit if and only if it has no more than one partial limit.

■ *Proof.*   We already know from Theorem 4.3.7 that no sequence which has a limit can have more than one partial limit. The *only if* part of the theorem is therefore clear. Suppose, now, that a given sequence (x_n) has no more than one partial limit. Then since Theorem 4.9.4 guarantees the existence of at least one partial limit of (x_n), we conclude that (x_n) has exactly one partial limit; we call it x. We need to show that $x_n \to x$, and to do so, we shall consider three cases.

Case 1: $x = \infty$. Since $-\infty$ is not a partial limit of (x_n), the sequence must be bounded below. Choose a lower bound α of (x_n). Let U be a neighborhood of ∞; and to obtain a contradiction suppose that (x_n) is frequently in $R \backslash U$. Choose a number w such that $(w, \infty) \subseteq U$. Then (x_n) is frequently in the closed, bounded set $[\alpha, w]$:

Therefore, by Theorem 4.9.3, (x_n) must have a partial limit in $[\alpha, w]$, which contradicts the fact that ∞ is the only partial limit of the sequence.

Case 2: $x = -\infty$. We leave this case as an exercise. You may use a proof which is analogous to the one we used in Case 1, or you may simply apply Case 1 to the sequence $(-x_n)$.

Case 3: $x \in R$. Since neither $-\infty$ nor ∞ can be a partial limit of (x_n), the sequence must be bounded. Choose a lower bound α and an upper bound β of (x_n). Let $\varepsilon > 0$; and to obtain a contradiction, we assume that (x_n) is frequently in $R \backslash (x - \varepsilon, x + \varepsilon)$. Then (x_n) is frequently in the closed, bounded set $[\alpha, \beta] \backslash (x - \varepsilon, x + \varepsilon)$; and by Theorem 4.9.3, (x_n) must have a partial limit in this set, contradicting the fact that x is the only partial limit of (x_n). ∎

4.10 LIMITS OF MONOTONE SEQUENCES

The message of Theorem 4.9.5 is that a sequence (x_n) will have a limit if it doesn't "oscillate" or "bounce around" too much. In this section we shall study one kind of sequence that doesn't oscillate at all, the monotone sequence. Not surprisingly, we shall see that every monotone sequence has a limit.

4.10.1 The Monotone Sequence Theorem.

(1) Every monotone sequence has a limit.

(2) A monotone sequence is convergent iff it is bounded.

(3) The limit of an increasing sequence is its supremum, and the limit of a decreasing sequence is its infimum.

■ *Proof.* Suppose that (x_n) is increasing, and define x to be the supremum of (x_n). We shall show that $x_n \to x$. Let U be a neighborhood of x. Choose a number $\alpha < x$ such that $(\alpha, x) \subseteq U$. Note that we can make this choice whether $x \in R$ or $x = \infty$.

Now using the fact that α is not an upper bound of (x_n), choose a natural number N such that $x_N > \alpha$. Then whenever $n \geq N$, we have $\alpha < x_N \leq x_n \leq x$; and therefore, whenever $n \geq N$, we have $x_n \in U$.

In the same way, one can prove that a decreasing sequence tends to its infimum. ∎

4.10.2 Some Examples.

In the examples that follow, we use Theorem 4.10.1 to discuss the limit behavior of some interesting monotone sequences.

(1) Suppose $c > 1$, and for each natural number n define $x_n = c^n$. The sequence (x_n) is clearly increasing, and from Theorem 4.10.1 we deduce that (x_n) has a limit. We shall prove that $x_n \to \infty$ as $n \to \infty$. We write $x = \lim x_n$; and to obtain a contradiction, we suppose that x is finite. Using the fact that $x/c < x$, choose n such that $x_n > x/c$. Then $x_{n+1} = cx_n > x$, contradicting part (3) of Theorem 4.10.1. [For another proof of this result, see Exercise 4.14(1).]

(2) Suppose $|c| < 1$, and for each natural number n define $x_n = c^n$. We shall show that $x_n \to 0$. Define $y_n = |x_n|$ for each n. Then in order to show that $x_n \to 0$, it is sufficient to show that $y_n \to 0$. The sequence (y_n) is clearly decreasing, and from Theorem 4.10.1 we deduce that (y_n) has a limit. Define $y = \lim y_n$; and to obtain a contradiction, assume that $y > 0$. Using the fact that $y/|c| > y$, choose n such that $y_n < y/|c|$. Then $y_{n+1} = |c|y_n < y$, contradicting part (3) of Theorem 4.10.1.

(3) Given any number $x \neq 1$ and any natural number n, the identity

$$1 + x + x^2 + \cdots + x^{n-1} = \frac{1 - x^n}{1 - x}$$

follows from the fact that

$$(1 - x)(1 + x + x^2 + \cdots + x^{n-1}) = 1 - x^n.$$

Using this identity, we see that for every natural number n the decimal expression

$$\frac{9}{10^1} + \frac{9}{10^2} + \frac{9}{10^3} + \cdots + \frac{9}{10^n}$$

is equal to $1 - 1/10^n$, which approaches 1 as $n \to \infty$.

(4) In this example we take a look at more general decimals. Suppose that (a_n) is a sequence in the set $\{0, 1, 2, 3, 4, 5, 6, 7, 8, 9\}$. For every natural number n we define

$$x_n = \frac{a_1}{10^1} + \frac{a_2}{10^2} + \frac{a_3}{10^3} + \cdots + \frac{a_n}{10^n}.$$

It is clear that (x_n) is an increasing sequence, and from example (3) we deduce that $x_n < 1$ for every n. The sequence (x_n) is therefore convergent, and its limit is the natural meaning of the **infinite decimal** whose nth digit is a_n for each n.

4.11 CAUCHY SEQUENCES

4.11.1 Introduction. The problem of determining whether or not a given sequence (x_n) converges to a given number x is quite different in character from the problem of testing a sequence (x_n) for convergence when we *don't* know what its limit might be. If all we want to know is whether (x_n) converges to a given number x, then the question before us is really one of inequalities; all we need to determine is whether or not the number $|x_n - x|$ can be made as small as we like by making n sufficiently large. On the other hand, if we want to know whether or not a given sequence (x_n) is convergent, and we have no idea what its limit might be, then the problem before us is considerably more difficult. Now we have to ask what sort of property the sequence must have in order to guarantee the *existence* of a number x such that $x_n \to x$. For monotone sequences the solution to this problem is provided by Theorem 4.10.1, which tells us that a monotone sequence is convergent if and only if it is bounded; but, in general, we need something more than just boundedness. In this section we discuss a criterion for convergence of sequences that was discovered by Cauchy. The sequences that satisfy Cauchy's criterion are called *Cauchy sequences;* and as we shall see in Theorem 4.11.3, the Cauchy sequences are precisely the sequences that converge. It is worth noting, however, that just as we needed the completeness property of $\mathbf{R}$ to prove the convergence of all bounded monotone sequences, we will need it again when we prove the convergence of all Cauchy sequences. Intuitively speaking, a Cauchy sequence is a sequence that "wants" to converge, and all it needs is a place to converge to. Had our number system been incomplete, there would have been Cauchy sequences that were divergent, simply because the points to which they "should" have converged were missing.

4.11.2 Definition of a Cauchy Sequence. A sequence (x_n) is said to be a **Cauchy sequence** if for every number $\varepsilon > 0$ there exists a natural number N such that for all natural numbers m and n satisfying $m \geq N$ and $n \geq N$, we have $|x_m - x_n| < \varepsilon$.

4.11.3 Theorem. A sequence (x_n) is convergent iff it is a Cauchy sequence.

■ *Proof.* First, we shall do Exercise 4.4(21) by showing that every convergent sequence must be a Cauchy sequence. Suppose that (x_n) is convergent with limit x. Let $\varepsilon > 0$. Choose a natural number N such that whenever $n \geq N$, we have $|x_n - x| < \varepsilon/2$. Then given any two natural numbers m and n such that $m \geq N$ and $n \geq N$, we have

$$|x_m - x_n| = |x_m - x + x - x_n| \leq |x_m - x| + |x - x_n| < \frac{\varepsilon}{2} + \frac{\varepsilon}{2} = \varepsilon.$$

Now in order to show that every Cauchy sequence must converge, assume that (x_n) is a Cauchy sequence. For convenience, we shall divide the proof that (x_n) converges into two steps.

Step 1.   We begin by showing that (x_n) is bounded. Choose a natural number N such that whenever $m \geq N$ and $n \geq N$, we have $|x_m - x_n| < 1$. Then for all $n \geq N$ we have $|x_n - x_N| < 1$, and it follows that for all $n \geq N$ we have $x_n \in (x_N - 1, x_N + 1)$. The range of (x_n) is therefore included in $\{x_n \mid n \leq N\}$ $\cup$ $(x_N - 1, x_N + 1)$, which, being the union of two bounded sets, must be bounded.

Step 2.   We now show that (x_n) cannot have more than one partial limit. To obtain a contradiction, assume that x and y are two partial limits of (x_n) and that $x < y$:

Choose two numbers α and β such that $x < \alpha < \beta < y$. Now choose a natural number N such that whenever $m \geq N$ and $n \geq N$, we have $|x_m - x_n| < \beta - \alpha$. Using the facts that $(-\infty, \alpha)$ is a neighborhood of x and that x is a partial limit of (x_n), choose a natural number $n \geq N$ such that $x_n < \alpha$. Now whenever $m \geq N$, it follows from the inequality $|x_m - x_n| < \beta - \alpha$ that $x_m < x_n + \beta - \alpha < \alpha + \beta - \alpha = \beta$; and therefore, (x_n) is not frequently in the neighborhood (β, ∞) of y. This contradicts our assumption that y is a partial limit of (x_n).

In view of Theorem 4.9.5, it follows at once from these two steps that the sequence (x_n) must be convergent.   ∎

4.12   THE SET OF PARTIAL LIMITS OF A SEQUENCE

What does the set of partial limits of a sequence look like? It isn't empty, of course, (as long as we count all points in $[-\infty, \infty]$), but what else can one say? If you look again at the examples of Section 4.3.6, you will see that the set of partial limits of a given sequence can be infinite and even unbounded; but as the next theorem shows, not every subset of $[-\infty, \infty]$ can be the set of partial limits of a sequence of real numbers.

4.12.1   Theorem.   The set of real partial limits of a sequence is closed.

∎ *Proof.*   Suppose (x_n) is a given sequence, and define A to be the set of its real partial limits. Let $y \in \bar{A}$. We need to show that $y \in A$, that is, that y is a partial limit of (x_n). Let $\varepsilon > 0$. Using the fact that $(y - \varepsilon, y + \varepsilon)$ is a neighborhood of y, choose a number x in $A \cap (y - \varepsilon, y + \varepsilon)$. Since $(y - \varepsilon, y + \varepsilon)$ is also a neighborhood of x, the sequence (x_n) must be frequently in $(y - \varepsilon, y + \varepsilon)$. This shows that y is a partial limit of (x_n).   ∎

4.12.2 Theorem. Every sequence has both a largest and a smallest partial limit.

■ *Proof.* We shall prove that every sequence has a largest partial limit. The proof that every sequence has a smallest partial limit is analogous and will be left as an exercise. Suppose (x_n) is any sequence. If ∞ happens to be a partial limit of (x_n), then ∞ is obviously the largest such partial limit; and the theorem is clear. From now on, we shall assume that ∞ is not a partial limit of (x_n)—in other words, that (x_n) is bounded above. There are now two possibilities: Either $x_n \to -\infty$, in which case the theorem is again obvious, or (x_n) has some real partial limits. In the latter case we define E to be the set of all real partial limits of (x_n), and notice that E is a nonempty set of real numbers and that E is bounded above. Since E is closed, we have sup $E \in E$, and therefore, sup E is the largest partial limit of (x_n).

■

4.12.3 Upper and Lower Limits of a Sequence. Given any sequence (x_n), the largest and smallest partial limits of (x_n) are called the **upper limit** and **lower limit** of (x_n) and are denoted as limsup x_n and liminf x_n, respectively. From Theorem 4.9.5 we see that a sequence has a limit if and only if its upper and lower limits are equal.

4.13 CANTOR'S INTERSECTION THEOREM

The Cantor intersection theorem is one more in a long line of theorems we have seen up to this point which make use of the completeness of our number system. Unlike some of the others, though, this result refers to a sequence of *sets* instead of a sequence of numbers. The theorem tells us that, under certain circumstances, if (A_n) is a sequence of nonempty sets, and if for every n we have $A_{n+1} \subseteq A_n$, then the intersection of the sets A_n will be nonempty.

4.13.1 Definition of a Contracting (or Nested) Sequence of Sets. A sequence (A_n) of sets is said to be **contracting** (or *nested*) if for every natural number n we have $A_{n+1} \subseteq A_n$.

4.13.2 Some Examples. The following are some simple examples of contracting sequences (A_n). Notice that although the sets A_n are nonempty in each of these examples, the set $\cap_{n=1}^{\infty} A_n$ is sometimes empty.

(1) For each natural number n we define $A_n = [0, 1/n]$. Clearly, the sequence (A_n) is contracting, and $\cap_{n=1}^{\infty} A_n = \{0\}$.

(2) For each natural number n we define $A_n = (0, 1/n)$. Again, (A_n) is a contracting sequence, but this time we see that $\cap_{n=1}^{\infty} A_n = \emptyset$.

(3) For each natural n we define $A_n = [n, \infty)$. Once again, the sequence (A_n) is contracting, and $\cap_{n=1}^{\infty} A_n = \emptyset$.

Notice that in example (2) the sets A_n are not closed and that in example (3) they are not bounded.

The following result is known as the **Cantor intersection theorem.** When the sets A_n are intervals, the theorem is commonly known as the **nested interval theorem**.

4.13.3 Theorem. Suppose that (A_n) is a contracting sequence of nonempty, closed, bounded subsets of R. Then there exists a real number x which belongs to every one of the sets A_n. In other words, the set $\bigcap_{n=1}^{\infty} A_n$ is not empty.

■ *Proof.* This proof is really a solution to Exercise 3.6(10). For each natural number n we define x_n to be the least member of the set A_n. The sequence (x_n) is obviously increasing and is bounded above by the greatest member of A_1. Define $x = \lim x_n$. We complete the proof by showing that x belongs to every one of the sets A_n. Let m be any natural; and to obtain a contradiction, assume that $x \notin A_m$. Using the fact that $R \backslash A_m$ is a neighborhood of x, choose a natural number N such that whenever $n \geq N$, we have $x_n \in R \backslash A_m$. Now choose a natural number n which is greater than both of the numbers N and m. Then since $n > N$, we have $x_n \in R \backslash A_m$; and since $n > m$, we have $x_n \in A_n \subseteq A_m$, which is a contradiction. ■

A sequence (A_n) of sets is said to be **expanding** if for every natural number n we have $A_n \subseteq A_{n+1}$. The next theorem is an analogue of Cantor's theorem for expanding sequences of open sets.

4.13.4 Theorem. Suppose that H is a closed, bounded subset of R and that (U_n) is an expanding sequence of open sets such that $H \subseteq \bigcup_{n=1}^{\infty} U_n$. Then it is possible to find a natural number N such that $H \subseteq U_N$.

■ *Proof.* For every natural number n we define $H_n = H \backslash U_n$. We see that each set H_n is closed and bounded and that for each n we have $H_{n+1} \subseteq H_n$. Now given any point $x \in H$, because $H \subseteq \bigcup_{n=1}^{\infty} U_n$, we know that there is a natural number n such that $x \in U_n$; and for this n we have $x \notin H_n$. Therefore, $\bigcap_{n=1}^{\infty} H_n = \emptyset$; and it follows from Theorem 4.13.3 that for some natural number N we must have $H_N = \emptyset$. In other words, for some N we must have $H \subseteq U_N$. ■

4.14 EXERCISES

1. In Section 4.10.2 we used the monotone sequence theorem to show that if $c > 1$ and $x_n = c^n$ for each natural number n, then $x_n \to \infty$ as $n \to \infty$. In this exercise we shall explore an alternative proof that does not make use of the monotone sequence theorem. Define $\delta = c - 1$. Use Example 2.6.5(2) to show that $x_n \geq 1 + n\delta$ for each n. Deduce that $x_n \to \infty$.

2. Given that $0 < p < 1$ and that for each natural number n

$$x_n = 1 + p^1 + p^2 + \cdots + p^{n-1},$$

prove that the sequence (x_n) is increasing and bounded and, therefore, convergent. What is the limit of this sequence?

3. Suppose that (x_n) is a given sequence, that $x_1 = 0$, and that for every natural number n we have

$$8x_{n+1}^3 = 6x_n + 1.$$

   (a) Work out the first few terms of this sequence.
   (b) Using mathematical induction, prove that $x_n < 1$ for all n.
   (c) Using mathematical induction, prove that the sequence (x_n) is increasing.
   (d) Prove that the sequence (x_n) is convergent.
   *▪(e) Prove that $x_n \to \cos \pi/9$ as $n \to \infty$.

4. Given that x is a partial limit of (x_n) and that $y < x$, prove that there must be infinitely many natural numbers n for which $x_n > y$.

5. Given that $x = \limsup x_n$ and that $y > x$, prove that there can only be finitely many natural numbers n for which $x_n > y$. *Hint:* Say why (x_n) must be bounded above. Then show that if z is any upper bound of (x_n) and $z > y$, then (x_n) is not frequently in the set $[y, z]$.

6. Suppose that (x_n) is a given sequence, that $x \in [-\infty, \infty]$, and that the following two conditions hold:
   (i) For every number $y > x$ the set $\{n \in \mathbf{Z}^+ \mid x_n > y\}$ is finite.
   (ii) For every number $y < x$ the set $\{n \in \mathbf{Z}^+ \mid x_n > y\}$ is infinite.
   Prove that x is a partial limit of (x_n), and then proceed to show that $x = \limsup x_n$.

7. Suppose that (x_n) is a given sequence. For each $n \in \mathbf{Z}^+$ we define

$$y_n = \sup\{x_m \mid m \geq n\}.$$

   (a) Why is the sequence (y_n) decreasing?
   (b) We now define $y = \lim y_n$. Prove that $y = \limsup x_n$. *Hint:* Use exercise 6.

▪ 8. Given $H \subseteq \mathbf{R}$, prove that the following three conditions are equivalent.
   (a) H is closed and bounded.
   (b) Every sequence in H has a partial limit that belongs to H.
   (c) No sequence in H can have a limit outside H.
   Hint: Show that (a) $\Rightarrow$ (b), (b) $\Rightarrow$ (c), and (c) $\Rightarrow$ (a).
   After trying this exercise, turn ahead to Section 6.7.1, where you will find a discussion on the topic of closed, bounded sets. The properties of these sets will be very important to us in our study of continuous functions.

■ **9.** Given a contracting sequence (H_n) of closed, bounded sets, given $x_n \in H_n$ for every n, and given any partial limit x of (x_n), prove that $x \in \cap_{n=1}^{\infty} H_n$.

*■ **10.** Is it true that for every closed, bounded set H of real numbers, there exists a sequence (x_n) in H such that H is the set of partial limits of (x_n)?

4.15*   OPEN COVERS AND THE HEINE–BOREL THEOREM

This last section of the chapter may be used to motivate the notion of *compactness*, which plays an important role when the theory of limits and convergence is applied to more general spaces. This section is optional and may be omitted without loss of continuity.

4.15.1   Definition of a Cover.   A family $\mathscr{C}$ of subsets of $\boldsymbol{R}$ is said to be a *cover* of a given set A if A is included in the union of all the members of $\mathscr{C}$.

Equivalently, a family $\mathscr{C}$ of subsets of $\boldsymbol{R}$ is a cover of a given set A if for every point $x \in A$ there exists a member U of $\mathscr{C}$ such that $x \in U$. If $\mathscr{C}$ is a cover of A, then we sometimes say that the family $\mathscr{C}$ *covers* A. In this section we are particularly interested in families of *open* sets which cover a given set A. We call these families **open covers** of A.

4.15.2   Examples.   In each of the following examples $\mathscr{C}$ is an open cover of A.

(1) A is any set of real numbers, and $\mathscr{C} = \{\boldsymbol{R}\}$.

(2) $A = [0, 6]$, and $\mathscr{C} = \{(x - 1, x + 1) \mid 0 \le x \le 6\}$.

(3) $A = [0, 6]$, and $\mathscr{C} = \{(x - 1, x + 1) \mid x \in \{0, 1, 2, 3, 4, 5, 6\}\}$.

(4) $A = (0, 1]$, and $\mathscr{C} = \{(x/2, 2x) \mid x > 0\}$.

(5) A is the range of a strictly increasing sequence (x_n), and $\mathscr{C} = \{(x_{n-1}, x_{n+1}) \mid n \ge 2\} \cup \{(x_1 - 1, x_2)\}$.

4.15.3   The Heine-Borel Theorem.   Suppose H is any set of real numbers. The following two conditions are equivalent:

(1) H is closed and bounded.

(2) Given any open cover $\mathscr{C}$ of H, there exists a finite subfamily $\mathscr{D}$ of $\mathscr{C}$ such that $\mathscr{D}$ is also a cover of H.

■ ***Proof.***   To prove that $(2) \Rightarrow (1)$, assume that (2) holds. We need to prove that H is closed and bounded. In order to prove that H is bounded, we define $\mathscr{C} = \{(x - 1, x + 1) \mid x \in H\}$. Noting that $\mathscr{C}$ is an open cover of H, choose a finite subfamily $\mathscr{D}$ of $\mathscr{C}$ such that $\mathscr{D}$ covers H. So H is included in the union of finitely many bounded intervals, and it follows that H is bounded. Now to prove that H is closed, suppose that $\alpha \in \mathbf{R}\backslash H$. We must show that α does not lie in $\bar{H}$. We define

$$\mathscr{C} = \{(-\infty, \alpha - \delta) \cup (\alpha + \delta, \infty) \mid \delta > 0\}.$$

Noting that $\mathscr{C}$ is an open cover of H, choose a finite subfamily $\mathscr{D}$ of $\mathscr{C}$ such that $\mathscr{D}$ covers H. Define ε to be the least of the finitely many positive numbers δ for which the set $(-\infty, \alpha - \delta) \cup (\alpha + \delta, \infty)$ lies in $\mathscr{D}$. Since $H \subseteq (-\infty, \alpha - \varepsilon) \cup (\alpha + \varepsilon, \infty)$, we see that $(\alpha - \varepsilon, \alpha + \varepsilon)$ is a neighborhood of α which does not intersect with H; and therefore, $\alpha \notin \bar{H}$.

Now to prove that $(1) \Rightarrow (2)$, assume that (1) holds. Let $\mathscr{C}$ be an open cover of H; and to obtain a contradiction, assume that no finite subfamily of $\mathscr{C}$ covers H. Since H is a nonempty, closed and bounded set, H has a least member. Call this least member a. Define E to be the set of all those points x in H for which it is possible to find a finite subfamily $\mathscr{D}$ of $\mathscr{C}$ such that $\mathscr{D}$ covers $(-\infty, x] \cap H$. Now a lies in some member W of $\mathscr{C}$; and since $\{W\}$ is a finite subfamily of $\mathscr{C}$ and $\{W\}$ covers $(-\infty, a] \cap H$, we deduce that $a \in E$. Therefore, E is nonempty. Define $\alpha = \sup E$. Since $\alpha \in \bar{E}$ and $E \subseteq H$ and H is closed, we see that $\alpha \in H$. We now choose a member U of $\mathscr{C}$ such that $\alpha \in U$, and we choose $\delta > 0$ such that $(\alpha - \delta, \alpha + \delta) \subseteq U$. Using the fact that $\alpha - \delta$ is not an upper bound of E, choose a number $b \in E$ such that $b > \alpha - \delta$. Now choose a finite subfamily $\mathscr{D}$ of $\mathscr{C}$ which covers the set $(-\infty, b] \cap H$, and observe that the finite family $\mathscr{D} \cup \{U\}$ covers $(-\infty, \alpha + \delta) \cap H$:

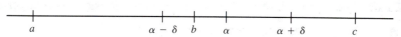

Since the finite family $\mathscr{D} \cup \{U\}$ does not cover H, it follows that H must contain some points $x \geq \alpha + \delta$. Define c to be the least member of the set $H \cap [\alpha + \delta, \infty)$. Choose a member V of $\mathscr{C}$ such that $c \in V$, and observe that the finite family $\mathscr{D} \cup \{U, V\}$ covers $(-\infty, c] \cap H$. But this fact implies that $c \in E$, contradicting the fact that $\alpha = \sup E$. ∎

Chapter 5

Limits of Functions

5.1 INTRODUCTION

The subject of limits of functions is the starting point of almost all elementary calculus courses, for it is limits of functions that we use when we define derivatives. In this book, however, in which our approach to calculus is somewhat more advanced, we have taken things a little more slowly, delaying limits of functions until after we have looked at some facts about the real number system, including its topological properties, and the theory of sequences of real numbers. Now that this has been done, we are ready to take a closer look at limits of functions than would have been possible in an elementary calculus course. Let us begin by recalling the notation traditionally employed for functions. If S is a set of real numbers and f is a real-valued function with domain S, then we write $f: S \to \boldsymbol{R}$ to mean that the real number $f(x)$ (called the *value of f at x*) is defined for every point $x \in S$. A more precise approach to the concept of a function can be found in Appendix A.

Now roughly speaking, the condition $f(x) \to \alpha$ as $x \to a$ means that we can make $f(x)$ as close as we like to α by making $x \neq a$ and x sufficiently close to a.

Notice how this idea makes no demand at all about the value of $f(x)$ when $x = a$. All the definition demands is that $f(x)$ should be as close as we like to α for values of x in the domain of f, *unequal* to a and lying close enough to a. This means that $f(a)$ itself might be very far away from α, and in fact, there is no reason why $f(a)$ should even be defined. To understand why we are so insistent that the number $f(a)$ should play no role at all in the limit concept, $f(x) \to \alpha$ as $x \to a$, let us take a brief look at the important limit on which the definition of a derivative is based.

In order to define the derivative $f'(a)$ of a function f at a point a in the usual form,

$$f'(a) = \lim_{x \to a} \frac{f(x) - f(a)}{x - a},$$

we are really defining a new function q by the equation

$$q(x) = \frac{f(x) - f(a)}{x - a}$$

and then saying that $q(x) \to f'(a)$ as $x \to a$. Notice that the one value of x at which this function q is definitely not defined is $x = a$.

So to repeat what we said above: When we take a limit saying that $f(x) \to \alpha$ as $x \to a$, we don't make any demands about the value (if any) of the number $f(a)$. If we denote the domain of f by S, then in order to say that $f(x) \to \alpha$ as $x \to a$, we do not need to assume that $a \in S$. Roughly speaking, the condition $f(x) \to \alpha$ as $x \to a$ should say that given any neighborhood V of α, we can make $f(x) \in V$ for all points $x \in S \backslash \{a\}$ which lie sufficiently close to a.

5.2 AN UNSUCCESSFUL ATTEMPT AT DEFINING THE LIMIT OF A FUNCTION

Suppose that $S \subseteq R$, that $f: S \to R$, and that a and α are given numbers. In view of the introduction, it seems natural to define a limit by saying that $f(x) \to \alpha$ as $x \to a$ when for every neighborhood V of α there exists a neighborhood U of a such that for every point $x \in U \cap S \backslash \{a\}$ we have $f(x) \in V$. This definition is pretty good, but it is not quite good enough. To see what can go wrong, look at the following example.

5.2.1 Example. We take $S = [0, 1] \cup \{2\}$:

And we take *any* function $f: S \to R$. We shall now prove that according to the previous definition, $f(x) \to 17$ as $x \to 2$. Let V be a neighborhood of 17. Define $U = (1, \infty)$, and notice that U is a neighborhood of 2. Furthermore, the set $S \cap U \backslash \{2\}$ is empty, and it therefore follows at once that for every point $x \in U \cap S \backslash \{2\}$ we have $f(x) \in V$. In case you find this last claim confusing, let us follow our practice of writing the denial of a statement in order to help us understand what the statement is really saying: The denial of the statement that $f(x) \in V$ for every point $x \in U \cap S\backslash\{2\}$ says that there should exist a point $x \in U \cap S\backslash\{2\}$ such that $f(x) \notin V$. But this denial would require, in particular, that there should exist a point $x \in U \cap S\backslash\{2\}$; and as we have said, the set $U \cap S\backslash\{2\}$ is empty. So we have proved that $f(x) \rightarrow 17$ as $x \rightarrow 2$. Surely, it can't be a satisfactory state of affairs if no matter what function f we take on this set S, we have $f(x) \rightarrow 17$ as $x \rightarrow 2$. Now let us go a little further and observe that this proof that $f(x) \rightarrow 17$ as $x \rightarrow 2$ has nothing to do with the number 17. We could just as well have proved that $f(x) \rightarrow -3026$ as $x \rightarrow 2$. We therefore come to the absurd conclusion that given any function f whatsoever defined on this set S, and given any number α, we have $f(x) \rightarrow \alpha$ as $x \rightarrow 2$.

This example should serve as a warning that given $f: S \rightarrow R$ and given numbers a and α, before we speak about the notion $f(x) \rightarrow \alpha$ as $x \rightarrow a$, we should make sure that there isn't a neighborhood U of a for which the set $U \cap S\backslash\{a\}$ is empty. In other words, we need to know that for every neighborhood U of a we have $U \cap S\backslash\{a\} \neq \emptyset$, and this is precisely the condition that a should be a limit point of the set S.

5.3   A SUCCESSFUL DEFINITION OF LIMIT OF A FUNCTION

5.3.1   The Definition.   Suppose that $f: S \rightarrow R$, where $S \subseteq R$; suppose that a and α are points in $[-\infty, \infty]$; and suppose that the point a is a limit point of S.[1] Then we say that $f(x) \rightarrow \alpha$ as $x \rightarrow a$ if for every neighborhood V of α there exists a neighborhood U of a such that whenever $x \in U \cap S\backslash\{a\}$, we have $f(x) \in V$.

5.3.2   Some Equivalent Forms of the Definition.   The definition given in Section 5.3.1 is a fairly general formulation of the definition of a limit and includes a number of important special cases. For example, a and α might both be real numbers, or one or both of them might be ∞ or $-\infty$. In this section we examine some of the cases that can occur, and in each of them we give a specific form of the definition which can be readily applied in that case.

(1)   Suppose that $f: S \rightarrow R$, where $S \subseteq R$; suppose that a is a limit point of S; and suppose that α is a real number. Then the following two conditions are equivalent:

(a) $f(x) \rightarrow \alpha$ as $x \rightarrow a$.

(b) For every number $\varepsilon > 0$ there exists a neighborhood U of a such that whenever $x \in U \cap S\backslash\{a\}$, we have $f(x) \in (\alpha - \varepsilon, \alpha + \varepsilon)$, that is, $|f(x) - \alpha| < \varepsilon$.

---

[1] Don't forget that ∞ is a limit point of S iff S is unbounded above and that $-\infty$ is a limit point of S iff S is unbounded below. Take another look at Sections 3.15.2 and 3.15.3 if you need a review.

■ *Proof.* To prove that (a) $\Rightarrow$ (b), suppose that $f(x) \to \alpha$ as $x \to a$, and let $\varepsilon > 0$. Then $(\alpha - \varepsilon, \alpha + \varepsilon)$ is a neighborhood of α, and it follows at once that a neighborhood U of a can be found satisfying the requirements of condition (b).

Now to prove that (b) $\Rightarrow$ (a), assume that (b) holds, and let V be a neighborhood of α. Choose $\varepsilon > 0$ such that $(\alpha - \varepsilon, \alpha + \varepsilon) \subseteq V$. Now choose a neighborhood U of a satisfying the requirements of condition (b), and observe that for every $x \in U \cap S\backslash\{a\}$ we must have $f(x) \in V$. ■

(2) Suppose that $f : S \to R$, where $S \subseteq R$, and that a is a limit point of S. Then the following two conditions are equivalent:

(a) $f(x) \to \infty$ as $x \to a$.

(b) For every number w there exists a neighborhood U of a such that for all points $x \in U \cap S\backslash\{a\}$ we have $f(x) > w$.

■ *Proof.* To prove that (a) $\Rightarrow$ (b), assume that (a) holds, and let w be any real number. Then (w, ∞) is a neighborhood of ∞, and it follows at once that a neighborhood U of a can be found satisfying the requirements of condition (b).

Now to prove that (b) $\Rightarrow$ (a), assume that (b) holds; and let V be a neighborhood of ∞. Choose a real number w such that $(w, \infty) \subseteq V$. Now using condition (b), choose a neighborhood U of a such that for all points $x \in U \cap S\backslash\{a\}$ we have $f(x) > w$; and note that for all points $x \in U \cap S\backslash\{a\}$ we must have $f(x) \in V$. ■

(3) Suppose that $f : S \to R$, where $S \subseteq R$; suppose that a is a real number which is a limit point of S; and suppose that $\alpha \in [-\infty, \infty]$. Then the following two conditions are equivalent:

(a) $f(x) \to \alpha$ as $x \to a$.

(b) For every neighborhood V of α there exists a number $\delta > 0$ such that whenever $x \in (a - \delta, a + \delta) \cap S\backslash\{a\}$, we have $f(x) \in V$.

Sometimes, it is convenient to write condition (b) in the following alternative way:

(b) For every neighborhood V of α there exists a number $\delta > 0$ such that whenever $x \in S\backslash\{a\}$ and $|x - a| < \delta$ we have $f(x) \in V$.

■ *Proof.* To prove that (a) $\Rightarrow$ (b), assume that (a) holds, and let V be a neighborhood of α. Choose a neighborhood U of a such that for all points $x \in U \cap S\backslash\{a\}$ we have $f(x) \in V$. Now choose a number $\delta > 0$ such that $(a - \delta, a + \delta) \subseteq U$. This number δ obviously satisfies the requirements of condition (b).

Now to prove that (b) $\Rightarrow$ (a), assume that (b) holds, and let V be a neighborhood of α. Choose a number $\delta > 0$ such that for all points $x \in (a - \delta, a + \delta) \cap S\backslash\{a\}$ we have $f(x) \in V$. Now define $U = (a - \delta, a + \delta)$. Then U is a neighborhood of a, and for all points $x \in U \cap S\backslash\{a\}$ we have $f(x) \in V$. ■

The following results are similar to the first three, and we leave their proofs as exercises.

(4) Suppose that $f: S \to R$, where $S \subseteq R$; suppose that ∞ is a limit point of S; and suppose that $\alpha \in [-\infty, \infty]$. Then the following two conditions are equivalent:

(a) $f(x) \to \alpha$ as $x \to \infty$.

(b) For every neighborhood V of α there exists a number w such that whenever $x \in (w, \infty) \cap S$, we have $f(x) \in V$.

(5) Suppose that $f: S \to R$, where $S \subseteq R$, and suppose that $-\infty$ is a limit point of S. Then the following two conditions are equivalent:

(a) $f(x) \to \infty$ as $x \to -\infty$.

(b) For every number w there exists a number u such that whenever $x \in (-\infty, u) \cap S$, we have $f(x) > w$.

(6) Suppose that $f: S \to R$, where $S \subseteq R$; suppose that a and α are real numbers; and suppose that the number a is a limit point of S. Then the following two conditions are equivalent:

(a) $f(x) \to \alpha$ as $x \to a$.

(b) For every number $\varepsilon > 0$ there exists a number $\delta > 0$ such that whenever $x \in S \backslash \{a\}$ and $|x - a| < \delta$, we have $|f(x) - \alpha| < \varepsilon$.

5.3.3 Some Examples. The examples of this section illustrate the idea of a limit. At the same time, they will acquaint you with some important techniques which will help you to understand the ideas contained in the coming chapters.

(1) Define $f(x) = x^2 + 1$ for all $x \in R$. We shall show that $f(x) \to 5$ as $x \to 2$. We shall use the version of the definition of a limit which is given in Section 5.3.2(6). Let $\varepsilon > 0$. We need to find a number $\delta > 0$ such that whenever $|x - 2| < \delta$ and $x \neq 2$, we have $|f(x) - 5| < \varepsilon$. Now given any number x,

$$|f(x) - 5| = |x^2 + 1 - 5| = |(x - 2)(x + 2)|.$$

Roughly speaking, the reason why $|f(x) - 5|$ is small whenever x is close to 2 is that $|x - 2|$ is small and $|x + 2|$ is not too large. In fact, if $|x - 2| < 1$, then $1 < x < 3$:

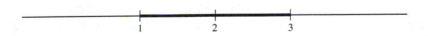

and it is clear that $|x + 2| < 5$.

This tells us how to complete the proof. We define δ to be the smaller of the numbers 1 and $\varepsilon/5$. Then whenever $|x - 2| < \delta$ and $x \neq 2$, we have

$$|f(x) - 5| = |(x - 2)(x + 2)| < \frac{5\varepsilon}{5} = \varepsilon.$$

(2) Define

$$f(x) = \frac{x - 1}{x^2 + 3}$$

for all $x \in \mathbf{R}$. We shall show that $f(x) \to \frac{1}{7}$ as $x \to 2$. Let $\varepsilon > 0$. Now for any number x we have

$$\left| f(x) - \frac{1}{7} \right| = \left| \frac{x - 1}{x^2 + 3} - \frac{1}{7} \right| = \frac{|(x - 2)(x - 5)|}{7(x^2 + 3)} \leq \frac{|(x - 2)(x - 5)|}{21}.$$

Roughly speaking, the reason why $|f(x) - \frac{1}{7}|$ is small when x is close to 2 is that $|x - 2|$ is small and $|x - 5|$ is not too large. In fact, if $|x - 2| < 1$, then it is easy to see that $|x - 5| < 4$, and this tells us how to complete the proof. We define δ to be the smaller of the numbers 1 and $\varepsilon/4$. Then whenever $x \neq 2$ and $|x - 2| < \delta$, we have

$$\left| f(x) - \frac{1}{7} \right| \leq \frac{|(x - 2)(x - 5)|}{21} < \frac{\varepsilon}{21} < \varepsilon.$$

(3) Define $f(x) = x/|x|$ for all numbers $x \neq 0$. In other words, $f(x) = 1$ whenever $x > 0$, and $f(x) = -1$ whenever $x < 0$. We shall show that f has no limit at 0. To obtain a contradiction, suppose that for some α we have $f(x) \to \alpha$ as $x \to 0$. Now α cannot be equal to both of the numbers 1 and -1. Assume for the moment that $\alpha \neq 1$, and define $V = \mathbf{R} \backslash \{1\}$. Then although V is a neighborhood of α, it is clear that for every number $\delta > 0$ there are numbers $x \in (0 - \delta, 0 + \delta) \backslash \{0\}$ for which $f(x) = 1$, and for all such numbers x we cannot have $f(x) \in V$:

This gives us the required contradiction in the case $\alpha \neq 1$. An analogous proof may be applied to the case $\alpha = 1$, taking $V = \mathbf{R} \backslash \{-1\}$. We leave the details as an exercise.

(4) Suppose f is the function of example (3), and define g to be the restriction of f to the interval $(0, \infty)$; in other words, $g(x) = f(x) = 1$ for all $x > 0$. Even though f has no limit at 0, it is clear that $g(x) \to 1$ as $x \to 0$. Now let us define h to be the restriction of f to the interval $(-\infty, 0)$; in other words, $h(x) = f(x) = -1$ for all $x < 0$. We see that $h(x) \to -1$ as $x \to 0$. This example suggests the idea of *limits from the left* and *limits from the right,* which we shall define in Section 5.3.4.

(5) Define $f(x) = x/(x - 3)^2$ for all $x \neq 3$. We shall show that $f(x) \to \infty$ as $x \to 3$. We shall use the version of the definition of a limit which is given in Section 5.3.2(2). Let w be any real number. We need to find a number $\delta > 0$ such that whenever $x \neq 3$ and $|x - 3| < \delta$, we have $f(x) > w$. Roughly speaking, the reason $f(x)$ is large when x is close to 3 is that $1/(x - 3)^2$ is large and x is not too small. In fact, if $|x - 3| < 1$, then we have $x > 2$, and this tells us how to complete the proof. We define $\delta = 1/(1 + |w|)$. Then whenever $x \neq 3$ and $|x - 3| < \delta$, we have

$$f(x) = \frac{x}{(x - 3)^2} > \frac{2}{\delta^2} = 2(1 + |w|)^2 > w.$$

(6) Define $f(x) = x/(x - 3)^2$ for all $x \neq 3$. We shall show that $f(x) \to 0$ as $x \to \infty$. To do this, we observe first that for each $x \neq 3$ we have

$$f(x) = \frac{x - 3 + 3}{(x - 3)^2} = \frac{1}{x - 3} + \frac{3}{(x - 3)^2}.$$

And roughly speaking, the reason why $f(x)$ is small for x large is that both terms on the right side of the latter identity are small. Let $\varepsilon > 0$. In order to guarantee that $1/(x - 3) < \varepsilon/2$, we require $x > 3 + 2/\varepsilon$; and in order to guarantee that $3/(x - 3)^2 < \varepsilon/2$, we require $(x - 3)^2 > 6/\varepsilon$. Now as long as $x \geq 4$, the latter inequality will hold when $x - 3 > 6/\varepsilon$; and this tells us how to complete the proof. We define w to be the larger of the two numbers 4 and $3 + 6/\varepsilon$. Then whenever $x > w$, we have

$$f(x) = \frac{1}{x - 3} + \frac{3}{(x - 3)^2} < \frac{\varepsilon}{2} + \frac{3\varepsilon}{6} = \varepsilon.$$

5.3.4   One-Sided Limits.

In examples (3) and (4) of the previous section we saw that if $f(x) = x/|x|$ for all $x \neq 0$, then f has no limit at 0, even though its restrictions to $(-\infty, 0)$ and $(0, \infty)$ do have limits. We saw, in fact, that if we define $g(x) = f(x)$ for $x > 0$, then $g(x) \to 1$ as $x \to 0$; and because of this, it is natural to say that $f(x) \to 1$ as $x \to 0$ *from the right*. Similarly, if we define $h(x) = f(x)$ for $x < 0$, then $h(x) \to -1$ as $x \to 0$; and it is therefore natural to say that $f(x) \to -1$ as $x \to 0$ *from the left*.

This discussion suggests the following precise definitions of one-sided limits.

Limits from the Left.   Suppose that $f: S \to R$, where $S \subseteq R$, and suppose that a and α are points in $[-\infty, \infty]$ such that a is a limit point of $S \cap (-\infty, a)$. Suppose that g is the function from $S \cap (-\infty, a)$ to R defined by $g(x) = f(x)$ for all $x \in S \cap (-\infty, a)$. In the event that $g(x) \to \alpha$ as $x \to a$, we say that $f(x) \to \alpha$ as $x \to a$ *from the left*; and we write $f(x) \to \alpha$ as $x \to a-$. See Exercise 5.4(6) for another way of looking at limits from the left.

Limits from the Right.   Suppose that $f: S \to R$, where $S \subseteq R$, and suppose that a and α are points in $[-\infty, \infty]$ such that a is a limit point of $S \cap (a, \infty)$. Suppose that g is the function from $S \cap (a, \infty)$ to R defined by $g(x) = f(x)$ for all $x \in S \cap (a, \infty)$. In the event that $g(x) \to \alpha$ as $x \to a$, we say that $f(x) \to \alpha$ as $x \to a$ *from the right*; and we write $f(x) \to \alpha$ as $x \to a+$.

5.3.5   Some Examples of Functions Which Fail to Have One-Sided Limits.

Roughly speaking, a function can fail to have a one-sided limit if it "wobbles" too much. Here are two examples of this type of behavior.

(1) We define $f(x) = 1$ if $x \in Q$ and $f(x) = -1$ if $x \in R \backslash Q$. By an argument similar to the one we used in example (3) of Section 5.3.3, we see that f does not have a limit, not even a one-sided limit, at any number.

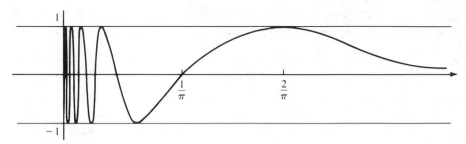

Figure 5.1

(2) See Figure 5.1. We define $f(x) = \sin(1/x)$ for all $x > 0$. Given any natural number n, we note that if $x = 1/n\pi$, then $f(x) = 0$; and if $x = 1/(n\pi + \pi/2)$, then $f(x) = \pm 1$. Since there are numbers x of both these forms in every neighborhood of 0, it follows once again by the sort of argument we used in Example 5.3.3(3) that f has no limit (from the right) at 0.

5.4 EXERCISES

■ **1.** Working directly from the definition of a limit, prove each of the following statements.

(a) $x^3 - 3x \to 2$ as $x \to -1$.

(b) $\dfrac{1}{x} \to \dfrac{1}{3}$ as $x \to 3$.

(c) $\dfrac{x^3 - 8}{x^2 + x - 6} \to \dfrac{12}{5}$ as $x \to 2$.

(d) $\dfrac{x^3 - 8}{x^2 + x - 6} \to \infty$ as $x \to \infty$.

(e) $\dfrac{x}{x - 2} \to \infty$ as $x \to 2+$.

(f) $\dfrac{3x^2 + x - 1}{5x^2 + 4} \to \dfrac{3}{5}$ as $x \to \infty$.

2. Given $f(x) = x/(x - 2)$ for $x \neq 2$, prove that f has no limit at 2.

3. Given $f(x) = x$ for $0 < x < 1$, and $f(x) = x^2$ for $x > 1$, prove that $f(x) \to 1$ as $x \to 1$.

4. Given that $f(x) = x$ for all $x \in Q$ and that $f(x) = x^2$ for all $x \in R\backslash Q$, prove the following:

(a) $f(x) \to 0$ as $x \to 0$.

(b) $f(x) \to 1$ as $x \to 1$.

(c) f does not have a limit at 2, not even a one-sided limit at 2.

5. Given $f(x) = (x^2 - 9)/|x - 3|$ whenever $x \neq 3$, prove that f has limits both from the left and from the right at 3 but that f has no limit at 3.

■ 6. Given that $f: S \to R$, that $\alpha \in [-\infty, \infty]$, and that a is a real number which is a limit point of the set $(-\infty, a) \cap S$, prove that the following two conditions are equivalent.
   (a) $f(x) \to \alpha$ as $x \to a-$.
   (b) For every neighborhood V of α there exists a number $\delta > 0$ such that whenever $x \in S$ and $0 < a - x < \delta$, we have $f(x) \in V$.

7. Given that $f: S \to R$, that $\alpha \in [-\infty, \infty]$, and that a is a real number which is a limit point of both of the sets $(-\infty, a) \cap S$ and $S \cap (a, \infty)$, prove that the following two conditions are equivalent.
   (a) $f(x) \to \alpha$ as $x \to a$.
   (b) $f(x) \to \alpha$ as $x \to a-$ and $f(x) \to \alpha$ as $x \to a+$.

■ 8. Suppose that $f: S \to R$, that $\alpha \in [-\infty, \infty]$, and that a is a real number which is a limit point of $S \cap (a, \infty)$ but is not a limit point of the set $(-\infty, a) \cap S$. Prove that the condition $f(x) \to \alpha$ as $x \to a$ is equivalent to the condition $f(x) \to \alpha$ as $x \to a+$.

9. Suppose that $f: S \to R$ and that $f(x) \to \alpha \in R$ as $x \to a$. Prove that if $\varepsilon > 0$, then it is possible to find a neighborhood U of a such that for all points x and t in the set $U \cap S \backslash \{a\}$ we have $|f(x) - f(t)| < \varepsilon$.

10. Given that $f(x) \to \alpha \in R$ as $x \to a$, prove that $|f(x)| \to |\alpha|$ as $x \to a$. *Hint:* Use Exercise 2.1.3(5).

■ 11. Suppose that $a < b$ and that the function f is increasing on (a, b); in other words, suppose that whenever $a < t \leq x < b$, we have $f(t) \leq f(x)$. Prove that if $\alpha = \sup\{f(x) \mid a < x < b\}$, then $f(x) \to \alpha$ as $x \to b$. A solution to this exercise will be provided in Theorem 6.10.6.

12. Given that $f: S \to R$, that a and α are real numbers, and that a is a limit point of S, complete the following sentence: "f fails to have a limit α at a when there exists a number $\varepsilon > 0$ such that for every $\delta > 0, \ldots$."

5.5 THE RELATIONSHIP BETWEEN LIMITS OF FUNCTIONS AND LIMITS OF SEQUENCES

We shall see in this section that the concepts of limit of a function and limit of a sequence are closely related; and later on, we shall use this relationship to obtain a free ride through much of the theory of limits of functions, using the results which we developed in Chapter 4. The key to the relationship between limits of functions and limits of sequences is contained in Section 4.7; therefore, you may wish to review this section now. At the same time, you might take a look at Exercise 4.8(9).

5.5.1 Theorem. Suppose that $S \subseteq R$, that $f: S \to R$, that a and α are points in $[-\infty, \infty]$, and that a is a limit point of S. Then the following two conditions are equivalent:

(1) $f(x) \to \alpha$ as $x \to a$.

(2) For every sequence (x_n) in the set $S \backslash \{a\}$ satisfying $x_n \to a$, we have $f(x_n) \to \alpha$.

■ **Proof.** To prove that (1) ⇒ (2), assume that (1) holds, and let (x_n) be a sequence in $S \backslash \{a\}$ such that $x_n \to a$. To show that $f(x_n) \to \alpha$, let V be a neighborhood of α. Choose a neighborhood U of a such that for every point $x \in U \cap S \backslash \{a\}$ we have $f(x) \in V$. Now using the fact that $x_n \to a$, choose a natural number N such that whenever $n \geq N$, we have $x_n \in U$. Then for all natural numbers $n \geq N$ we have $x_n \in U \cap S \backslash \{a\}$, and therefore, we have $f(x_n) \in V$.

Now to prove that (2) ⇒ (1), assume that (2) holds; and to obtain a contradiction, assume that condition (1) is false. Choose a neighborhood V of α such that for every neighborhood U of a there are points x in the set $U \cap S \backslash \{a\}$ for which $f(x) \notin V$. At this stage we need to separate the proof into three cases.

Case 1: Suppose that $a \in R$. For each natural number n, using the fact that $(a - 1/n, a + 1/n)$ is a neighborhood of a, choose a point, which we shall call x_n, in $(a - 1/n, a + 1/n) \cap S \backslash \{a\}$ such that $f(x_n) \notin V$. We notice that (x_n) is a sequence in $S \backslash \{a\}$ and that $x_n \to a$, and from condition (2) we deduce that $f(x_n) \to \alpha$. Therefore, the sequence $(f(x_n))$ is eventually in V even though it is *never* in V, and this contradiction completes the proof in this case.

Case 2: Suppose that $a = \infty$. For every natural number n, using the fact that (n, ∞) is a neighborhood of ∞, choose a point, which we shall call x_n, in $(n, \infty) \cap S$ such that $f(x_n) \notin V$. We notice that (x_n) is a sequence in $S \backslash \{\infty\}$, in other words, in S, and that $x_n \to a$; and from condition (2) we deduce that $f(x_n) \to \alpha$. Just as in Case 1, this is a contradiction.

Case 3: Suppose that $a = -\infty$. This case is similar to Case 2, and we leave it as an exercise. ■

5.6 A FREE RIDE THROUGH THE THEORY OF LIMITS OF FUNCTIONS

5.6.1 The Uniqueness of Limits of Functions. Suppose that $f: S \to R$, where $S \subseteq R$, and that a is a limit point of S. Then f has at most one limit at a.

■ **Proof.** Suppose α and β are both limits of f at a. Using the fact that a is a limit point of S, choose a sequence (x_n) in $S \backslash \{a\}$ such that $x_n \to a$.[2] From Theorem 5.5

---

[2] In view of Section 5.2, you should not be surprised that the proof of this uniqueness theorem uses the fact that a is a limit point of S.

we deduce that the sequence $(f(x_n))$ has both α and β as limits, and it follows from the theorem of uniqueness of limits of sequences that $\alpha = \beta$. ∎

5.6.2 Limit Notation for Functions. (Compare this section with Section 4.3.9.) If $f(x) \to \alpha$ as $x \to a$, then since α is the only limit of the function f at a, we can give this limit a name. We call it $\lim_{x \to a} f(x)$; or alternatively, since the symbol "x" is quite superfluous here, we may refer to the limit more simply (and more precisely) as $\lim_a f$. In the same way, if $f(x) \to \alpha$ as $x \to a-$, then we write

$$\alpha = \lim_{x \to a-} f(x) = \lim_{a-} f;$$

and we employ a similar notation for limits from the right.

5.6.3 The Sandwich Theorem for Functions. Suppose $S \subseteq R$ and a is a limit point of S. Suppose that f, g, and h are functions from S to R, and that for every point $x \in S$ we have $f(x) \le g(x) \le h(x)$. Suppose $\alpha \in [-\infty, \infty]$ and that $f(x) \to \alpha$ as $x \to a$ and $h(x) \to \alpha$ as $x \to a$. Then $g(x) \to \alpha$ as $x \to a$.

■ **Proof.** Let (x_n) be a sequence in $S \backslash \{a\}$ such that $x_n \to a$. We need to show that $g(x_n) \to \alpha$. But we know that $f(x_n) \to \alpha$ and that $h(x_n) \to \alpha$; and we know that $f(x_n) \le g(x_n) \le h(x_n)$ for every n. Therefore, the fact that $g(x_n) \to \alpha$ follows at once from the sandwich theorem for sequences (Theorem 4.3.11 and Exercise 4.4.18). ∎

5.6.4 The Arithmetical Rules for Limits of Functions. Suppose that $S \subseteq R$, that a is a limit point of S, that f and g are functions defined on S, and that α and β are points in $[-\infty, \infty]$. And suppose finally that $f(x) \to \alpha$ as $x \to a$ and $g(x) \to \beta$ as $x \to a$. Then each of the following holds provided that the appropriate limit value is defined:

(1) $f(x) + g(x) \to \alpha + \beta$ as $x \to a$.

(2) $f(x) - g(x) \to \alpha - \beta$ as $x \to a$.

(3) $f(x) \cdot g(x) \to \alpha\beta$ as $x \to a$.

(4) $f(x)/g(x) \to \alpha/\beta$ as $x \to a$.

■ **Proof.** Let (x_n) be a sequence in $S \backslash \{a\}$ such that $x_n \to a$. Then $f(x_n) \to \alpha$ and $g(x_n) \to \beta$; and therefore, whenever the appropriate limit values are defined, we have

$$f(x_n) + g(x_n) \to \alpha + \beta, \qquad f(x_n) - g(x_n) \to \alpha - \beta,$$

$$f(x_n)g(x_n) \to \alpha\beta, \qquad \frac{f(x_n)}{g(x_n)} \to \frac{\alpha}{\beta}.$$

The result therefore follows at once from Theorem 5.5. ∎

5.7 THE COMPOSITION THEOREM

As you know, the composition $g \circ f$ of two functions f and g is defined as long as $f(x)$ is in the domain of g for every point x in the domain of f, and for each such point x we define $g \circ f(x) = g(f(x))$. [See also Appendix A, Exercise A.4(3).] The **composition theorem for limits** says roughly that if $f(x) \to \alpha$ as $x \to a$ and $g(y) \to \beta$ as $y \to \alpha$, then $g \circ f(x) \to \beta$ as $x \to a$. Roughly speaking, if x is close to a, then $f(x)$ is a number y close to α; and so $g(f(x))$ is close to β. There is, however, one thing that can go wrong, and that is the fact that we have no knowledge about $g(f(x))$ whenever $f(x) = \alpha$. The precise statement of the theorem must take this into account.

5.7.1 Theorem.   Suppose that $f : S \to \mathbf{R}$, that for all points $x \in S$ the number $f(x)$ lies in a set T, and that $g : T \to \mathbf{R}$. Suppose also that a is a limit point of S, that α is a limit point of T, that $f(x) \to \alpha$ as $x \to a$, and that $g(y) \to \beta$ as $y \to \alpha$. Then $g \circ f(x) \to \beta$ as $x \to a$ if and only if at least one of the following two conditions holds:

(1) There exists a neighborhood U_1 of a such that for all $x \in U_1 \cap S \backslash \{a\}$ we have $f(x) \neq \alpha$.

(2) $\alpha \in T$ and $g(\alpha) = \beta$.

■ *Proof.*   We separate the proof into three cases.

Case 1: Suppose that condition (1) holds.   We need to show that $g \circ f(x) \to \beta$ as $x \to a$. Let W be a neighborhood of β. Using the fact that $g(y) \to \beta$ as $y \to \alpha$, choose a neighborhood V of α such that for all $y \in V \cap T \backslash \{\alpha\}$ we have $g(y) \in W$. Now using the fact that $f(x) \to \alpha$ as $x \to a$, choose a neighborhood U of a such that for all points $x \in U \cap S \backslash \{a\}$ we have $f(x) \in V$. Using condition (1), choose a neighborhood U_1 of a such that for all points x in $U_1 \cap S \backslash \{a\}$ we have $f(x) \neq \alpha$. Then $U \cap U_1$ is a neighborhood of a, and for all points $x \in (U \cap U_1) \cap S \backslash \{a\}$ we have $f(x) \in V \cap T \backslash \{\alpha\}$; and so $g \circ f(x) \in W$.

Case 2: Suppose that condition (2) holds.   Once again, we need to show that $g \circ f(x) \to \beta$ as $x \to a$. Let W be a neighborhood of β. Using the fact that $g(y) \to \beta$ as $y \to \alpha$, choose a neighborhood V of α such that for all points $y \in V \cap T \backslash \{\alpha\}$ we have $g(y) \in W$. Now using the fact that $f(x) \to \alpha$ as $x \to a$, choose a neighborhood U of a such that for all points $x \in U \cap S \backslash \{a\}$ we have $f(x) \in V$. We shall now see that given any point $x \in U \cap S \backslash \{a\}$, we must have $g(f(x)) \in W$. Let $x \in U \cap S \backslash \{a\}$. There are two cases: Either $f(x) = \alpha$, in which case $g(f(x)) = \beta \in W$; or $f(x) \neq \alpha$, in which case the fact that $g(f(x)) \in W$ follows from our choice of U and V.

Case 3: Suppose that neither condition (1) nor condition (2) holds.   This time we must show that it is *false* that $g(f(x)) \to \beta$ as $x \to a$. We observe first that the failure of condition (1) guarantees the existence of points x for which $f(x) = \alpha$, and it follows that $\alpha \in T$. Therefore, since condition (2) also fails, we must have $g(\alpha) \neq \beta$. Choose a neighborhood W of β such that $g(\alpha) \notin W$. (We could, for example, have taken $W = \boldsymbol{R} \backslash \{g(\alpha)\}$.) Now we see that for every neighborhood U of a there must be a point $x \in U \cap S \backslash \{a\}$ such that $g(f(x)) = g(\alpha) \notin W$; and this shows that we can't have $g(f(x)) \to \beta$ as $x \to a$. ∎

5.8 THE "RULER FUNCTION": A FUNCTION THAT IS POSITIVE AT EVERY RATIONAL BUT THAT HAS A ZERO LIMIT EVERYWHERE

The purpose of this section is to show you an example of a function f defined on $[0, 1]$ which has the property that $f(x) > 0$ at every rational number x in the interval, and yet $f(x) \to 0$ as $x \to a$ for every number a in $[0, 1]$. Before describing the example, we shall take a look at a simple lemma that should make the example easier to understand.

5.8.1 Lemma. Suppose that $f : [0, 1] \to [0, \infty)$, and suppose that for every number $\varepsilon > 0$ the set $\{x \in [0, 1] \mid f(x) \geq \varepsilon\}$ is finite. Then for every $a \in [0, 1]$ we have $f(x) \to 0$ as $x \to a$.

■ **Proof.** Let $a \in [0, 1]$, and let $\varepsilon > 0$. We need to find a neighborhood U of a such that whenever $x \in U \cap [0, 1] \backslash \{a\}$, we will have $f(x) \in (-\varepsilon, \varepsilon)$. Since $\{x \in [0, 1] \mid f(x) \geq \varepsilon\}$ is finite, it follows that the smaller set $H = \{x \in [0, 1] \mid f(x) \geq \varepsilon\} \backslash \{a\}$ is finite. But every finite set is closed. Therefore, H is closed, and so $\boldsymbol{R} \backslash H$ is open. It follows that $\boldsymbol{R} \backslash H$ is a neighborhood of a. We now complete the proof by observing that for every point $x \in (\boldsymbol{R} \backslash H) \cap [0, 1] \backslash \{a\}$ we have $f(x) < \varepsilon$, and consequently, $f(x) \in (-\varepsilon, \varepsilon)$. ∎

5.8.2 The Example. See Figure 5.2. For every irrational number $x \in [0, 1]$ we define $f(x) = 0$. Now given any rational number x, one can write x uniquely in its so-called reduced form m/n, where m and n are integers with no common factor and $n > 0$. (The reduced form of the number 0 is understood to be $0/1$.) For every rational number $x \in [0, 1]$ with reduced form m/n, we define $f(x) = 1/n$. To see that this function f has the properties assumed in the lemma, all we have to notice is that if $\varepsilon > 0$, then there are only finitely many rational numbers with reduced form m/n in the interval $[0, 1]$ for which $1/n \geq \varepsilon$.

5.9 EXERCISES

1. Prove that $x \sin(1/x) \to 0$ as $x \to 0$. *Hint:* Use the sandwich theorem (Theorem 5.6.3).

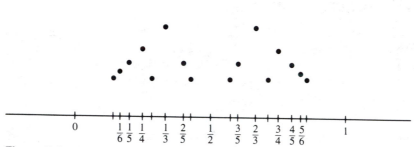

Figure 5.2

2. Given that $f: S \to R$, that a is a limit point of S, and that $\alpha \in [-\infty, \infty]$, prove that the following two conditions are equivalent:

 (a) $f(x) \to \alpha$ as $x \to a$.

 (b) For every sequence (x_n) in $S\setminus\{a\}$ which has a as a partial limit, α will be a partial limit of the sequence $(f(x_n))$.

3. Prove that Lemma 5.8.1 has a converse. In other words, prove that if $f: [0, 1] \to R$, and if for every $a \in [0, 1]$, $f(x) \to 0$ as $x \to a$, then for every number $\varepsilon > 0$ the set $\{x \in [0, 1] \mid f(x) \ge \varepsilon\}$ must be finite.

*4. State and prove a converse to Exercise 5.4(9). *Hint:* Take another look at the section on Cauchy sequences (Section 4.11). The converse you are looking for should depend upon the completeness of the real number system.

Chapter 6

Continuity

In Chapter 5, when we described the notion $f(x) \to \alpha$ as $x \to a$, we went to great lengths to point out that the number $f(a)$ plays no role at all in the definition. We said that in order for us to have $f(x) \to \alpha$ as $x \to a$, we do not have to know that a lies in the domain of f; and even if it does, the number $f(a)$ does not have to be close to α. But there are occasions when the number $f(a)$ does exist and is exactly equal to α, and it is this extra bit of good behavior that distinguishes the idea of continuity from the limit concept of Chapter 5. We begin with the precise definition.

6.1 DEFINITION OF CONTINUITY

6.1.1 Continuity at a Point. Suppose $S \subseteq R$ and $f: S \to R$, and suppose $a \in S$.[1] We say that the function f is **continuous at the point** a if for every neighborhood V of $f(a)$ there exists a neighborhood U of a such that whenever $x \in U \cap S$, we have $f(x) \in V$.

---

[1] Notice that we do not require a to be a limit point of S. Notice also that since $a \in S$, we must have $a \in R$; in other words, a can't be $\pm\infty$.

6.1.2 Continuity on a Set. Suppose $S \subseteq R$ and $f: S \to R$. We say that f is **continuous on** S if for every point $a \in S$ the function f is continuous at a.

6.2 A COMPARISON BETWEEN LIMITS AND CONTINUITY

It is important to notice that in the definition of continuity we speak about "all points $x \in U \cap S$" and do not slash out the singleton $\{a\}$, as we did in the definition of a limit. The following theorem will help us understand how this removal or nonremoval of the point a from $U \cap S$ embodies much of the distinction between the ideas of limit at a and continuity at a.

6.2.1 Theorem. Suppose $f: S \to R$ and $a \in S$. Suppose $\alpha \in [-\infty, \infty]$, and suppose that for every neighborhood V of α there exists a neighborhood U of a such that for all points $x \in U \cap S$ we have $f(x) \in V$. Then $\alpha = f(a)$, and f is continuous at a.

■ *Proof.* Obviously, once we have shown that $\alpha = f(a)$, the given information will at once tell us that f is continuous at a. Now to obtain a contradiction, assume that $\alpha \neq f(a)$. Choose a neighborhood V of α such that $f(a) \notin V$. For example, we could choose $V = R\backslash\{f(a)\}$. Now using the given information, choose a neighborhood U of a such that for every point $x \in U \cap S$ we have $f(x) \in V$. But $a \in U \cap S$, and therefore, $f(a) \in V$, contradicting the choice of V. ■

It is worth noticing that even though we only assumed $\alpha \in [-\infty, \infty]$ in the statement of the theorem, the fact that $\alpha = f(a)$ guarantees that α is a real number.

6.2.2 Theorem. Suppose that $f: S \to R$ and that $a \in S \cap \mathcal{L}(S)$. Then the following two conditions are equivalent:

(1) f is continuous at a.

(2) $f(x) \to f(a)$ as $x \to a$.

■ *Proof.* To prove that (1) $\Rightarrow$ (2), assume that (1) holds. Let V be a neighborhood of $f(a)$; and using the fact that f is continuous at a, choose a neighborhood U of a such that for all points $x \in U \cap S$ we have $f(x) \in V$. Then certainly, for all points x in the smaller set $U \cap S\backslash\{a\}$ we have $f(x) \in V$.

Now to prove that (2) $\Rightarrow$ (1), assume that (2) holds. Let V be a neighborhood of $f(a)$; and using the fact that $f(x) \to f(a)$ as $x \to a$, choose a neighborhood U of a such that for all points $x \in U \cap S\backslash\{a\}$ we have $f(x) \in V$. Now the only point of $U \cap S$ which does not lie in the set $U \cap S\backslash\{a\}$ is a itself, and we certainly have $f(a) \in V$. Therefore, for all points $x \in U \cap S$ we have $f(x) \in V$. ■

6.2.3 Theorem. Suppose that $S \subseteq R$ and that $a \in S$ but that a is not a limit point of S. Then given any function $f: S \to R$, f is automatically continuous at a.

■ ***Proof.***   Let V be a neighborhood of $f(a)$. Choose a neighborhood U of a such that $U \cap S = \{a\}$. Then since the only point in the set $U \cap S$ is a itself, it is clear that for every point $x \in U \cap S$ we have $f(x) \in V$.   ■

6.2.4   Theorem.   Suppose that $f: S \to \mathbf{R}$ and that $a \in S$. Then the following two conditions are equivalent:

(1) f is continuous at a.

(2) For every number $\varepsilon > 0$ there exists a number $\delta > 0$ such that whenever $x \in S$ and $|x - a| < \delta$, we have $|f(x) - f(a)| < \varepsilon$.

■ ***Proof.***   We shall prove that $(1) \Rightarrow (2)$ and leave the proof that $(2) \Rightarrow (1)$ as an exercise. Assume, then, that (1) holds. Let $\varepsilon > 0$. Using the fact that f is continuous at a and that $(f(a) - \varepsilon, f(a) + \varepsilon)$ is a neighborhood of $f(a)$, choose a neighborhood U of a such that for all points $x \in U \cap S$ we have $f(x) \in (f(a) - \varepsilon, f(a) + \varepsilon)$. Now choose a number $\delta > 0$ such that $(a - \delta, a + \delta) \subseteq U$, and observe that δ satisfies the requirements of condition (2).   ■

6.2.5   Failure of Continuity.   With Theorem 6.2.4 in mind, we observe that if $f: S \to \mathbf{R}$ and $a \in S$, then a condition for f to be discontinuous at a is that there should exist a number $\varepsilon > 0$ such that for every number $\delta > 0$ there is at least one number $x \in (a - \delta, a + \delta) \cap S$ such that $|f(x) - f(a)| \geq \varepsilon$.

6.2.6   Some Examples.   The following examples illustrate the notion of continuity.

(1) Define $f(x) = x^2 + 1$ for all $x \in \mathbf{R}$. In Example 5.3.3(1) we showed that $f(x) \to 5$ as $x \to 2$. Since $f(2) = 5$, we may conclude that f is continuous at 2. We leave it as an exercise to show that for every number a the function f is continuous at a.

(2) Define $f(x) = (x - 1)/(x^2 + 3)$ for all $x \in \mathbf{R}$. Our conclusion in Example 5.3.3(2) shows that f is continuous at 2. One may show similarly that f is continuous at any real number a; in other words, f is continuous on $\mathbf{R}$.

(3) Define $f(0) = 0$; and for each $x \neq 0$, define $f(x) = x \sin(1/x)$. From the fact that $|f(x)| \leq |x|$ for every x, it is easy to show that $f(x) \to 0$ as $x \to 0$. This shows that f is continuous at 0.

6.3   THE RELATIONSHIP BETWEEN CONTINUITY AND LIMITS OF SEQUENCES

6.3.1   Theorem.   Suppose $f: S \to \mathbf{R}$ and $a \in S$. Then the following two conditions are equivalent:

(1) f is continuous at a.

(2) For every sequence (x_n) in S converging to a, we have $f(x_n) \to f(a)$.

■ **Proof.**   To prove that $(1) \Rightarrow (2)$, assume that (1) holds. Let (x_n) be a sequence in S converging to a; and in order to show that $f(x_n) \to f(a)$, let V be a neighborhood of $f(a)$. Using the fact that f is continuous at a, choose a neighborhood U of a such that for all points $x \in U \cap S$ we have $f(x) \in V$. Now using the fact that $x_n \to a$, choose a natural number N such that whenever $n \geq N$, we have $x_n \in U$. Then whenever $n \geq N$, we have $x_n \in U \cap S$, and consequently, $f(x_n) \in V$. This shows that $f(x_n) \to f(a)$.

Now to prove that $(2) \Rightarrow (1)$, assume that (2) holds; and to obtain a contradiction, assume that f is not continuous at a. Choose a number $\varepsilon > 0$ such that for every positive number δ there exists at least one number $x \in (a - \delta, a + \delta) \cap S$ for which $f(x) \notin (f(a) - \varepsilon, f(a) + \varepsilon)$. For every natural number n, choose a number $x_n \in (a - 1/n, a + 1/n) \cap S$ such that $f(x_n) \notin (f(a) - \varepsilon, f(a) + \varepsilon)$. We see that (x_n) is a sequence in S and $x_n \to a$. Therefore, $f(x_n) \to f(a)$. Using the fact that $(f(a) - \varepsilon, f(a) + \varepsilon)$ is a neighborhood of $f(a)$, choose a natural number N such that whenever $n \geq N$, we have $f(x_n) \in (f(a) - \varepsilon, f(a) + \varepsilon)$. This gives us an obvious contradiction.   ■

6.3.2 Corollary.   Given $S \subseteq R$ and $f: S \to R$, the following two conditions are equivalent:

(1) f is continuous on S.

(2) For every sequence (x_n) in S which converges to a point x of S, the sequence $(f(x_n))$ converges to $f(x)$.

6.4 COMBINATIONS OF CONTINUOUS FUNCTIONS

6.4.1 Theorem.   Suppose that $S \subseteq R$, that $a \in S$, that f and g are functions with domain S, and that f and g are continuous at a. Then the functions $f + g, f - g$, and fg must be continuous at a; and provided that $g(a) \neq 0$, the function f/g will also be continuous at a.

This theorem follows simply from Theorem 6.3.1 and its proof will be left as an exercise.

6.4.2 Corollary.   Suppose that $S \subseteq R$, that f and g are functions with domain S, and that f and g are continuous on S. Then the functions $f + g, f - g$, and fg must be continuous on S; and provided that $g(x) \neq 0$ for every $x \in S$, the function f/g will also be continuous on S.

6.4.3 Corollary. Polynomials are continuous at every point in R. Rational functions are continuous at every point in R at which their denominators are not zero.

6.4.4 The Composition Theorem. Suppose that $f: S \to R$ and that f is continuous on S. Suppose that T is a set which includes the range of f and that g is continuous on T. Then the composition $g \circ f$ is continuous on S.

▪ **Proof.** Let (x_n) be a sequence in S converging to a point $x \in S$. We need to show that $g(f(x_n))$ converges to $g(f(x))$. Now by the continuity of f at x, we have $f(x_n) \to f(x)$. Therefore, since g is continuous at $f(x)$, we have $g(f(x_n)) \to g(f(x))$, as required. ▪

Notice how much easier the composition theorem for continuity is than the composition theorem for limits in Chapter 5.

6.5 EXERCISES

1. Prove that every function $f: Z \to R$ is continuous on Z.

2. Given that $S = \{1/n \mid n \in Z^+\}$, prove that every function $f: S \to R$ must be continuous on S.

3. Suppose that $S = \{1/n \mid n \in Z^+\}$ and that for every natural number n we have $f(1/n) = (-1)^n$. Why doesn't it contradict Theorem 6.3.1 to say that $1/n \to 0$ and that f is continuous on S but that the sequence $(f(1/n))$ diverges?

4. Suppose that $S = \{1/n \mid n \in Z^+\} \cup \{0\}$, that $f(1/n) = (-1)^n/n$ for every natural number n, and that $f(0) = 0$. Prove that f is continuous on S.

5. Give an example of a function $f: R \to R$ which is continuous at every point except 2.

6. Given that $S \subseteq R$ and that $\mathcal{L}(S) \subseteq R\backslash S$, prove that every function $f: S \to R$ must be continuous on S. Give an example of a set S such that $\mathcal{L}(S) \neq \emptyset$ and $\mathcal{L}(S) \subseteq R\backslash S$.

▪ 7. Given that S is a closed subset of R, that f is continuous on S, and that (x_n) is a convergent sequence in S, prove that the sequence $(f(x_n))$ is convergent. Take another look at exercise 3 to make sure you understand that somewhere in your proof you have to use the fact that S is closed.

▪ 8. Given that S is an open subset of R, that $a \in S$, that $f: S \to R$, and that the function f is continuous at a, prove that for every number $\varepsilon > 0$ there exists a number $\delta > 0$ such that for all numbers x satisfying $|x - a| < \delta$, we have $|f(x) - f(a)| < \varepsilon$. *Note:* If you didn't use the fact that S is open, then your proof is *wrong*.

9. Suppose that a and b are real numbers, that $a < b$, and that f is a function defined on $[a, b]$. Prove that the condition that f be continuous at a is equivalent to the condition that for every neighborhood V of $f(a)$ there exists a number $\delta > 0$ such that for all numbers x satisfying $a < x < a + \delta$, we have $f(x) \in V$.

10. Given a, b, and f as in exercise 9, state and prove an analogue of exercise 9 for continuity of f at b.

11. Suppose that $S \subseteq R$, that $f: S \to R$, that f is continuous on S, that $x \in S$, and that (x_n) is a sequence in S with x as a partial limit. Prove that $f(x)$ is a partial limit of the sequence $(f(x_n))$.

■ 12. Suppose $S \subseteq R$ and $a \in S$. Suppose that f and g are functions defined on S and that f is continuous at a but that g is not continuous at a.
(a) Can one conclude that $f + g$ is not continuous at a?
(b) Can one conclude that fg is not continuous at a?
In each case, if you answer yes, support your answer with a proof; if you answer no, support your answer with an example.

13. Repeat exercise 12, but assume this time that both f and g are discontinuous at a.

14. Suppose that $f: S \to R$ and that f is continuous on S. Suppose that T is a set which includes the range of f and that $g(x)$ is defined for all $x \in T$ but that g is not continuous on T. Can one conclude that the composition $g \circ f$ is not continuous on S? What if neither f nor g is continuous?

15. Suppose that f is the ruler function of Section 5.8. Prove that f is continuous at every irrational number in $[0, 1]$ and discontinuous at every rational. *Hint:* Look at the limit properties of the ruler function that were derived in Section 5.8.

16. Given that $f: S \to R$ and that f is continuous on S, prove that $|f|$ is continuous on S.

17. Suppose that $S \subseteq R$, that f and g are functions defined on S, and that g is continuous on S; and suppose that for all points x and t of S we have $|f(x) - f(t)| \leq |g(x) - g(t)|$. Prove that f must be continuous on S.

18. Suppose that $f: R \to R$ and that f is continuous on R. Suppose that $E \subseteq R$, that $a \in \bar{E}$, and that $P = \{f(x) \mid x \in E\}$. Prove that $f(a) \in \bar{P}$.

19. Suppose that $f: R \to R$ and that f is continuous on R. Prove that if V is an open subset of R, then the set $\{x \mid f(x) \in V\}$ must also be open. Given an open subset U of R, can one conclude that the set $\{f(x) \mid x \in U\}$ is open?

■ 20. Suppose that $f: R \to R$ and that f is continuous on R. Prove that if H is a closed subset of R, the set $\{x \mid f(x) \in H\}$ must also be closed. Given a closed subset H of R, can one conclude that the set $\{f(x) \mid x \in H\}$ is closed?

21. Suppose that $S \subseteq \mathbf{R}$ and that f is continuous on S. Suppose that $E \subseteq S$ and that the function $g : E \to \mathbf{R}$ is defined by $g(x) = f(x)$ for every point $x \in E$. Prove that g is continuous on E.

22. Suppose that f is continuous on the interval $[a, b]$, where $a \le b$. Suppose that the function g is defined by $g(x) = f(a)$ for every $x < a$, by $g(x) = f(x)$ for every $x \in [a, b]$, and by $g(x) = f(b)$ for every $x > b$. Prove that g is continuous on $\mathbf{R}$.

23. Given that f is continuous on $\mathbf{R}$, prove that the set $\{x \mid f(x) = 0\}$ is closed. Having done this, prove that if f and g are continuous on $\mathbf{R}$, then the set $\{x \mid f(x) = g(x)\}$ must be closed.

24. Given that f and g are continuous on $\mathbf{R}$ and that $f(x) = g(x)$ for every $x \in \mathbf{Q}$, prove that $f = g$.

*25. Suppose that $f : \mathbf{Q} \to \mathbf{R}$, that $f(1) = 1$, and that for all rational numbers x and t we have $f(x + t) = f(x) + f(t)$. Prove that $f(x) = x$ for every $x \in \mathbf{Q}$. *Hint:* This question is pure algebra! First, show that $f(2) = 2$. Then use mathematical induction to show that $f(n) = n$ for all $n \in \mathbf{Z}^+$. Finally, extend the identity to negative integers and rationals.

*26. Suppose that $f : \mathbf{R} \to \mathbf{R}$, that f is continuous on $\mathbf{R}$, that $f(1) = 1$, and that for all rational numbers x and t we have $f(x + t) = f(x) + f(t)$. Prove that $f(x) = x$ for all $x \in \mathbf{R}$.

*27. Suppose that $f : \mathbf{R} \to \mathbf{R}$, that f is an increasing function, that $f(1) = 1$, and that for all rational numbers x and t we have $f(x + t) = f(x) + f(t)$. Prove that $f(x) = x$ for all $x \in \mathbf{R}$.[2]

*28. Suppose that $f : \mathbf{R} \to \mathbf{R}$ and that for all numbers x and t we have $|f(x) - f(t)| \le |x - t|^2$. Prove that the function f must be constant. A much easier proof will be available after Theorem 7.12.6.

6.6* ADDITIONAL EXERCISES: DISTANCE FUNCTION TO A SET

Given a nonempty set A of real numbers, we define the function ρ_A by

$$\rho_A(x) = \inf\{|x - a| \mid a \in A\} \qquad \text{for all} \qquad x \in \mathbf{R}.$$

For every number x we call $\rho_A(x)$ the *distance from x to the set A*.

---

[2] Notice how the fact that this function f is increasing makes f automatically continuous. This exercise is actually the simplest in a long line of interesting theorems that say, roughly, that if a function f satisfies the identity $f(x + t) = f(x) + f(t)$, then either f is continuous, or it is very badly behaved. See Boas [5] for some more detailed results of this type. Some abstract results of this type can be found in Lewin [9].

*1. Suppose A is a nonempty set of real numbers.

  (a) Prove that $\rho_A(x) = 0$ iff $x \in \bar{A}$.

  (b) Prove that for every point $a \in A$ and all numbers x and t, we have $\rho_A(x) \leq |x - t| + |t - a|$.

  (c) Prove that for all numbers x and t we have

$$\rho_A(x) \leq |x - t| + \rho_A(t).$$

  (d) Prove that for all numbers x and t we have

$$|\rho_A(x) - \rho_A(t)| \leq |x - t|.$$

  (e) Prove that the function ρ_A is continuous on $\boldsymbol{R}$.

*2. This exercise refers to the concept of separated sets, which was introduced in Exercise 3.9(7). Prove that if two sets A and B are separated, then there exist two mutually disjoint open sets U and V such that $A \subseteq U$ and $B \subseteq V$. *Hint:* Define U to be the set of numbers x for which $\rho_A(x) - \rho_B(x) < 0$.

*3. We say that two sets A and B are *strongly separated* when $\bar{A} \cap \bar{B} = \emptyset$. Suppose that A and B are given sets and that there exists a function f which is continuous on $\boldsymbol{R}$ and which has the property that $f(x) = 0$ for every $x \in A$ and $f(x) = 1$ for every $x \in B$. Prove that the sets A and B must be strongly separated.

*4. Given two strongly separated sets A and B, prove that there exists a function f continuous on $\boldsymbol{R}$ such that $0 \leq f(x) \leq 1$ for every $x \in \boldsymbol{R}$, that $f(x) = 0$ for every $x \in A$, and that $f(x) = 1$ for every $x \in B$. *Hint:* Look at $\rho_A/(\rho_A + \rho_B)$.

6.7 THE BEHAVIOR OF CONTINUOUS FUNCTIONS ON CLOSED, BOUNDED SETS

We come, at last, to those important properties of continuity that depend on the completeness of the number system. In this section we shall be studying the special behavior of functions which are continuous on closed, bounded sets; and in order to do this, we shall use some of the special properties of closed, bounded sets (properties which depend on completeness), which we investigated in Chapter 4. We begin with a quick review of the properties of closed, bounded sets that we need.

6.7.1 A Review of Closed, Bounded Sets

  (1) The central theorem of Section 4.9 is Theorem 4.9.3, which tells us that if a set H is closed and bounded, then every sequence which is frequently in H must have a partial limit in H. (Take another look at the proof of that theorem and see again how it uses the Bolzano-Weierstrass theorem.)

(2) Theorem 4.7.4 tells us that if a set H is unbounded above, then one can find a sequence (x_n) in H such that $x_n \to \infty$; and such a sequence certainly doesn't have a partial limit in H. Similarly, if H is unbounded below, there must be a sequence in H which has no partial limit in H.

(3) Finally, we can deduce from Theorem 4.7.3 that if a set H is not closed, then there has to be a sequence in H which converges to a point in $R \backslash H$; and again, we find a sequence in H which has no partial limit in H. In Exercise 4.14(8) we summed up all this information as follows: Given any subset H of R, the following two conditions are equivalent:
   (a) H is closed and bounded.
   (b) Every sequence in H must have a partial limit in H.

6.7.2 The Main Result. Suppose that H is a closed, bounded set of real numbers and that f is continuous on H. Then the range of f is also closed and bounded.

■ *Proof.* We shall write the range of f as K. In order to show that K is closed and bounded, we shall show that every sequence in K must have a partial limit in K. Let (y_n) be a sequence in K. For every natural number n, choose a point $x_n \in H$ such that $f(x_n) = y_n$. This defines a sequence (x_n) in H. Using the fact that H is closed and bounded, choose a partial limit x of (x_n) such that $x \in H$. We now complete the proof by showing that $f(x)$ is a partial limit of (y_n). Let V be a neighborhood of $f(x)$. Using the fact that f is continuous at x, choose a neighborhood U of x such that for all points $t \in U \cap H$ we have $f(t) \in V$. Since x is a partial limit of (x_n), there are infinitely many natural numbers n for which $x_n \in U$; and for all such n we have $f(x_n) \in V$. ■

6.7.3 Corollary. Suppose that H is a nonempty, closed, bounded subset of R, that $f : H \to R$, and that f is continuous on H. Then f has a maximum value and a minimum value. More precisely, there exist points x_1 and x_2 in H such that for every point $x \in H$ we have $f(x_1) \le f(x) \le f(x_2)$.

■ *Proof.* Define K to be the range of f. Then by Theorem 6.7.2, the set K must be closed and bounded; and since H is nonempty, so is K. Therefore, since sup K and inf K belong to K, we see that K has both a largest and a smallest member. ■

6.8 BOLZANO'S INTERMEDIATE VALUE THEOREM

The intermediate value theorem says:

   If a function is continuous on an interval, then its range is an interval.

Roughly speaking, this theorem tells us that if the domain of a continuous function is an interval, then one can draw its graph without lifting one's pencil from the paper.

Now the precise statement of the theorem talks about *intervals,* and so we need to know something about intervals. What we need to know about them appeared in Exercise 2.5(17), and since this exercise has become very important to us, we shall repeat most of it here and provide a solution. If you don't need to see this solution, skip it and go on to Theorem 6.8.1.

A set A of real numbers is said to be **convex** if whenever $x < y < z$ and both x and z belong to A, the number y must also belong to A.

(2) Prove that if A is convex and unbounded both above and below, then $A = \mathbf{R}$.

(3) Prove that if A is convex, bounded below, and unbounded above and if inf $A = \alpha$, then A is one of the sets $[\alpha, \infty)$, (α, ∞).

(4) Prove that if a set A is convex, then A must have one of the following forms: $\emptyset, \mathbf{R}, (-\infty, \beta), (-\infty, \beta], (\alpha, \infty), [\alpha, \infty), (\alpha, \beta), [\alpha, \beta), (\alpha, \beta], [\alpha, \beta]$. Note that a set which has one of these ten forms is called an **interval**. Therefore, in this exercise you are showing that every convex set has to be an interval.

Solution to part (2).   Assume that A is convex and unbounded both above and below. We must show that every real number belongs to A. Let $x \in \mathbf{R}$. Using the fact that x is neither an upper bound nor a lower bound of A, choose two members p and q of A such that $p < x < q$. It now follows at once from the convexity of A that $x \in A$.

Solution to part (3).   Assume that A is convex, bounded below, and unbounded above and that $\alpha = \inf A$. Obviously, no number which is less than α can be a member of A; and therefore, in order to show that A is one of the sets $[\alpha, \infty)$, (α, ∞), we need only show that for every real number $x > \alpha$ we have $x \in A$. Let $x > \alpha$. Then since A is unbounded above, x is not an upper bound of A; and since $x > \alpha$, x is not a lower bound of A. Therefore, as in part (2), we can choose two members p and q of A such that $p < x < q$. And this guarantees that $x \in A$.

Solution to part (4).   All we have to do is note that if $A \neq \emptyset$, then either A is unbounded both above and below, in which case we use part (2); or A is bounded below but not above, in which case we use part (3); or A is bounded above but not below, or both above and below. In the latter two cases we use obvious analogues of parts (2) and (3).

6.8.1   Bolzano's Intermediate Value Theorem.   If a function is continuous on an interval, then its range is an interval.

■ *Proof.*   Suppose that f is continuous on an interval S and that T is the range of f. We need to show that T is convex. To obtain a contradiction, assume that T is not convex, and choose two points p and q in T and a number $y \notin T$ such that $p < y < q$. Choose a and b in S such that $p = f(a)$ and $q = f(b)$. Now either $a < b$ or $a > b$;

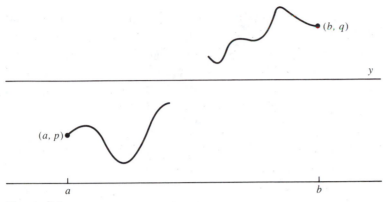

(b, q)

y

(a, p)

a

b

Figure 6.1

and since these two cases require similar arguments, we shall assume that $a < b$. See Figure 6.1. From this point there are several ways in which the desired contradiction is commonly reached. Each of the following three approaches is interesting enough to be worth knowing, and we cannot resist showing them all. Note that our assumption guarantees that there is no point x in $[a, b]$ such that $f(x) = y$.

Method 1. Define $A = \{x \in [a, b] \mid f(x) < y\}$ and $B = \{x \in [a, b] \mid f(x) > y\}$, and observe that $A \cup B = [a, b]$. Since $a \in A$, we know that $A \neq \emptyset$; and A is certainly bounded above. Define $c = \sup A$. There are now two possibilities: Either $c \in A$ or $c \in B$, and we shall show that each of them leads to a contradiction.

a   c   b

Suppose, first, that $c \in A$. This means that $f(c) < y$; and since $f(b) > y$, we see that in this case $c \neq b$. Using the facts that $(-\infty, y)$ is a neighborhood of $f(c)$ and that f is continuous at c, choose a number $\delta > 0$ such that for all points $x \in (c - \delta, c + \delta) \cap [a, b]$, we have $f(x) < y$:

a   c − δ   c   x   c + δ   b

Now choose a number x lying between c and the smaller of the two numbers $c + \delta$ and b. Then $x \in [a, b]$, and we must have $f(x) < y$ even though $x > c$. This contradicts the fact that c is an upper bound of A.

Second, suppose that $c \in B$. This means that $f(c) > y$; and since $f(a) < y$, we see that in this case $c \neq a$. Using the facts that (y, ∞) is a neighborhood of $f(c)$ and that f is continuous at c, choose a number $\delta > 0$ such that for all points $x \in (c - \delta, c + \delta) \cap [a, b]$, we have $f(x) > y$. Define d to be the larger of the two numbers $c - \delta$ and a. Then $d < c$; and since $f(x) > y$ for every point $x \in (d, c)$, it follows that d is an upper bound of A, contradicting the fact that c is the *least* upper bound of A. This completes Method 1.

Method 2. Method 2 is really Method 1 all over again; but this time, we let sequences do the work for us. As before, define $A = \{x \in [a, b] \mid f(x) < y\}$ and $B = \{x \in [a, b] \mid f(x) > y\}$, and define $c = \sup A$. Using the fact that $c \in \bar{A}$, choose a sequence (x_n) in A which converges to c. Since f is continuous at c, we have $f(x_n) \rightarrow f(c)$; and therefore, since $f(x_n) < y$ for every natural number n, we conclude that $f(c) \leq y$. Since $f(b) > y$, we deduce that $c < b$. Choose a sequence (t_n) in $(c, b]$ which converges to c. Then $f(t_n) \rightarrow f(c)$; and because $f(t_n) > y$ for every natural number n, we deduce that $f(c) \geq y$. Therefore, $f(c) = y$, contradicting our assumption that y is not in the range of f.

Method 3. We define a function $g : \boldsymbol{R} \rightarrow \boldsymbol{R}$ as follows:

$$g(x) = f(a) \quad \text{for} \quad x < a,$$

$$g(x) = f(x) \quad \text{for} \quad a \leq x \leq b,$$

and

$$g(x) = f(b) \quad \text{for} \quad x > b.$$

In Exercise 6.5(22) we saw that this function g is continuous on $\boldsymbol{R}$. Define $A = \{x \in \boldsymbol{R} \mid f(x) < y\}$. From Exercise 6.5(19) we deduce that A is open. Also, since y does not lie in the range of f, we have $\boldsymbol{R} \backslash A = \{x \in \boldsymbol{R} \mid f(x) > y\}$, which is also open. We now have a contradiction to Theorem 3.7, because A is both open and closed, $A \neq \emptyset$ (because $a \in A$), and $A \neq \boldsymbol{R}$ (because $b \notin A$). ∎

6.8.2 Corollary: The Existence of *n*th Roots. Given any positive number y and any natural number n, there is a unique positive number x such that $x^n = y$.

∎ ***Proof.*** Define $f(x) = x^n$ for every $x \in \boldsymbol{R}$. Then f, being a polynomial, is continuous on $\boldsymbol{R}$. Choose a number b which is greater than the larger of the two numbers 1 and y. Since $f(b) = b^n > b \geq y$ and $f(0) = 0 < y$, we deduce from Theorem 6.8.1 that there must exist a number $x \in [0, b]$ such that $f(x) = y$. Obviously, $x > 0$, and so the existence part of the theorem is proved. Now whenever $0 < t < x$, we have $t^n < x^n$; and whenever $t > x$, we have $t^n > x^n$; and therefore, x is the only positive nth root of y. ∎

6.9 EXERCISES

1. Suppose that $S \subseteq \boldsymbol{R}$, that $a \in \mathcal{L}(S) \backslash S$, and that for every point $x \in S$ we have $f(x) = |x - a|$. Prove that f is continuous on S (this is meant to be easy) but that f has no minimum.

2. Suppose that $S \subseteq \boldsymbol{R}$, that $a \in \mathcal{L}(S) \backslash S$, and that for every point $x \in S$ we have $f(x) = 1/|x - a|$. Prove that f is continuous on S (this is meant to be easy) but that f has no maximum. Show that, in fact, f is not even bounded.

3. Given that $S \subseteq \boldsymbol{R}$ and that every continuous function on S has a maximum, prove that S must be closed.

4. Given that $S \subseteq \boldsymbol{R}$ and that every continuous function on S is bounded, prove that S must be closed.

5. Suppose that S is an unbounded set of real numbers, and that for every $x \in S$ we have $f(x) = x$. Prove that f is continuous on S but that f is not bounded. This exercise is meant to be trivial.

6. Given $S \subseteq \boldsymbol{R}$, prove that the following three conditions are equivalent:
   (a) S is closed and bounded.
   (b) Every continuous function on S is bounded.
   (c) Every continuous function on S has a maximum.

• 7. Suppose that $S \subseteq \boldsymbol{R}$ and that for every continuous function f on S which satisfies $f(x) > 0$ for every point $x \in S$, the function f must have a minimum. Can one conclude from this that S is closed and bounded? If you answer *yes*, give a proof; if you say *no*, give an example.

8. Suppose that f is continuous on $\boldsymbol{R}$, that $f(0) = 1$, and that whenever $|x| > 6$, we have $f(x) > 1$. Prove that f has a minimum.

9. Suppose that f is a quartic polynomial of the form

   $$f(x) = x^4 + ax^3 + bx^2 + cx + d \qquad \text{for all} \quad x \in \boldsymbol{R}.$$

   Prove that $f(x) \to \infty$ as $x \to \infty$ and also as $x \to -\infty$. Now prove that f has a minimum.

10. Can you extend exercise 9 to more general polynomials of even degree?

11. Given that $f(x) = x^5 - 4x + 2$ for all $x \in \boldsymbol{R}$, prove that there are at least three real numbers x such that $f(x) = 0$. *Hint:* Use Bolzano's theorem.

12. Show that if $f(x) = \sqrt{x}$ for all $x \geq 0$, then f is continuous on the set $[0, \infty)$.

6.10 ONE–ONE FUNCTIONS AND MONOTONE FUNCTIONS

We begin by reviewing some definitions.

A function $f: S \to \boldsymbol{R}$ is said to be **increasing** on S if given any points x and t in S such that $x < t$, we have $f(x) \leq f(t)$; and we say that f is **strictly increasing** on S if given any points x and t in S such that $x < t$, we have $f(x) < f(t)$. *Decreasing* and *strictly decreasing functions* are defined similarly. A function which is either increasing or decreasing is said to be **monotone**, and a function which is either strictly increasing or strictly decreasing is said to be **strictly monotone**. A function $f: S \to \boldsymbol{R}$ is said to be **one–one** on S if given any two points x and t in S such that $x \neq t$, we have $f(x) \neq f(t)$. (Some further information about one–one functions can be found in Appendix A, Section A.3.4.)

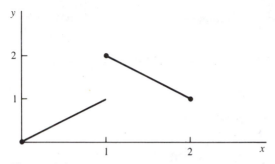

Figure 6.2

It goes without saying that every strictly monotone function must be one–one, but it is by no means true that every one–one function has to be strictly monotone. The following example shows what can go wrong.

6.10.1 Example. See Figure 6.2. We define $f: [0, 2] \to \boldsymbol{R}$ by $f(x) = x$ for $0 \le x < 1$, and $f(x) = 3 - x$ for $1 \le x \le 2$.

This function is obviously one–one, and both its domain and its range are equal to [0, 2]. But f is not monotone, and f is not continuous (because it isn't continuous at 1). So even when a function is one–one and both its domain and its range are intervals, the function doesn't have to be monotone or continuous. In this section you will see, however, that a one–one *continuous* function on an interval has to be monotone. You will also see that a monotone function on an interval will be continuous if and only if its range is an interval.

Roughly speaking, when a function is not monotone, either it must rise and then fall, or it must fall and then rise. The following useful definition of a *switch* will help us say this precisely and will simplify the proof of Theorem 6.10.5.

6.10.2 Definition of a Switch. Suppose that $f: S \to \boldsymbol{R}$. A **switch** of the function f consists of three points a, b, and c of S such that $a < b < c$ and either one of the following two conditions holds:

(1) $f(a) < f(b)$ and $f(b) > f(c)$.

(2) $f(a) > f(b)$ and $f(b) < f(c)$.

Figure 6.3 illustrates the two ways in which a switch can occur.

6.10.3 Lemma. A function with no switch must be monotone.

■ *Proof.* To obtain a contradiction, assume that $f: S \to \boldsymbol{R}$ and that f has no switch in S but that f fails to be monotone on S. Choose points a and b in S such that $a < b$ and $f(a) < f(b)$, and choose points c and d in S such that $c < d$ and

Figure 6.3

$f(c) > f(d)$. We now complete the proof by looking at each of the six cases specified in Figure 6.4.

For example, in the first case, since $f(a) < f(b)$ and f has no switch, we see that $f(b) \leq f(c)$. Therefore, $f(a) < f(c)$; and because f has no switch, $f(c) \leq f(d)$, a contradiction. The other five cases are similar.   ∎

6.10.4   Lemma.   Suppose that S is an interval, that $f: S \to R$, and that f is continuous and one–one on S. Then f has no switch in S.

■ Proof.   To obtain a contradiction, assume that f does have a switch in S. From Definition 6.10.2, f must have a switch either of type (1) or of type (2). Let us assume that it is of type (1); the other case is similar. Choose points a, b, and c in S such that $a < b < c$, $f(a) < f(b)$, and $f(b) > f(c)$. See Figure 6.5.

Choose a number y which is less than $f(b)$ but is greater than the larger of the two numbers $f(a)$ and $f(c)$. We now apply the intermediate value theorem to f on each of the intervals $[a, b]$ and $[b, c]$, and we choose a number $t \in (a, b)$ such that $f(t) = y$ and a number $x \in (b, c)$ such that $f(x) = y$. Then $f(x) = f(t)$ even though $t \neq x$, and this contradicts the fact that f is one–one.   ■

The following theorem follows at once from the two lemmas.

6.10.5   Theorem.   A one–one continuous function on an interval must be strictly monotone.

We now take a closer look at monotone functions. First, we shall show that unlike the "oscillating" functions that were discussed in Section 5.3.5, monotone functions always have one-sided limits. Then we shall keep the promise we made earlier and show that if the range of a monotone function is an interval, then the

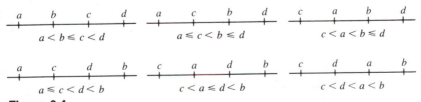

Figure 6.4

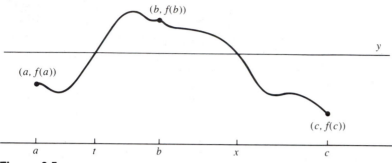

Figure 6.5

function must be continuous. Roughly speaking, the reason why this holds is that the only way a monotone function can fail to be continuous at a given point is that the function has a "jump" there. See Figure 6.6.

The following theorem provides a solution to Exercise 5.4(11).

6.10.6 Theorem: One-Sided Limits of an Increasing Function. Suppose that S is an interval, that $f: S \to R$, that f is increasing on S, and that $a \in S$.

(1) If a is not the left endpoint of S, then f has a limit from the left at a; and this limit is $\sup\{f(x) \mid x \in S$ and $x < a\}$.

(2) If a is not the right endpoint of S, then f has a limit from the right at a; and this limit is $\inf\{f(x) \mid x \in S$ and $x > a\}$.

■ **Proof.** We shall prove (1) and leave the proof of (2) as an exercise.

Assume that a is not the left endpoint of S, and define $\alpha = \sup\{f(x) \mid x \in S$ and $x < a\}$. We want to show that f has a limit from the left at a and that this limit equals α. Let V be a neighborhood of α. Choose a number $w < \alpha$ such that $(w, \alpha] \subseteq V$. Now using the fact that w is not an upper bound of the set $\{f(x) \mid x \in S$ and $x < a\}$, choose a point $c \in S$ such that $c < a$ and $f(c) > w$. Then for every point $x \in (c, a)$ we have $w < f(c) \le f(x) \le \alpha$, and this means that for every point $x \in (c, a)$ we have $f(x) \in (w, \alpha] \subseteq V$. ■

Figure 6.6

6.10.7 Theorem: One-Sided Limits of a Decreasing Function. Suppose that S is an interval, that $f: S \to R$, that f is decreasing on S, and that $a \in S$.

(1) If a is not the left endpoint of S, then f has a limit from the left at a; and this limit is $\inf\{f(x) \mid x \in S \text{ and } x < a\}$.

(2) If a is not the right endpoint of S, then f has a limit from the right at a; and this limit is $\sup\{f(x) \mid x \in S \text{ and } x > a\}$.

It might be a good exercise to write out the proof of at least one part of this theorem.

6.10.8 Theorem: Converse of Bolzano's Theorem for Monotone Functions. Suppose that S is an interval, that $f: S \to R$, that f is monotone, and that the range of f is an interval. Then f must be continuous on S.

■ **Proof.** We shall assume that f is increasing. The case in which f is decreasing can be treated in an analogous way, or one may simply apply the first case to $-f$. Now to obtain a contradiction, let us assume that f is not continuous on S. Choose a point $a \in S$ such that f is not continuous at a. There are two cases.

Case 1. The number a is not the left endpoint of S, and $f(a)$ is not the limit from the left of f at a. It follows from part (1) of Theorem 6.10.6 that $f(a) \neq \sup\{f(x) \mid x \in S \text{ and } x < a\}$, and this means that $f(a) > \sup\{f(x) \mid x \in S \text{ and } x < a\}$. Choose a number y such that $f(a) > y > \sup\{f(x) \mid x \in S \text{ and } x < a\}$. Then y is flanked on either side by points in the range of f, but obviously, y cannot belong to the range of f. This contradicts the fact that the range of f is an interval.

Case 2. The number a is not the right endpoint of S, and $f(a)$ is not the limit from the right of f at a. The proof in this case is similar to the proof given for Case 1 and will be omitted. ■

6.11 INVERSE FUNCTIONS

6.11.1 Definition of an Inverse Function. Suppose that f is a one–one function whose domain is S and whose range is T. For every point $y \in T$ there is one and only one point $x \in S$ such that $y = f(x)$. The **inverse function** f^{-1} of f is defined to be the function from T to S whose value $f^{-1}(y)$ at each point y of T is the unique point $x \in S$ for which $y = f(x)$.

Notice that f^{-1} is a one–one function whose domain is T and whose range is S, and that the inverse function of f^{-1} is f. Like many of the concepts that have been introduced in this book, the idea of an inverse function is probably not new to you. As you know, the functions exp and log are inverses of each other. So are sin (restricted to $[-\pi/2, \pi/2]$) and arcsin, and there are many other such pairs of functions that played a role in your elementary calculus courses. You will see inverse functions in a number of places in this book. For example, in Appendix A the concept of inverse function is a useful tool in our study of sets, and in Chapter 7 we are concerned with derivatives of inverse functions. In this section we are concerned with their continuity. Specifically, we shall see that if a function is one–one and continuous on an interval, then its inverse function is also continuous. This important result is called the (one-dimensional) inverse function theorem.

6.11.2 The Inverse Function Theorem. Suppose that S is an interval, that $f: S \to R$, and that f is one–one and continuous on S. Suppose that T is the range of f. Then the inverse function f^{-1} of f is continuous on T.

■ *Proof.* It follows at once from Bolzano's intermediate value theorem that T is an interval. Therefore, since both the domain and range of f^{-1} are intervals and f^{-1} is one–one, the continuity of f^{-1} will follow from Theorem 6.10.8 as soon as we have shown that this function is monotone. Now from Theorem 6.10.5 we deduce that f is strictly monotone. There are two cases to consider: Either f is increasing, or f is decreasing. We shall assume that f is increasing and leave the other case as an easy exercise. To show that f^{-1} is also increasing, suppose that y_1 and y_2 are any two points of T and that $y_1 < y_2$. Let us write $f^{-1}(y_1) = x_1$ and $f^{-1}(y_2) = x_2$. We need to see that $x_1 < x_2$. Clearly, $x_1 \neq x_2$; and since $y_1 = f(x_1)$ and $y_2 = f(x_2)$, and since f is increasing, it is impossible to have $x_1 > x_2$. Therefore, $x_1 < x_2$, as required. ■

6.12 UNIFORM CONTINUITY

The condition that a function f be continuous on a set S is too weak for some of the important theorems in the theory of integration. In order to arrive at these theorems, we need a condition that will tell us that a given function f is continuous at "about the same rate" throughout the set S, and for this purpose we introduce the concept of uniform continuity.

6.12.1 Definition of Uniform Continuity. Given $S \subseteq R$ and $f: S \to R$, we say that f is **uniformly continuous** on S if for every number $\varepsilon > 0$ there exists a number $\delta > 0$ such that for all points x and t in S satisfying $|x - t| < \delta$, we have $|f(x) - f(t)| < \varepsilon$.

6.12.2 A Comparison Between Uniform Continuity and Continuity. Suppose $f: S \to R$. The condition that f be continuous on S says that for every point $x \in S$, f is continuous at x. This is equivalent to saying the following:

> For every point $x \in S$ and for every number $\varepsilon > 0$ there exists a number $\delta > 0$ such that for all points $t \in S$ satisfying $|x - t| < \delta$, we have $|f(x) - f(t)| < \varepsilon$.

You should notice that in this statement of continuity of f on S, both the symbols x and ε were introduced with the quantifier $\forall$, and it therefore doesn't matter in which order these two symbols appear. We could just as well have stated the condition of continuity of f on S as follows:

> For every number $\varepsilon > 0$ and for every point $x \in S$, there exists a number $\delta > 0$ such that for all points $t \in S$ satisfying $|x - t| < \delta$, we have $|f(x) - f(t)| < \varepsilon$.

A comparison of this condition with the definition of uniform continuity may be reminiscent of some of the exercises of Chapter 1 which contain several pairs of similar-looking sentences that don't say the same thing. Notice that in both the previous condition and in the definition of uniform continuity, we want to guarantee the inequality $|f(x) - f(t)| < \varepsilon$, where ε is some given positive number. In both cases we are saying that we can guarantee this inequality provided that $|x - t|$ is small enough, smaller than an appropriate positive number δ. But the two concepts differ significantly in the way the number δ appears in each of them. In order for f to be uniformly continuous on S, we need to know that once ε has been prescribed, we must be able to find a single number $\delta > 0$ which will guarantee that for *all* pairs of points x and t in the set S satisfying $|x - t| < \delta$, we have $|f(x) - f(t)| < \varepsilon$. But on the other hand, in order for f to be just plain continuous on S, we only need to know how to find δ *after* we have already said at what point x the continuity of f is being investigated.

6.12.3 Some Examples of Continuous Functions Which Are Not Uniformly Continuous. The examples which follow will help us to appreciate the distinction between continuity and uniform continuity.

(1) We define $f(x) = x^2$ for every real number x. Of course, f, being a polynomial, must be continuous on $\mathbf{R}$. In order to show that f is not uniformly continuous on $\mathbf{R}$, we shall show that it is impossible to find a number $\delta > 0$ such that the inequality $|f(x) - f(t)| < 1$ holds for all numbers x and t which satisfy $|x - t| < \delta$. Let δ be any positive number. We shall find two numbers x and t such that $|x - t| < \delta$ and $|x^2 - t^2| > 1$. To do this, we shall choose two numbers x and t such that $x = t + \delta/2$ and such that x and t are large enough to ensure that $x^2 - t^2 > 1$. What we need is $(t + \delta/2)^2 - t^2 > 1$, that is, $\delta t + \delta^2/4 > 1$; and we shall therefore choose $t = 1/\delta$ and $x = 1/\delta + \delta/2$.

(2) [You should compare this example with Example 5.3.5(2).] We define $f(x) = \sin(1/x)$ for all $x > 0$. Because of the continuity of rational functions and our assumption of continuity of sin, it follows from the composition theorem that f is continuous on $(0, \infty)$. To show that f fails to be uniformly continuous on $(0, \infty)$, let $\delta > 0$. We now want to find two points x and t in $(0, \infty)$ such that $|x - t| < \delta$ and $|f(x) - f(t)| = 1$. To do this, we define $x = 1/n\pi$ and $t = 1/(n\pi + \pi/2)$, where n is a natural number chosen large enough to ensure that both x and t are less than δ:

These two examples show the two typical ways in which a continuous function might fail to be uniformly continuous. In example (1) the function grows too quickly, while in example (2) the function, though bounded, oscillates too much. Notice that in example (2) the trouble (in other words, the oscillating) occurs near 0, and the function would still fail to be uniformly continuous even if we were to restrict it to a smaller domain like (0, 1). Now we look at another oscillating function.

(3) We define $S = \{1/n \mid n \in \mathbf{Z}^+\}$, and we define $f(1/n) = (-1)^n$ for every $n \in \mathbf{Z}^+$. Since no limit point of S belongs to S, we know from Exercise 6.5(6) that f must be continuous on S. Now to show that f is not uniformly continuous on S, let $\delta > 0$. We want to find two points x and t in S such that $|x - t| < \delta$ and $|f(x) - f(t)| = 2$. To do this, choose a natural number $n > 1/\delta$, and define $x = 1/n$ and $t = 1/(n + 1)$:

Since one of the two natural numbers n and $n + 1$ must be even and the other must be odd, our choice of x and t obviously works.

(4) Our last example in this section is another fast grower. We define $f(x) = 1/x$ for every point $x \in (0, 1)$. The function f is certainly continuous on $(0, 1)$. Now let $\delta > 0$. Choose $x \in (0, \delta)$. Now using the fact that $f(t) \to \infty$ as $t \to 0$, choose a number $t \in (0, x)$ such that $f(t) > 1 + f(x)$. Then even though $|t - x| < \delta$, we have $|f(x) - f(t)| > 1$. Therefore, f cannot be uniformly continuous on $(0, 1)$.

6.12.4 Some Examples of Uniformly Continuous Functions

(1) Suppose that S is a bounded set of real numbers and that $f(x) = x^2$ for every point $x \in S$. Before we show that f is uniformly continuous on S, we shall use the boundedness of S to choose a number k such that for every point $x \in S$ we have $|x| < k$. Notice that for all points x and t in S we have

$$|x^2 - t^2| \le |x + t|\,|x - t| \le 2k|x - t|.$$

Now to show that f is uniformly continuous on S, let $\varepsilon > 0$. Define $\delta = \varepsilon/2k$. We see at once that for all points x and t in S satisfying $|x - t| < \delta$, we must have $|f(x) - f(t)| < \varepsilon$.

Example (1) shows that x^2 is well behaved if we keep x away from $\pm\infty$. In the next example we shall see that $1/x$ gives no trouble if we keep x away from 0.

(2) Suppose that $S \subseteq \mathbf{R}$ and that $0 \notin \bar{S}$, and define $f(x) = 1/x$ for every point $x \in S$. Before we show that f is uniformly continuous on S, we shall use the fact that $0 \notin \bar{S}$ and choose a number $\alpha > 0$ such that $(-\alpha, \alpha) \cap S = \emptyset$. Define $k = 1/\alpha^2$, and notice that for all points x and t in S we have $|f(x) - f(t)| \le k|x - t|$. Now to show that f is uniformly continuous on S, let $\varepsilon > 0$. Define $\delta = \varepsilon/k$. We see at once that for all points x and t in S satisfying $|x - t| < \delta$, we must have $|f(x) - f(t)| < \varepsilon$.

6.12.5 The Principal Theorem on Uniform Continuity. Suppose S is a closed, bounded set of real numbers. Then every function which is continuous on S will be uniformly continuous on S.

■ *Proof.* To obtain a contradiction, suppose that the theorem is false. Choose a function f continuous on S such that f is not uniformly continuous on S. Choose a number $\varepsilon > 0$ such that for every number $\delta > 0$ there are points x and t in S such that $|x - t| < \delta$, but $|f(x) - f(t)| \geq \varepsilon$.

For every natural number n, using the fact that $1/n > 0$, choose two points of S, which we shall call x_n and t_n, such that $|x_n - t_n| < 1/n$, but $|f(x_n) - f(t_n)| \geq \varepsilon$. Using Theorem 4.9.3, choose a partial limit x of (x_n) such that $x \in S$. Now we use the fact that f is continuous at x. Choose a number $\delta > 0$ such that for all points $t \in (x - \delta, x + \delta) \cap S$ we have $|f(t) - f(x)| < \varepsilon/2$. What we are going to do now is find a natural number n such that both x_n and t_n lie in the interval $(x - \delta, x + \delta)$. This will lead to an immediate contradiction, because in spite of the fact that $|f(x_n) - f(t_n)| \geq \varepsilon$, we must also have

$$|f(x_n) - f(t_n)| \leq |f(x_n) - f(x)| + |f(x) - f(t_n)| < \frac{\varepsilon}{2} + \frac{\varepsilon}{2} = \varepsilon.$$

So we need to find this number n. Using the fact that x is a partial limit of (x_n), choose $n > 2/\delta$ such that $|x_n - x| < \delta/2$:

Now we have $|t_n - x| \leq |t_n - x_n| + |x_n - x| < 1/n + \delta/2 < \delta/2 + \delta/2 = \delta$, and we deduce that both x_n and t_n lie in $(x - \delta, x + \delta)$, as required. ■

6.13 EXERCISES

1. Given that f is monotone on an open interval S and that $a \in S$, prove that f is continuous at a if and only if $\lim_{a-} f = \lim_{a+} f$.

2. Suppose that f is an increasing function on an interval $[a, b]$ and that $\delta > 0$. Prove that the number of points $x \in [a, b]$ at which $\lim_{x+} f - \lim_{x-} f > \delta$ does not exceed $[f(b) - f(a)]/\delta$.

*■ 3. (You should attempt this exercise only if you have read Appendix A.) Prove that if a function is monotone on an interval, then the set of points at which the function is discontinuous is countable.

4. Suppose that S is an interval, that $f: S \to R$, that f is one–one, and that whenever a and b are two points of S and $a < b$, the set $\{f(x) \mid a \leq x \leq b\}$ is convex. Prove that f has no switch, and deduce from this that f must be continuous.

5. Complete the following sentence: "Given $f: S \to R$, the condition that f should fail to be uniformly continuous on S is that there should exist a number $\varepsilon > 0$ such that for every $\delta > 0$ . . . ." Compare this exercise with Example 1.3.4(21).

6. If $f: S \to R$, then we say that f is **lipschitzian** if there exists a number k such that for all points x and t in S we have $|f(x) - f(t)| \le k|x - t|$. Prove that a lipschitzian function must be uniformly continuous.

7. Prove that the functions in Examples 6.12.4(1) and 6.12.4(2) are lipschitzian.

■ 8. Given that $f(x) = \sqrt{x}$ for all $x \in [0, 1]$, prove that f is uniformly continuous on $[0, 1]$ but not lipschitzian.

■ 9. (a) Given that $f: S \to R$ and that f is unbounded above, prove that there exists a sequence (x_n) in S such that for every $n \in Z^+$ we have $f(x_{n+1}) \ge 1 + f(x_n)$.
   (b) Given that $f: S \to R$, that the set S is bounded, and that f is uniformly continuous on S, prove that f must be bounded.

10. Given that f is uniformly continuous on S and that (x_n) is a Cauchy sequence in S, prove that the sequence $(f(x_n))$ is a Cauchy sequence. Take another look at Exercise 6.5(3) to make sure you understand why uniform continuity of f is really needed here.

11. (Go back and take a second look at Exercises 6.5(3) and 6.5(7) before you begin this exercise.) Suppose that f is uniformly continuous on S, that (x_n) is a sequence in S, and that (x_n) converges to some real number x. Prove that the sequence $(f(x_n))$ is convergent. *Hint:* Use exercise 10. You may *not* assume that $x \in S$.

12. Prove that if f is uniformly continuous on (a, b), then f has a limit from the right at a and from the left at b.

6.14* ADDITIONAL EXERCISES: UNIFORM CONTINUITY AND THE SMALL-DISTANCE PROPERTY

The small-distance property was defined in Section 3.12.

*1. Prove that if A and B are mutually disjoint, closed, bounded sets, then

$$\inf\{|x - t| \mid x \in A \text{ and } t \in B\} > 0.$$

*2. In exercise 1, what if only one of the two sets is bounded? What if neither of the two sets is bounded?

*3. Given that $E \subseteq R$ and that E has the small-distance property, prove that for every number a the set $E\backslash\{a\}$ must have the small-distance property. Prove that if F is any finite set, then $E\backslash F$ must have the small-distance property.

*4. Given that $E \subseteq \boldsymbol{R}$ and that E has the small-distance property, show that it is possible to find two mutually disjoint subsets A and B of E, of the form $A = \{a_n \mid n \in \boldsymbol{Z}^+\}$ and $B = \{b_n \mid n \in \boldsymbol{Z}^+\}$, such that for every $n \in \boldsymbol{Z}^+$ we have $|a_n - b_n| < 1/n$.

*5. Prove the following sharper version of Theorem 6.12.5: Suppose that a set S of real numbers has the property that for every subset A of S which has the small-distance property, the set A must have a limit point in S. Then every function which is continuous on S is uniformly continuous on S.

*6. In this exercise we obtain a converse to exercise 5. Suppose that $S \subseteq \boldsymbol{R}$, and suppose that S has a subset E which has the small-distance property but which has no limit point that lies in S. Prove that there exists a function $f: S \rightarrow [0, 1]$ which is continuous on S but not uniformly continuous on S. *Hint:* First use exercise 4. Then apply the technique of Exercise 6.5(4).

*7. Prove that if S is a set of real numbers and if every subset E of S which has the small-distance property must have a limit point in S, then it must be possible to write S as the union of a closed, bounded set and a set which does not have the small-distance property.

Chapter 7

Differentiation

7.1 INTRODUCTION

When you studied elementary calculus, you probably acquired only the vaguest notion of what limits and continuity are all about; but nevertheless, you still developed a good intuitive feel for those two great central themes of calculus, the derivative and the integral. As you know, the derivative $f'(x)$ of a function f at a point x is defined by

$$f'(x) = \lim_{h \to 0} \frac{f(x + h) - f(x)}{h},$$

or equivalently, by

$$f'(x) = \lim_{t \to x} \frac{f(t) - f(x)}{t - x}.$$

You know a variety of rules for working out derivatives; and you have been taught how to apply differentiation to a whole host of related topics such as curve sketching and maxima and minima, and perhaps even to science and economics.

In short, you have mastered the derivative of Newton and Leibniz, and you are ready now to see the derivative of Bolzano, Cauchy, and Weierstrass. In this chapter on differentiation you will use your newly acquired understanding of limits and continuity. First, you will develop a proper understanding of the definition of a derivative. Then you will see how, starting from the definition of a derivative, we can derive and *understand* all the usual rules for differentiation, like the product rule, the quotient rule, and the chain rule. And finally, we shall begin to look at the stuff that really matters: the theorems that depend upon the completeness of our number system. We shall ask, for example, why a function with an identically zero derivative has to be constant. On the face of it, this statement seems to be obvious, but it is not obvious at all; and as a matter of fact, it depends directly on the completeness of the number system R.

In order to gain some insight that will help us understand the role played by completeness in this innocent-looking fact about derivatives, let us digress for a moment and imagine that the number system with which we are working is the rational number system Q. In Q one can define neighborhoods, limits, and continuity, just as we defined them in R; and many of the basic theorems can be proved just as we proved them in R. On the other hand, if we are working in Q, then we have to exclude any theorem that depends on the existence of a supremum, which means that just about every result of any real interest will be lost. With this in mind, let us look at the following example of a function defined on Q:

Given any $x \in Q$, if $x \leq 0$ or $x^2 < 2$, we define $f(x) = 1$; and if $x > 0$ and $x^2 > 2$, we define $f(x) = 6$.

Notice that at every (rational) number x we have $f'(x) = 0$, even though f certainly isn't constant. Look at Figure 7.1. Does the graph of f jump anywhere? If you say yes, then your eyes are playing tricks on you. Your eyes are refusing to look at the rational number system Q and are, instead, looking at a solid line that includes the "forbidden" number $\sqrt{2}$, where from the standpoint of the real number system the graph of f has its jump. Inside Q, f is a perfectly smooth function whose derivative is zero everywhere, and yet f is not constant! This might serve as a warning that in the calculus of the number system R a logical path must be found from the definition of a derivative to the fact that a function with a zero derivative must be constant. This path is by no means trivial, and somewhere in it, the completeness of the real number system has to be used.

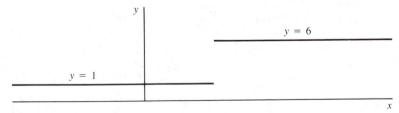

Figure 7.1

7.2 DEFINITION OF A DERIVATIVE; DIFFERENTIABILITY

Suppose that S is a set of real numbers and that $f: S \to R$; and suppose that $x \in S \cap \mathcal{L}(S)$. For each point $t \in S \setminus \{x\}$, define

$$q(t) = \frac{f(t) - f(x)}{t - x}.$$

If $\lim_{t \to x} q(t)$ exists, then we say that f is **differentiable** at x; and we define the **derivative** $f'(x)$ of f at x to be $\lim_{t \to x} q(t)$. In other words, we define

$$f'(x) = \lim_{t \to x} \frac{f(t) - f(x)}{t - x}.$$

The following three cases are of particular interest to us.

(1) S is an interval and x is not an endpoint of S. In this case $f'(x)$ is called the **two-sided derivative** of f at x.

(2) S is an interval and x is the left endpoint of S. In this case $f'(x)$ is called the derivative of f **from the right** at x.

(3) S is an interval and x is the right endpoint of S. In this case $f'(x)$ is called the derivative of f **from the left** at x.

Finally, if f is differentiable at every point of S, then we say that f is *differentiable on* S.

7.3 EXAMPLES

The following examples illustrate the definition of a derivative.

(1) Suppose $c \in R$, and define $f(t) = c$ for all $t \in R$. Then given any number x, defining

$$q(t) = \frac{f(t) - f(x)}{t - x}$$

for all $t \neq x$, we see that for all such t we have

$$q(t) = \frac{c - c}{t - x} = 0;$$

and it follows that $f'(x) = \lim_{t \to x} q(t) = 0$. Therefore, the derivative of a constant function is everywhere zero. Do not confuse this easy fact with the substantially more difficult result that we mentioned in the introduction (Section 7.1), namely, that a function with a zero derivative has to be constant.

(2) Define $f(t) = t^2$ for all $t \in \mathbf{R}$. Then given any number x, defining

$$q(t) = \frac{f(t) - f(x)}{t - x}$$

for all $t \neq x$, we see that for all such t we have

$$q(t) = \frac{t^2 - x^2}{t - x} = t + x;$$

and it follows that $f'(x) = \lim_{t \to x} (t + x) = 2x$.

(3) Suppose n is any natural number, and define $f(t) = t^n$ for all $t \in \mathbf{R}$. Then given any number x, defining

$$q(t) = \frac{f(t) - f(x)}{t - x}$$

for all $t \neq x$, we see that for all such t we have

$$q(t) = \frac{t^n - x^n}{t - x} = \frac{(t - x)(t^{n-1} + t^{n-2}x + t^{n-3}x^2 + \cdots + x^{n-1})}{t - x};$$

and it follows simply that $f'(x) = nx^{n-1}$. You may recall that this result was used to illustrate the principle of mathematical induction in Example 2.6.5(3). The induction proof makes use of the product rule for differentiation, which we shall prove as Theorem 7.9.3. In Section 7.8 we shall produce the familiar extension of the formula to more general values of n.

(4) Define $f(t) = t^{3/2}$ for all $t \geq 0$. Then given any $x \geq 0$, defining

$$q(t) = \frac{f(t) - f(x)}{t - x}$$

for all $t \in [0, \infty) \setminus \{x\}$, we have

$$q(t) = \frac{t^{3/2} - x^{3/2}}{t - x} = \frac{(t^{3/2} - x^{3/2})(t^{3/2} + x^{3/2})}{(t - x)(t^{3/2} + x^{3/2})}$$

$$= \frac{t^2 + tx + x^2}{t^{3/2} + x^{3/2}} \to \frac{3\sqrt{x}}{2} \qquad \text{as} \qquad t \to x.$$

Note that if $x = 0$, then this limit is automatically a derivative from the right; while if $x > 0$, we are looking at a two-sided derivative.

(5) Define $f(t) = t^2$ for all $t \in Q$, and define $f(t) = 0$ for all $t \in R\backslash Q$. Then for all $t \neq 0$ we have

$$\left| \frac{f(t) - f(0)}{t - 0} \right| \leq |t|;$$

and it follows from the sandwich theorem that $f'(0) = 0$.

(6) Define $f(t) = t$ for all $t \in Q$, and define $f(t) = 0$ for all $t \in R\backslash Q$. Then given $t \neq 0$, we have

$$\frac{f(t) - f(0)}{t - 0} = 1 \qquad \text{for} \qquad t \in Q,$$

and

$$\frac{f(t) - f(0)}{t - 0} = 0 \qquad \text{for} \qquad t \in R\backslash Q.$$

Since every neighborhood of 0 contains both rationals and irrationals, it follows that f is not differentiable at 0.

(7) See Figure 7.2. For every nonzero number t we define $f(t) = t \sin(1/t)$, and we define $f(0) = 0$. Then whenever $t \neq 0$, we have

$$\frac{f(t) - f(0)}{t - 0} = \sin \frac{1}{t}.$$

Now given any neighborhood U of 0, the numbers $1/n\pi$ and $1/(\pi/2 + n\pi)$ will lie in U whenever n is a sufficiently large natural number; and it follows simply that f is not differentiable at 0.

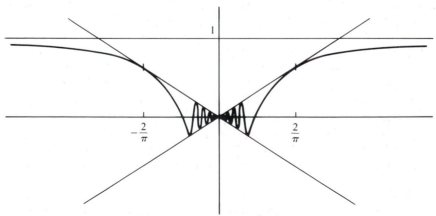

Figure 7.2

7.4   DIFFERENTIABILITY AND CONTINUITY

7.4.1   Theorem.   If a function f is differentiable at a number x, then f is continuous at x.

■ ***Proof.***   Suppose that $f: S \to \mathbf{R}$ and that $f'(x)$ exists. Since for every number $t \in S \backslash \{x\}$ we have

$$f(t) = \frac{f(t) - f(x)}{t - x} (t - x) + f(x),$$

it follows from the arithmetical rules for limits that

$$\lim_{t \to x} f(t) = f'(x) \cdot 0 + f(x);$$

and the continuity of f at x now follows at once from Theorem 6.2.2.   ■

7.5   THE RELATIONSHIP BETWEEN DIFFERENTIABILITY AND LIMITS OF SEQUENCES

7.5.1   Theorem.   Suppose $f: S \to \mathbf{R}$ and $x \in S \cap \mathcal{L}(S)$. Then the following two conditions are equivalent:

(1) f is differentiable at x.

(2) There exists a number w such that for every sequence (x_n) in the set $S \backslash \{x\}$ which converges to x, we have

$$\frac{f(x_n) - f(x)}{x_n - x} \to w \quad \text{as} \quad n \to \infty.$$

Furthermore, if condition (2) holds, then the number w must be $f'(x)$.

This theorem follows at once from Theorem 5.5.1 which relates limits of functions and limits of sequences. In the following useful variation of the theorem we allow the possibility $x_n = x$.

7.5.2   Theorem.   Suppose that $f: S \to \mathbf{R}$ and that f is differentiable at a point x of S.[1] Suppose that (x_n) is a sequence in S which converges to x, and suppose that the sequence (α_n) is defined as follows:

$$\alpha_n = \frac{f(x_n) - f(x)}{x_n - x} \quad \text{whenever} \quad n \in \mathbf{Z}^+ \quad \text{and} \quad x_n \neq x;$$

and

$$\alpha_n = f'(x) \quad \text{whenever} \quad n \in \mathbf{Z}^+ \quad \text{and} \quad x_n = x.$$

Then the sequence (α_n) converges to $f'(x)$.

[1] This automatically includes the assumption that $x \in S \cap \mathcal{L}(S)$.

■ **Proof.**   Let V be a neighborhood of $f'(x)$, and choose a neighborhood U of x such that for every point $t \in U \cap S\backslash\{x\}$ we have $[f(t) - f(x)]/(t - x) \in V$. Now using the fact that $x_n \to x$, choose a natural number N such that whenever $n \geq N$, we have $x_n \in U$. Now given any natural number $n \geq N$, there are two possibilities: Either $x_n = x$, or $x_n \neq x$; and in both of these cases it is obvious that $\alpha_n \in V$. ■

7.6   THE CHAIN RULE: THE COMPOSITION THEOREM FOR DERIVATIVES

7.6.1   Theorem.   Suppose that S and T are subsets of $\boldsymbol{R}$, that $f: S \to T$, and that $g: T \to \boldsymbol{R}$. Suppose that f is differentiable at a given point $x \in S$ and that g is differentiable at the point $f(x)$. Then the function $g \circ f$ must be differentiable at x, and we have

$$(g \circ f)'(x) = g'(f(x)) \cdot f'(x).$$

Even though you must have used the chain rule hundreds of times in your first calculus course, it might be a good idea to illustrate it once more with a simple example before we finally prove it. We shall take $f(x) = x^3 + 1$ for all $x \in \boldsymbol{R}$ and $g(u) = u^2$ for all $u \in \boldsymbol{R}$. Then for every $x \in \boldsymbol{R}$ we see that $g \circ f(x) = (x^3 + 1)^2 = x^6 + 2x^3 + 1$, and $(g \circ f)'(x) = 6x^5 + 6x^2$. At the same time,

$$g'(f(x)) \cdot f'(x) = 2f(x) \cdot 3x^2 = 2(x^3 + 1) \cdot 3x^2 = 6x^5 + 6x^2.$$

■ **Proof.**   Let (x_n) be any sequence in $S\backslash\{x\}$ which converges to x. We need to show that

$$\frac{g \circ f(x_n) - g \circ f(x)}{x_n - x} \to g'(f(x)) \cdot f'(x) \qquad \text{as} \qquad n \to \infty.$$

Note that

$$\frac{f(x_n) - f(x)}{x_n - x} \to f'(x) \qquad \text{as} \qquad n \to \infty,$$

and since f, being differentiable at x, must be continuous at x, we have $f(x_n) \to f(x)$. With Theorem 7.5.2 in mind we now define a sequence (α_n) as follows:

$$\alpha_n = \frac{g(f(x_n)) - g(f(x))}{f(x_n) - f(x)} \qquad \text{whenever} \qquad f(x_n) \neq f(x);$$

and

$$\alpha_n = g'(f(x)) \qquad \text{when} \qquad f(x_n) = f(x).$$

From Theorem 7.5.2 we deduce that $\alpha_n \to g'(f(x))$. Now for every natural number n [regardless of whether or not $f(x_n) = f(x)$], we have

$$\frac{g \circ f(x_n) - g \circ f(x)}{x_n - x} = \alpha_n \left[\frac{f(x_n) - f(x)}{x_n - x} \right],$$

and so the result follows at once, letting $n \to \infty$. ■

7.7 DIFFERENTIATION OF INVERSE FUNCTIONS

Before you begin this section, you should make sure that you have a proper understanding of Section 6.11 where we discussed the continuity of inverse functions. In particular, you should understand that a continuous one–one function on an interval has to be strictly monotone and that its inverse function must also be continuous.

In order to motivate the following theorem, we observe that if f is a one–one differentiable function on an interval S, and if the inverse function g of f is differentiable at the point $f(x)$ for each $x \in S$, then an application of the chain rule to the identity $x = g(f(x))$ yields $1 = g'(f(x))f'(x)$.

7.7.1 Theorem. Suppose that f is a one–one continuous function on an interval S and that g is the inverse function of f. Suppose that f is differentiable at a given point $x \in S$ and that $f'(x) \neq 0$. Then g must be differentiable at the point $f(x)$, and we have

$$g'(f(x)) = \frac{1}{f'(x)}.$$

■ **Proof.** We shall write the range of f as T. Then T is an interval, and g is continuous on T. Now let (y_n) be a sequence in $T \setminus \{f(x)\}$, which converges to $f(x)$. We need to show that

$$\frac{g(y_n) - g(f(x))}{y_n - f(x)} \to \frac{1}{f'(x)} \qquad \text{as} \qquad n \to \infty.$$

For each $n \in \mathbf{Z}^+$, define $x_n = g(y_n)$. Then (x_n) is a sequence in $S \setminus \{x\}$; and by the continuity of g we have $x_n \to x$. It follows from the fact that f is differentiable at x that

$$\frac{g(y_n) - g(f(x))}{y_n - f(x)} = \frac{x_n - x}{f(x_n) - f(x)} = \left[\frac{f(x_n) - f(x)}{x_n - x} \right]^{-1} \to \frac{1}{f'(x)},$$

as required. ■

7.8 THE POWER RULE

Roughly speaking, the power rule says that given any number α, if $f(t) = t^\alpha$ whenever t^α is defined, then for any number x for which $f(x)$ is defined, f must be differentiable at x, and $f'(x) = \alpha x^{\alpha-1}$. The easiest case of the power rule occurs when α is a natural number; and as we saw in Example 7.3(3), the rule holds in this case without any restrictions on x. We also saw the rule in Example 7.3(4), where we took $\alpha = \frac{3}{2}$, and in this case we had to have $x \geq 0$. Unfortunately, the most general form of the rule is not available to us in this chapter because we are not yet in a position to give a definition of t^α if α is irrational. This will be done in Section 9.2. When α is rational, however, we can appeal to Corollary 6.8.2 and give the usual definition of t^{α} for $t > 0$, this being $(t^{1/n})^m$, where α has reduced form m/n. We shall restrict ourselves to rational exponents in this section and leave the most general case until after we have defined the exponential function in Chapter 9. In the first result of this section we look at the case $\alpha = -1$.

7.8.1 Lemma. Suppose $f(x) = 1/x$ for all $x \neq 0$. Then given any number $x \neq 0$, we have $f'(x) = -1/x^2$.

We leave the proof of this lemma as a simple exercise that you have probably done many times before.

7.8.2 Lemma. Suppose n is any integer, and for all $x \neq 0$, define $f(x) = x^n$. Then given any number $x \neq 0$, we have $f'(x) = nx^{n-1}$.

■ **Proof.** In Example 7.3(3) we saw that this theorem holds for $n > 0$, and the theorem is trivial if $n = 0$. Assume now that $n < 0$. Define $m = -n$, and define $g(x) = 1/x$ for every nonzero number x. We see now that for every number $x \neq 0$ we have $f(x) = (g(x))^m$; and it follows at once from the chain rule that $f'(x) = m(g(x))^{m-1}g'(x) = m(1/x)^{m-1}(-1/x^2) = nx^{n-1}$. ■

7.8.3 Lemma. Suppose n is any natural number; and for all positive[2] numbers x, define $f(x) = x^{1/n}$. Then given any number $x > 0$ we have $f'(x) = (1/n)x^{(1/n)-1}$.

■ **Proof.** Define g to be the inverse function of f; in other words, $g(y) = y^n$ for every $y > 0$. Now let $x > 0$, and define $y = f(x)$. By Theorem 7.7 we have

$$f'(x) = \frac{1}{g'(y)} = \frac{1}{ny^{n-1}} = \left(\frac{1}{n}\right)x^{(1/n)-1}.$$
■

7.8.4 Theorem. Suppose α is any rational number; and for all $x > 0$, define $f(x) = x^{\alpha}$. Then for every $x > 0$ we have $f'(x) = \alpha x^{\alpha-1}$.

■ **Proof.** Suppose α has reduced form m/n. Then for every $x > 0$ we have $f(x) = (x^{1/n})^m$, and the result follows at once from the chain rule. ■

7.8.5 Theorem. Suppose that $f: S \to R$, that f is differentiable at a point x of S, and that $f(t) > 0$ for all $t \in S$. Suppose α is any rational number, and define $h(t) = [f(t)]^{\alpha}$ for all $t \in S$. Then we have

$$h'(x) = \alpha[f(x)]^{\alpha-1}f'(x).$$

■ **Proof.** In view of Theorem 7.8.4, this theorem follows at once from the chain rule. ■

---

[2] In the event that n is odd, we can relax this requirement a little and assume merely that $x \neq 0$.

7.9   THE ARITHMETICAL RULES FOR DIFFERENTIATION

7.9.1   The Sum Rule.   Suppose that both f and g are defined on a set S and are differentiable at a point $x \in S$. Then $f + g$ is differentiable at x, and we have $(f + g)'(x) = f'(x) + g'(x)$.

■ **Proof.**   For every $t \in S\backslash\{x\}$ we have

$$\frac{(f + g)(t) - (f + g)(x)}{t - x} = \frac{f(t) - f(x)}{t - x} + \frac{g(t) - g(x)}{t - x}$$

and the result follows at once by letting $t \to x$.   ■

We won't bother even to state the rule for differences of functions.

7.9.2   The Constant-Multiple Rule.   (A special case of the product rule.) Suppose that $f: S \to R$ and that f is differentiable at a point x of S. Suppose $c \in R$, and define $h(t) = cf(t)$ for every $t \in S$. Then h is differentiable at x, and we have $h'(x) = cf'(x)$.

We leave the simple proof of this rule as an exercise.

7.9.3   The Product Rule.   Suppose that both f and g are defined on a set S and are differentiable at a point $x \in S$. Then fg is differentiable at x, and we have $(fg)'(x) = f'(x)g(x) + f(x)g'(x)$.

■ **First Proof.**   (We use the preceding arithmetical rules and the power rule.) We observe that

$$fg = \tfrac{1}{2}(f + g)^2 - \tfrac{1}{2}f^2 - \tfrac{1}{2}g^2,$$

and it therefore follows at once that

$$(fg)'(x) = \tfrac{1}{2} \cdot 2[f(x) + g(x)][f'(x) + g'(x)]$$

$$- \tfrac{1}{2} \cdot 2f(x)f'(x) - \tfrac{1}{2} \cdot 2g(x)g'(x)$$

$$= f'(x)g(x) + f(x)g'(x).$$

■ **Second Proof.**   (This is the proof you know from elementary calculus.) For every point $t \in S\backslash\{x\}$ we have

$$\frac{f(t)g(t) - f(x)g(x)}{t - x} = \frac{f(t)g(t) - f(x)g(t) + f(x)g(t) - f(x)g(x)}{t - x}$$

$$= g(t)\left[\frac{f(t) - f(x)}{t - x}\right] + f(x)\left[\frac{g(t) - g(x)}{t - x}\right],$$

and the result follows by letting $t \to x$. Be careful not to overlook the fact that as we do this, we use the continuity of the function g at x.   ■

7.9.4 The Quotient Rule. Suppose f and g are defined on a set S and are differentiable at a point x of S; and suppose $g(t) \neq 0$ for all $t \in S$. Then f/g is differentiable at x, and we have

$$\left(\frac{f}{g}\right)'(x) = \frac{g(x)f'(x) - f(x)g'(x)}{[g(x)]^2}.$$

■ *Proof.* When f is just the constant function 1, this theorem follows from Lemma 7.8.1 and the chain rule. In general, we can write $f/g = (f)(1/g)$, and the theorem follows at once from the product rule. Of course, if you want to, you can prove the quotient rule directly. ■

7.10 HIGHER–ORDER DERIVATIVES

Suppose that $f: S \to R$ and that f is differentiable at every point of S. Then f' is also a function defined on S, and it may happen that f' is itself differentiable at a given point $x \in S$. If so, then the derivative of f' at x is (of course) called the **second derivative** of f at x and is written as $f''(x)$. The third and subsequent derivatives are defined similarly; and as you know, an alternative notation for the second derivative of f is $f^{(2)}$, and in general, the notation for the nth derivative of f is $f^{(n)}$. In keeping with this notation, the symbol $f^{(0)}$ stands for the function f differentiated no times—in other words, for f itself.

7.11 EXERCISES

1. Given $f(x) = |x|$ for all x, prove that $f'(0)$ does not exist.

2. Given $f(x) = |x|$ for all $x \in [-2, -1] \cup [0, 1]$, prove that $f'(0)$ does exist.

3. Given $f(x) = x^2 \sin(1/x)$, draw a rough sketch of the graph of f (especially near the origin), and prove that f is differentiable on R but that f' is not continuous at 0. (You may assume the usual formulas for the derivatives of the trigonometric functions.)

4. Given $f(x) = x^3 \sin(1/x)$, draw a rough sketch of the graph of f (especially near the origin), and prove that f' is continuous at 0 but that f' is not differentiable at 0.

■ 5. Given that $f: S \to R$ and that for some given point $x \in S$ we have $f'(x) > 0$, prove that there exists a neighborhood U of x such that for all points $t \in U \cap S \setminus \{x\}$, if $t < x$, then we have $f(t) < f(x)$; and if $t > x$, then we have $f(t) > f(x)$.

6. State an analogue of exercise 5 in which $f'(x) < 0$.

7. Find an example of a strictly increasing polynomial whose derivative is not always positive. Why doesn't this example contradict exercises 5 and 6?

7.12 THE MEAN VALUE THEOREMS

We are ready now for the most important, the most powerful, and the most interesting properties of derivatives, these being the properties that depend on the completeness of our number system. The main results of this section stem from two important theorems called the *intermediate value theorem for derivatives* and *Rolle's theorem,* both of which tell us that if a function satisfies certain conditions, then its derivative has to vanish somewhere. As we shall see in a moment, the derivative of a function has to be zero at points where the function is a maximum or a minimum; and what we therefore need, if we are to find points where the derivative of a given function vanishes, is a theorem that guarantees the existence of maximum and minimum points. This guarantee is provided by Corollary 6.7.3, and it is by making use of Corollary 6.7.3 that we invoke the completeness of the number system in this section. We begin by taking a careful look at Exercise 7.11(5) which contains the technical information that we need.

7.12.1 Solution of Exercise 7.11(5). For convenience, we state the exercise again:

> Given that $f: S \to R$ and that for some given point $x \in S$ we have $f'(x) > 0$, prove that there exists a neighborhood U of x such that for all points $t \in U \cap S \backslash \{x\}$, if $t < x$, then we have $f(t) < f(x)$; and if $t > x$, then we have $f(t) > f(x)$.

Using the fact that $(0, \infty)$ is a neighborhood of $f'(x)$, choose a neighborhood U of x such that for all points $t \in U \cap S \backslash \{x\}$ we have $[f(t) - f(x)]/(t - x) > 0$. Thus $f(t) - f(x)$ and $t - x$ have the same sign, and so U is the neighborhood we were looking for.

7.12.2 Lemma. Suppose that S is an interval, that $f: S \to R$, and that f is differentiable at a point a of S.

(1) If $f(a)$ is the maximum value of f and a is not the left endpoint of S, then we have $f'(a) \geq 0$.

(2) If $f(a)$ is the minimum value of f and a is not the left endpoint of S, then we have $f'(a) \leq 0$.

(3) If $f(a)$ is the maximum value of f and a is not the right endpoint of S, then we have $f'(a) \leq 0$.

(4) If $f(a)$ is the minimum value of f and a is not the right endpoint of S, then we have $f'(a) \geq 0$.

(5) If a is not an endpoint of S and $f(a)$ is either the maximum or the minimum value of f, then $f'(a) = 0$.

■ *Proof.* To prove part (1), assume that $f(a)$ is the maximum value of f and that a is not the left endpoint of S. To obtain a contradiction, assume that $f'(a) < 0$.

Using Exercise 7.11(6), choose a neighborhood U of a such that whenever $x \in U \cap S$ and $x < a$, we have $f(x) > f(a)$. Since a is not the left endpoint of S, there are such numbers x; and this yields the desired contradiction. Parts (2), (3), and (4) follow similarly, and part (5) follows automatically from the first four. ∎

7.12.3 An Intermediate Value Theorem for Derivatives. Suppose that f is differentiable on $[a, b]$, where $a < b$, and that either of the following two conditions holds:

(1) $f'(a) < 0$ and $f'(b) > 0$.

(2) $f'(a) > 0$ and $f'(b) < 0$.

Then it is possible to find a number $c \in (a, b)$ such that $f'(c) = 0$.

■ **Proof.** We shall assume that $f'(a) < 0$ and $f'(b) > 0$, leaving the other case as a simple exercise. Since $a < b$, we make the obvious observation that a is not the right endpoint of the interval $[a, b]$, and b is not the left endpoint. It therefore follows from Lemma 7.12.2, parts (2) and (4), that the minimum value of f cannot occur at either of the two endpoints a and b. But we know from Corollary 6.7.3 that f does have a minimum, and the minimum therefore occurs at some point $c \in (a, b)$. By part (5) of Lemma 7.12.2, we have $f'(c) = 0$. ∎

The following theorem is an improvement on Theorem 7.12.3.

7.12.4 Darboux's Intermediate Value Theorem. Suppose S is an interval and f is differentiable on S. Then the range of f' is an interval.

■ **Proof.** We need to see, of course, that the range of f' is convex. Suppose, then, that a and b are points of S, and that $f'(a) < \alpha < f'(b)$. In order to find a number c between a and b such that $f'(c) = \alpha$, we shall assume that $a < b$. The case $b < a$ is analogous. For every point $x \in S$, define $g(x) = f(x) - \alpha x$. Since $g'(x) = f'(x) - \alpha$ for each x, we have $g'(a) < 0$ and $g'(b) > 0$. Therefore, by Theorem 7.12.3 there must be a number $c \in (a, b)$ such that $g'(c) = 0$. Of course, $f'(c) = \alpha$. ∎

7.12.5 Rolle's Theorem. Suppose a and b are real numbers and $a < b$. Suppose that f is differentiable on (a, b) and continuous on $[a, b]$,[3] and suppose that $f(a) = f(b)$. Then there is at least one number $c \in (a, b)$ such that $f'(c) = 0$.

---

[3] There is, of course, some overlap in these assumptions. Naturally, if f is differentiable on (a, b), then f is automatically continuous on (a, b); and our stated requirement that f be continuous on $[a, b]$ can therefore be interpreted to mean that f is continuous at the two endpoints. This means, of course, that f is continuous from the right at a and from the left at b.

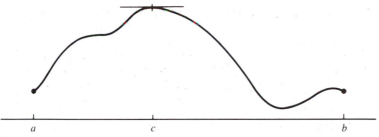

Figure 7.3

■ ***Proof.***   See Figure 7.3. If $f(a)$ is both the maximum and the minimum value of f on $[a, b]$, then f has to be constant on $[a, b]$; and we will have $f'(c) = 0$ for every $c \in [a, b]$. Otherwise, either the maximum or the minimum value of f must occur at a point $c \neq a$, not at a. Since $f(a) = f(b)$, it is clear that $c \neq b$; and it follows from Lemma 7.12.2, part (5), that $f'(c) = 0$.   ■

Our next theorem is the best known of all the mean value theorems, and it is sometimes known as *the* mean value theorem. We shall call it the *first mean value theorem* because, as you will see in the next section, the theorem is really just the first in a sequence of similar results.

7.12.6   The First Mean Value Theorem.   Suppose a and b are real numbers and $a < b$. Suppose f is differentiable on (a, b) and continuous on $[a, b]$. Then there must be at least one number $c \in (a, b)$ such that

$$f(b) = f(a) + (b - a)f'(c).$$

Before we prove the theorem, we might note that if we write the latter equation in the form

$$f'(c) = \frac{f(b) - f(a)}{b - a},$$

we can see a nice geometric interpretation of the first mean value theorem. See Figure 7.4. We can interpret $[f(b) - f(a)]/(b - a)$ to be the slope of the straight

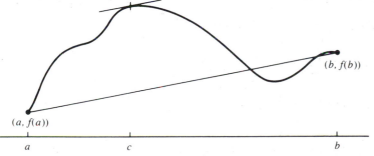

$(b, f(b))$

$(a, f(a))$

Figure 7.4

line that joins the two points $(a, f(a))$ and $(b, f(b))$, and the theorem can be seen as saying that there must be at least one point on the graph of f where the tangent line is parallel to the straight line that joins $(a, f(a))$ and $(b, f(b))$. Notice that in the event that $f(a) = f(b)$, the first mean value theorem reduces to Rolle's theorem.

▪ *Proof.* For each $x \in [a, b]$, define $h(x) = f(x) - f(a) - k(x - a)$, where the number k is chosen in such a way that $h(b) = 0$. As a matter of fact, this value of k is obviously $[f(b) - f(a)]/(b - a)$. Since h is differentiable on (a, b) and continuous on $[a, b]$, and $h(a) = h(b) = 0$, we can apply Rolle's theorem to h and choose a number $c \in (a, b)$ such that $h'(c) = 0$. But for every $x \in (a, b)$ we have $h'(x) = f'(x) - k$, and it follows that $f'(c) - k = 0$. This is the required result. ▪

7.12.7 The Cauchy Mean Value Theorem. Suppose a and b are real numbers and $a < b$. Suppose that f and g are differentiable on (a, b) and continuous on $[a, b]$, and that $g'(x) \neq 0$ for all $x \in (a, b)$. Then there must be at least one number $c \in (a, b)$ such that

$$\frac{f'(c)}{g'(c)} = \frac{f(b) - f(a)}{g(b) - g(a)}.$$

▪ *Proof.* Note first that since $g'(x) \neq 0$ for all $x \in (a, b)$, it follows from Rolle's theorem that $g(b) \neq g(a)$. For each $x \in [a, b]$, define

$$h(x) = f(x) - f(a) - k[g(x) - g(a)],$$

where the number k is chosen in such a way that $h(b) = 0$. The value of k is, of course, $[f(b) - f(a)]/[g(b) - g(a)]$. Now using the fact that $h(b) = h(a) = 0$, we apply Rolle's theorem to h on $[a, b]$ and choose a point $c \in (a, b)$ such that $h'(c) = 0$. But $h'(c) = f'(c) - kg'(c)$, and this gives us the desired result. ▪

It is worth noting that in the special case $g(x) = x$ for all x, the Cauchy mean value theorem reduces to the first mean value theorem. The Cauchy mean value theorem has some important applications, which will be explored in depth in Section 7.15, where we shall study L'Hospital's rule. The following simple form of L'Hospital's rule will serve as a preview to the theorems of Section 7.15.

7.12.8 A Simple Form of L'Hospital's Rule. Suppose a and b are real numbers and $a < b$. Suppose that f and g are differentiable on (a, b) and continuous on $[a, b]$, and that $g'(x) \neq 0$ for all $x \in (a, b)$. Suppose further that $f(b) = g(b) = 0$ and that $f'(x)/g'(x)$ approaches a limit α as $x \to b$. Then $f(x)/g(x) \to \alpha$ as $x \to b$.

▪ *Proof.* Let (x_n) be a sequence in $[a, b)$ such that $x_n \to b$. We need to show that $f(x_n)/g(x_n) \to \alpha$ as $n \to \infty$. For each n, using the Cauchy mean value theorem on the interval $[x_n, b]$, choose a number $t_n \in (x_n, b)$ such that

$$\frac{f'(t_n)}{g'(t_n)} = \frac{f(x_n) - f(b)}{g(x_n) - g(b)}.$$

Since $t_n \to b$ as $n \to \infty$, we see that

$$\frac{f(x_n)}{g(x_n)} = \frac{f(x_n) - f(b)}{g(x_n) - g(b)} = \frac{f'(t_n)}{g'(t_n)} \to \alpha \qquad \text{as} \qquad n \to \infty,$$

as required. ∎

7.13   THE MEAN VALUE THEOREMS OF HIGHER ORDER

7.13.1   Some Important Facts About Polynomials.

In this section we shall make the interesting observation that if f is a polynomial, for example, if we have

$$f(x) = c_0 + c_1 x + c_2 x^2 + \cdots + c_n x^n = \sum_{i=0}^{n} c_i x^i \qquad \text{for all } x,$$

then the coefficients $c_0, c_1, \ldots, c_n$ of f are determined uniquely by the behavior of f at the point 0.

In the case of c_0 this is obvious, because $c_0 = f(0)$. If we now differentiate f, we obtain

$$f'(x) = c_1 + 2c_2 x^1 + 3c_3 x^2 + \cdots + nc_n x^{n-1},$$

and from this it follows that $f'(0) = c_1$. Differentiating again, we obtain

$$f''(x) = 2 \cdot 1 c_2 + 3 \cdot 2 c_3 x + \cdots + n(n-1)c_n x^{n-2},$$

and this time we obtain $f''(0) = 2! c_2$. Continuing in this way, we see easily that for every $i = 0, 1, \ldots, n$ we have $c_i = f^{(i)}(0)/i!$.

As an application, let us consider the function f defined by $f(x) = (1 + x)^n$ for all $x \in \mathbf{R}$, where n is some natural number. It is clear that f is a polynomial of the form

$$f(x) = c_0 + c_1 x + c_2 x^2 + \cdots + c_n x^n,$$

and we can use the previous method to evaluate the coefficients. Clearly, $c_0 = f(0) = 1$, $c_1 = f'(0) = n$, $c_2 = f''(0)/2! = n(n-1)/2!$, and so forth; and continuing in this way, we obtain an easy proof of the **binomial theorem**:

$$(1 + x)^n = 1 + nx + \frac{n(n-1)}{2!} x^2 + \frac{n(n-1)(n-2)}{3!} x^3 + \cdots + \frac{n!}{n!} x^n.$$

7.13.2   An Extension of the Previous Idea.

Suppose f is a polynomial of degree not exceeding n, suppose a is any real number, and define $g(x) = f(x + a)$ for all x. If f has the form

$$f(x) = c_0 + c_1 x + c_2 x^2 + \cdots + c_n x^n \qquad \text{for all } x,$$

then for every x we have

$$g(x) = c_0 + c_1(x + a) + c_2(x + a)^2 + \cdots + c_n(x + a)^n,$$

and it follows that g, like f, is a polynomial of degree not exceeding n. It therefore follows from the previous section that for every x we have

$$g(x) = \sum_{i=0}^{n} \frac{g^{(i)}(0)}{i!} x^i.$$

Now since $g(x) = f(x + a)$ for all x, it follows from the chain rule that $g'(x) = f'(x + a)$ for all x; and in general, $g^{(i)}(x) = f^{(i)}(x + a)$ for every natural number i. This shows that for all x

$$f(x) = g(x - a) = \sum_{i=0}^{n} \frac{g^{(i)}(0)}{i!} (x - a)^i = \sum_{i=0}^{n} \frac{f^{(i)}(a)}{i!} (x - a)^i.$$

7.13.3 Taylor Polynomials.   Given a function f which has derivatives of all orders up to a natural number n at a number a, the **nth Taylor polynomial** *of f centered at a* means the polynomial p_n defined by

$$p_n(x) = \sum_{i=0}^{n} \frac{f^{(i)}(a)}{i!} (x - a)^i \qquad \text{for all } x \in \mathbf{R}.$$

From the two previous sections we see that in the event that f is itself a polynomial of degree not exceeding n, then $f = p_n$.

7.13.4 Some Examples

(1) Suppose that f is the polynomial defined by

$$f(x) = 2 - 4x + 3x^2 + 7x^3 \quad \text{for all } x \in \mathbf{R}.$$

Then for all x we have

$$p_0(x) = 2,$$
$$p_1(x) = 2 - 4x,$$
$$p_2(x) = 2 - 4x + 3x^2;$$

and for $n \geq 3$ we have $p_n = f$.

(2) The seventh Taylor polynomials centered at 0 of sin, cos, and exp are, respectively,

$$p_7(x) = x - \frac{x^3}{3!} + \frac{x^5}{5!} - \frac{x^7}{7!},$$

$$p_7(x) = 1 - \frac{x^2}{2!} + \frac{x^4}{4!} - \frac{x^6}{6!},$$

and

$$p_7(x) = 1 + x + \frac{x^2}{2!} + \frac{x^3}{3!} + \frac{x^4}{4!} + \frac{x^5}{5!} + \frac{x^6}{6!} + \frac{x^7}{7!}.$$

7.13.5 Lemma. Suppose that p_n is the nth Taylor polynomial of f centered at a. Then for every $i = 0, 1, 2, \ldots, n$ we have $p_n^{(i)}(a) = f^{(i)}(a)$.

■ **Proof.** By definition, we have

$$p_n(x) = f(a) + \frac{f'(a)}{1!}(x - a) + \frac{f''(a)}{2!}(x - a)^2 + \cdots + \frac{f^{(n)}(a)}{n!}(x - a)^n$$

for all x. From this we see at once that $p_n(a) = f(a)$, $p_n'(a) = f_n'(a)$; and just as we did in Section 7.13.1, we may continue in this way and show that for all $i = 0, 1, 2, \ldots, n$ we have $p_n^{(i)}(a) = f^{(i)}(a)$. ■

7.13.6 The Taylor Mean Value Theorem. Suppose that p_n is the nth Taylor polynomial of a function f centered at a. From the previous sections it seems reasonable to expect that p_n should be a pretty good approximation to f, at least near a. As we saw, not only must this polynomial agree with f at the point a, but we must also have $f'(a) = p_n'(a)$, and $f^{(2)}(a) = p_n^{(2)}(a)$, and $f^{(3)}(a) = p_n^{(3)}(a), \ldots,$ $f^{(n)}(a) = p_n^{(n)}(a)$. Furthermore, if f happens to be a polynomial of degree not exceeding n, then p_n will be exactly equal to f. The Taylor mean value theorem helps us understand just how well the polynomial p_n approximates f at any given number b. What it actually says is that the difference $f(b) - p_n(b)$ can be expressed in the form

$$\frac{f^{(n+1)}(c)}{(n + 1)!}(b - a)^{n+1}$$

where c is some (unknown) number lying between a and b:

It is interesting to notice that this expression for $f(b) - p_n(b)$ is very similar to the last term of $p_{n+1}(b)$, which is

$$\frac{f^{(n+1)}(a)}{(n + 1)!}(b - a)^{n+1}.$$

The precise statement of the Taylor mean value theorem is as follows:

Taylor Mean Value Theorem. Suppose a and b are real numbers and $a < b$. Suppose that n is a nonnegative integer, that f is a function defined on $[a, b]$, and that $f^{(n)}$ is differentiable on (a, b) and continuous on $[a, b]$. Then there must be at least one number $c \in (a, b)$ such that

$$f(b) = \sum_{i=0}^{n} \frac{f^{(i)}(a)}{i!}(b - a)^i + \frac{f^{(n+1)}(c)}{(n + 1)!}(b - a)^{n+1}.$$

We shall prove this theorem shortly, but first, we shall take a careful look at it for a few special values of n. Even in its simplest case, $n = 0$, the Taylor mean value theorem is interesting and useful, for in this case it reduces to the first mean value theorem. We shall now look at the case $n = 1$, which is called the second mean value theorem.

7.13.7 The Second Mean Value Theorem: Case $n = 1$ of Taylor's Theorem.

Suppose a and b are real numbers and $a < b$. Suppose that f is a function defined on $[a, b]$, and suppose that f' is differentiable on (a, b) and continuous on $[a, b]$. Then there must be at least one number $c \in (a, b)$ such that

$$f(b) = f(a) + (b - a)f'(a) + (b - a)^2 \frac{f''(c)}{2!}.$$

■ **Proof.** For each $x \in [a, b]$ we define

$$h(x) = f(x) - f(a) - f'(a)(x - a) - k(x - a)^2,$$

where the number k is chosen in such a way that $h(b) = 0$. Notice that what we have done is to subtract the first Taylor polynomial (centered at a) from f and then to subtract just one more term in $(x - a)^2$ in order to make $h(b) = 0$. The important thing to notice is that both $h(a)$ and $h'(a)$ have to be zero. We are now going to use Rolle's theorem twice. Using the fact that $h(a) = h(b) = 0$, we apply Rolle's theorem to h on $[a, b]$ and choose a number $c_1 \in (a, b)$ such that $h'(c_1) = 0$:

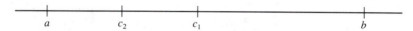

Now we use the fact that $h'(a) = h'(c_1) = 0$; and applying Rolle's theorem to h' on $[a, c_1]$, we choose a number $c_2 \in (a, c_1)$ such that $h''(c_2) = 0$. Differentiating h twice, we see that $h''(x) = f''(x) - 2!k$ for each $x \in (a, b)$; and substituting $x = c_2$, we obtain $k = f''(c_2)/(2!)$. Finally, putting this value of k into the definition of h and substituting $x = b$, we obtain

$$0 = f(b) - f(a) - (b - a)f'(a) - (b - a)^2 \frac{f''(c_2)}{2!}.$$

This is the desired result with $c = c_2$. ■

If you feel ready to read the proof of the general case of Taylor's theorem, skip Section 7.13.8, which deals with the third mean value theorem, and go straight to Section 7.13.9. If, however, you would prefer another stepping stone, then continue with Section 7.13.8.

7.13.8 The Third Mean Value Theorem: Case $n = 2$ of Taylor's Theorem.

Suppose a and b are real numbers and $a < b$. Suppose that f is a function defined on $[a, b]$, and suppose that $f^{(2)}$ is differentiable on (a, b) and continuous on $[a, b]$. Then there must be at least one number $c \in (a, b)$ such that

$$f(b) = f(a) + \frac{f^{(1)}(a)}{1!}(b - a) + \frac{f^{(2)}(a)}{2!}(b - a)^2 + \frac{f^{(3)}(c)}{3!}(b - a)^3.$$

■ **Proof.**   For each $x \in [a, b]$ we define

$$h(x) = f(x) - f(a) - \frac{f^{(1)}(a)}{1!}(x - a) - \frac{f^{(2)}(a)}{2!}(x - a)^2 - k(x - a)^3,$$

where the number k is chosen in such a way that $h(b) = 0$. Notice that what we have done is to subtract the second Taylor polynomial (centered at a) from f and then to subtract just one more term in $(x - a)^3$ in order to make $h(b) = 0$. The important thing to notice is that $h(a)$, $h^{(1)}(a)$, and $h^{(2)}(a)$ are all zero. We are now going to use Rolle's theorem three times. Using the fact that $h(a) = h(b) = 0$, we apply Rolle's theorem to h on $[a, b]$, and we choose a number $c_1 \in (a, b)$ such that $h'(c_1) = 0$:

Now we use the fact that $h'(a) = h'(c_1) = 0$; and applying Rolle's theorem to h' on $[a, c_1]$, we choose a number $c_2 \in (a, c_1)$ such that $h^{(2)}(c_2) = 0$. Finally, we use the fact that $h^{(2)}(a) = h^{(2)}(c_2) = 0$; and applying Rolle's theorem to $h^{(2)}$ on $[a, c_2]$, we choose a number c_3 (which we shall rename c) in the interval (a, c_2) such that $h^{(3)}(c) = 0$. By differentiating h three times, we see that for each point $x \in (a, b)$ we have $h^{(3)}(x) = f^{(3)}(x) - 3!k$; and on substituting $x = c$, we obtain $k = f^{(3)}(c)/3!$. Putting this value of k in the definition of h and substituting $x = b$, we obtain

$$0 = f(b) - f(a) - \frac{f^{(1)}(a)}{1!}(b - a) - \frac{f^{(2)}(a)}{2!}(b - a)^2 - \frac{f^{(3)}(c)}{3!}(b - a)^3,$$

and this gives the desired result.   ■

7.13.9   Proof of Taylor's Theorem.   For each $x \in [a, b]$ we define

$$h(x) = f(x) - \sum_{i=0}^{n} \frac{f^{(i)}(a)}{i!}(x - a)^i - k(x - a)^{n+1},$$

where the number k is chosen in such a way that $h(b) = 0$. Notice that what we have done is to subtract the nth Taylor polynomial (centered at a) from f and then to subtract just one more term in $(x - a)^{n+1}$ in order to make $h(b) = 0$. The important thing to notice is that all of the numbers $h^{(i)}(a)$ are zero for $i = 0$, $1, \ldots, n$. We are now going to use Rolle's theorem $n + 1$ times:

Using the fact that $h(a) = h(b) = 0$, we apply Rolle's theorem to h on $[a, b]$ and choose a number $c_1 \in (a, b)$ such that $h'(c_1) = 0$. Now we use the fact that $h'(a) = h'(c_1) = 0$; and applying Rolle's theorem to h' on $[a, c_1]$, we choose a number $c_2 \in (a, c_1)$ such that $h^{(2)}(c_2) = 0$. We continue this process until, at last, we have chosen a number c_{n+1} such that $h^{(n+1)}(c_{n+1}) = 0$; and we define $c = c_{n+1}$. Differentiating $n + 1$ times in the formula for h, we obtain

$$h^{(n+1)}(x) = f^{(n+1)}(x) - k(n + 1)!$$

for every $x \in (a, b)$; and therefore, substituting $x = c$, we obtain $k = f^{(n+1)}(c)/(n + 1)!$. Putting this value in the definition of h and substituting $x = b$, we obtain

$$0 = f(b) - \sum_{i=0}^{n} \frac{f^{(i)}(a)}{i!} (b - a)^i - \frac{f^{(n+1)}(c)}{(n + 1)!} (b - a)^{n+1},$$

and this gives the desired result. ∎

7.14 EXERCISES

- **1.** Given that $f: S \to R$, where S is an interval, and that $f'(x) = 0$ for all $x \in S$, prove that f must be constant on S.

2. Given that $f: S \to R$, where S is an interval, and that $f'(x) > 0$ for all $x \in S$, prove that f must be strictly increasing on S.

3. Given that $f: S \to R$, where S is an interval, and that $f'(x) < 0$ for all $x \in S$, prove that f must be strictly decreasing on S.

4. Given that $f: S \to R$, where S is an interval, and that $f'(x) \neq 0$ for all $x \in S$, prove that f must be one–one on S.

5. Suppose $f: R \to R$ and for all numbers x and t we have $|f(x) - f(t)| \leq |x - t|^2$. Prove that f is constant. Note that this problem also appeared as Exercise 6.5(28).

6. Given that f is differentiable on an interval S and that f' is bounded on S, prove that f must be lipschitzian on S.

7. Given that f is differentiable on $(0, \infty)$, that $\alpha \in [-\infty, \infty]$, and that $f'(x) \to \alpha$ as $x \to \infty$, prove that $f(x + 1) - f(x) \to \alpha$ as $x \to \infty$.

- **8.** Given that f is continuous on $[a, b]$ and differentiable on (a, b), and given that $w \in R$ and that $f'(x) \to w$ as $x \to a$, prove that $f'(a) = w$.

- **9.** Given that $a < b$, that $f''(x) < 0$ for all $x \in (a, b)$, and that f' is continuous on $[a, b]$, prove that the point $(b, f(b))$ must lie under the tangent line to the graph of f at the point $(a, f(a))$. *Hint:* Use the second mean value theorem.

- **10.** Given that $a < b$, that $f''(x) < 0$ for all $x \in (a, b)$, and that f' is continuous on $[a, b]$, prove that the straight-line segment joining the points $(a, f(a))$ and $(b, f(b))$ lies under the graph of f between a and b.

11. A function $f: S \to R$ is said to have a *local minimum* at a point $a \in S$ if there exists a neighborhood U of a such that for every $x \in U \cap S$ we have $f(x) \geq f(a)$. *Local maxima* are defined similarly. State and prove an analogue of Lemma 7.12.2 for local maxima and minima. Your proof should be very short.

12. Suppose that f is differentiable on an interval S and that $a \in S$, $f'(a) = 0$, and

$f''(a) > 0$. Prove that f must have a local minimum at a. *Hint:* Apply Exercise 7.11(5) to the function f', and then apply the first mean value theorem to f.

13. Suppose that f is $n + 1$ times differentiable on an interval S, that $a \in S$, that $f^{(i)}(a) = 0$ for $1 \le i \le n$, and that $f^{(n+1)}(a) \ne 0$. Apply Exercises 7.11(5) and 7.11(6) to the function $f^{(n)}$ and then apply Taylor's theorem to f; and prove the following assertions:
    (a) If n is even and a is not an endpoint of S, then f has neither a local maximum nor a local minimum at a.
    (b) If n is odd and $f^{(n+1)}(a) > 0$, then f has a local minimum at a.
    (c) If n is odd and $f^{(n+1)}(a) < 0$, then f has a local maximum at a.

■ 14. Given real numbers a and b such that $a > b$, prove the following variation of the Taylor mean value theorem: If n is a nonnegative integer, and f is a function defined on $[b, a]$, and $f^{(n)}$ is differentiable on (b, a) and continuous on $[b, a]$, then there must be at least one number $c \in (b, a)$ such that

$$f(b) = \sum_{i=0}^{n} \frac{f^{(i)}(a)}{i!} (b - a)^i + \frac{f^{(n+1)}(c)}{(n + 1)!} (b - a)^{n+1}.$$

15. As you know, if a function is differentiable on an interval and its derivative does not vanish, then the function has to be one–one on the interval. Now we look at "two–one" functions: We say that a function f is *two–one* on a set S if, given any number y, there are at most two points $x \in S$ such that $y = f(x)$. Prove that if f is twice differentiable on an interval S and f'' does not vanish in S, then f must be two–one on S. *Hint:* Use Rolle's theorem twice.

16. State and prove a generalization of exercise 15 for three–one functions.

17. Given two functions f and g which are differentiable on $\mathbf{R}$ and which satisfy $f' = g$ and $g' = -f$, prove that $f^2 + g^2$ (the sum of the squares of f and g) must be constant. Explain also why the nth derivatives $f^{(n)}$ and $g^{(n)}$ must exist for every natural number n and why $f'' = -f$.

18. Suppose that a and b are real numbers and $a < b$, that f and g are twice differentiable functions on (a, b), that f' and g' are continuous on $[a, b]$, and that $g''(x) \ne 0$ for all $x \in (a, b)$. Apply Rolle's theorem twice to the function h defined by

$$h(x) = f(x) - f(a) - (x - a)f'(a) - k[g(x) - g(a) - (x - a)g'(a)]$$

for all $x \in [a, b]$, where the number k is chosen in such a way that $h(b) = 0$; and obtain a "second Cauchy mean value theorem." Can you obtain an nth Cauchy mean value theorem?

■ 19. Suppose f is differentiable on $[0, \infty)$, that $f(0) = 0$, and that f' is increasing on $[0, \infty)$. For every $x > 0$ we define $g(x) = f(x)/x$. Prove that g is increasing on $(0, \infty)$.

20. Suppose that a function f has a positive derivative at a number a. In the light of Exercise 7.11(5), one might believe that f must be increasing in some neighborhood of a. By considering the function f defined by $f(x) = x/2 + x^2\sin(1/x)$ for $x \neq 0$ and $f(0) = 0$, and taking $a = 0$, prove that such belief is unfounded.

*21. The purpose of this exercise is to give you an opportunity to understand some of the theory behind **Newton's method** for approximating roots of equations. We shall assume that f is continuous on an interval $[a, b]$ and differentiable on (a, b), that f' is increasing on (a, b), and that $f(a) < 0$ and $f(b) > 0$. See Figure 7.5. From Bolzano's intermediate value theorem we know that there is a number $\alpha \in (a, b)$ such that $f(\alpha) = 0$. Newton's method gives a way of finding successive approximations to such a number α. The idea of the method is to let $x_1 = b$; and for each natural number n, if x_n has been defined, then we define x_{n+1} in such a way that the tangent line to the graph of f at the point $(x_n, f(x_n))$ meets the x-axis at the point $(x_{n+1}, 0)$. Since the equation of this tangent line is

$$y - f(x_n) = f'(x_n)(x - x_n),$$

we see that this line meets the x-axis when

$$x = x_n - \frac{f(x_n)}{f'(x_n)};$$

and therefore, for each natural number n we define

$$x_{n+1} = x_n - \frac{f(x_n)}{f'(x_n)}.$$

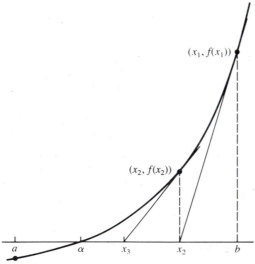

$(x_1, f(x_1))$

$(x_2, f(x_2))$

a α x_3 x_2 b

Figure 7.5

(a) Prove that there is exactly one number $\alpha \in (a, b)$ such that $f(\alpha) = 0$ and that whenever $\alpha \leq x < b$, we have $f'(x) > 0$.

(b) Prove that for each natural number n we have $\alpha < x_{n+1} < x_n$.

(c) Prove that if $x_n \to \beta$ as $n \to \infty$, then $f(\beta) = 0$; and deduce that $x_n \to \alpha$ as $n \to \infty$.

Some further information about Newton's method and some applications may be found on pages 161–166 of Thomas and Finney [14]. For an estimate of the error involved in replacing α by x_n, see page 52 of Atkinson [2].

7.15   L'HOSPITAL'S RULE

We know from Theorem 5.6.4 that if $f(x) \to \alpha$ and $g(x) \to \beta$ as $x \to a$ and if α/β exists, then $f(x)/g(x) \to \alpha/\beta$ as $x \to a$. Even when α/β does not exist, however, the limit $\lim_{x \to a} f(x)/g(x)$ might still exist. Typically, this happens when $\alpha = \beta = 0$ or when $\alpha = \beta = \pm\infty$, and in this event the expression $f(x)/g(x)$ is said to be an **indeterminate form** *in* x *as* $x \to a$. The following two limits are limits of indeterminate forms, and can be shown to be equal to $\frac{1}{6}$ and 0, respectively:

(1) $\lim\limits_{x \to 0} \left(\dfrac{x - \sin x}{x^3} \right).$ **(2)** $\lim\limits_{x \to \infty} \left\{ \dfrac{x^{100}}{\exp[(\log x)^2]} \right\}.$

Clearly, these limits cannot be evaluated by a simple application of Theorem 4.5.2. Some special techniques are needed, and among these is the theorem that has come to be known as L'Hospital's rule.[4] In this final section on differentiation we shall give a precise statement of L'Hospital's rule and a proof of the theorem. Then we shall explore some of the applications of the theorem to the evaluation of limits of indeterminate forms. In the course of these applications we shall evaluate some limits that will be very useful to us in our study of infinite series in Chapter 10.

7.15.1   L'Hospital's Rule Stated for Limits from the Left.   Suppose that $-\infty < a \leq \infty$ and that the functions f and g are defined on an interval S which has a as a right endpoint, and suppose that $f'(x)/g'(x)$ approaches a limit α as $x \to a$.[5] Suppose that one of the following two conditions holds:

(1) $f(x) \to 0$ as $x \to a$ and $g(x) \to 0$ as $x \to a$.

(2) $|g(x)| \to \infty$ as $x \to a$.

Then we have $f(x)/g(x) \to \alpha$ as $x \to a$.

■ ***Proof.***   The fact that $f'(x)/g'(x) \to \alpha$ as $x \to a$ guarantees that $g'(x) \neq 0$ for all x sufficiently close to a. Choose a number $c < a$ such that $g'(x) \neq 0$ for all points

---

[4] Although first published by L'Hospital, this theorem was in fact discovered by Johann Bernoulli, who communicated it to L'Hospital in a letter.

[5] We have not assumed that the limit α is finite.

$x \in (c, a)$. We deduce from Rolle's theorem that there cannot be more than one number x in the interval (c, a) at which $g(x) = 0$; and therefore, by making c larger, if necessary, we can guarantee that $g(x) \neq 0$ for all points $x \in (c, a)$.

Now to obtain a contradiction, let us assume that $f(x)/g(x)$ does not tend to α as $x \to a$. Using Theorem 5.5.1, choose a sequence (x_n) in S such that $x_n < a$ for all $n \in \mathbf{Z}^+$ and $x_n \to a$, and such that the sequence $(f(x_n)/g(x_n))$ does not have limit α. Choose a partial limit p of $(f(x_n)/g(x_n))$ such that $p \neq \alpha$. We may assume without loss of generality that $p < \alpha$:

Choose a number q such that $p < q < \alpha$. Using the fact that (q, ∞) is a neighborhood of α, choose a number $b < a$ such that whenever $b < x < a$, we have $f'(x)/g'(x) > q$:

Now if x and t are any two different points in (b, a), then the Cauchy mean value theorem (Theorem 7.12.7) guarantees that it is possible to find a point u between x and t such that

$$\frac{f(x) - f(t)}{g(x) - g(t)} = \frac{f'(u)}{g'(u)};$$

and it follows that whenever x and t are two different points in (b, a), we have

$$\frac{f(x) - f(t)}{g(x) - g(t)} > q.$$

Choose a natural number N such that whenever $n \geq N$, we have $x_n > b$. We shall now obtain the desired contradiction by looking separately at the two cases (1) and (2) in the statement of the theorem.

First case: Suppose that (1) holds. For every $n \geq N$ we have

$$\frac{f(x_n)}{g(x_n)} = \lim_{t \to a} \left[\frac{f(x_n) - f(t)}{g(x_n) - g(t)} \right] \geq q,$$

and this contradicts the fact that p is a partial limit of the sequence $(f(x_n)/g(x_n))$.

Second case: Suppose that (2) holds. Since $|g(x_n)| \to \infty$, it is clear that for n sufficiently large we have $g(x_n) \neq g(x_N)$. For all such n we have

$$\frac{f(x_n)}{g(x_n)} = \left[\frac{f(x_n) - f(x_N)}{g(x_n) - g(x_N)} \right] \left[1 - \frac{g(x_N)}{g(x_n)} \right] + \frac{f(x_N)}{g(x_n)},$$

and since $1 - g(x_N)/g(x_n) \to 1$ and $f(x_N)/g(x_n) \to 0$ as $n \to \infty$, it follows that the two sequences

$$\left(\frac{f(x_n)}{g(x_n)} \right) \qquad \text{and} \qquad \left(\frac{f(x_n) - f(x_N)}{g(x_n) - g(x_N)} \right)$$

must have the same partial limits. [See Exercises 4.6(14) and 4.6(15).] But as in the first case, no partial limit of the latter sequence can be less than q, contradicting the fact that p is a partial limit of the sequence $(f(x_n)/g(x_n))$.   ∎

7.15.2   L'Hospital's Rule Stated for Limits from the Right.

This version of L'Hospital's rule is analogous to the version just given for limits from the left. We leave as an exercise the task of stating it precisely and proving it.

The two-sided limit form of the theorem, which can be obtained at once by combining the two one-sided forms, can be stated as follows.

7.15.3   L'Hospital's Rule Stated for Two-Sided Limits.

Suppose that $-\infty < a < \infty$, that the functions f and g are defined on an interval S of which a is not an endpoint, and that $f'(x)/g'(x)$ approaches a limit α as $x \to a$. Suppose that one of the following two conditions holds:

(1) $f(x) \to 0$ as $x \to a$ and $g(x) \to 0$ as $x \to a$.

(2) $|g(x)| \to \infty$ as $x \to a$.

Then we have $f(x)/g(x) \to \alpha$ as $x \to a$.

7.15.4   Some Applications of L'Hospital's Rule

(1) We shall use the rule to evaluate

$$\lim_{x \to 0} \frac{\tan x - x}{x - \sin x}.$$

Since both the numerator and the denominator of this fraction approach 0 as $x \to 0$, we have

$$\lim_{x \to 0} \frac{\tan x - x}{x - \sin x} = \lim_{x \to 0} \frac{\sec^2 x - 1}{1 - \cos x} = \lim_{x \to 0} \frac{1 - \cos^2 x}{(\cos^2 x)(1 - \cos x)}$$

$$= \lim_{x \to 0} \frac{1 + \cos x}{\cos^2 x} = 2.$$

(2) We shall evaluate $\lim_{x \to \pi/2}(\sin x)^{\tan x}$. Define $f(x) = (\sin x)^{\tan x}$ for $0 < x < \pi$. Then for each such x we have

$$\log f(x) = (\tan x)(\log \sin x) = \frac{(\sin x)(\log \sin x)}{\cos x}.$$

Now since $\sin x \to 1$ as $x \to \pi/2$, we have

$$\lim_{x \to \pi/2} \log f(x) = \lim_{x \to \pi/2} \frac{\log \sin x}{\cos x}$$

provided that the latter limit exists. But

$$\lim_{x\to\pi/2} \frac{\log \sin x}{\cos x} = \lim_{x\to\pi/2} \frac{\cot x}{-\sin x} = 0,$$

and it follows that $f(x) \to e^0 = 1$ as $x \to \pi/2$.

(3) Suppose q is any number; and for all $x > 2$, define $f(x) = (x - 1)$ $\cdot [\log(x - 1)]^q$. We shall show that

$$x \log x \left[1 - \frac{1}{x} - \frac{f(x)}{f(x + 1)} \right] \to q \qquad \text{as} \qquad x \to \infty.$$

First of all, we notice that an easy application of L'Hospital's rule shows that $\log(x - 1)/\log x \to 1$ as $x \to \infty$. Therefore, since

$$x \log x \left[1 - \frac{1}{x} - \frac{f(x)}{f(x + 1)} \right] = \frac{1 - \left[\dfrac{\log(x - 1)}{\log x} \right]^q}{[(x - 1)(\log x)]^{-1}},$$

an application of L'Hospital's rule gives

$$\lim_{x\to\infty} x \log x \left[1 - \frac{1}{x} - \frac{f(x)}{f(x + 1)} \right]$$

$$= \lim_{x\to\infty} \frac{-q \left[\dfrac{\log(x - 1)}{\log x} \right]^{q-1} \dfrac{\left[\dfrac{\log x}{x - 1} - \dfrac{\log(x - 1)}{x} \right]}{(\log x)^2}}{-[(x - 1)(\log x)]^{-2} \left[\log x + \dfrac{x - 1}{x} \right]}$$

provided that the latter limit exists. Now once again, we use the fact that $\log(x - 1)/\log x \to 1$ as $x \to \infty$. This allows us to simplify the preceding limit and obtain

$$\lim_{x\to\infty} \frac{q(x - 1)[x \log x - (x - 1) \log (x - 1)]}{x \log x + x - 1}$$

$$= \lim_{x\to\infty} q \left(\frac{x - 1}{x} \right) \frac{[x \log x - (x - 1) \log (x - 1)]}{\log x \left[1 + \dfrac{1}{\log x} - \dfrac{1}{x \log x} \right]}.$$

Since both of the expressions $(x - 1)/x$ and $[1 + 1/\log x - 1/(x \log x)]$ approach 1 as $x \to \infty$, the preceding limit may be written in the form

$$\lim_{x\to\infty} q \left[\frac{x \log x - (x - 1) \log (x - 1)}{\log x} \right],$$

and applying L'Hospital's rule twice more, we obtain

$$\lim_{x\to\infty} q \left[\frac{\log x - \log (x - 1)}{1/x} \right] = \lim_{x\to\infty} q \left[\frac{1/x - 1/(x - 1)}{-1/x^2} \right] = q,$$

as required.

7.16 EXERCISES

Evaluate the limits specified in exercises 1–4.

▪ **1.** $\lim_{x \to 0} \left(\dfrac{x - \sin x}{x^3} \right)$

▪ **2.** $\lim_{x \to 0} \dfrac{\tan x - \sin x}{x^3}$

▪ **3.** $\lim_{x \to 0} \dfrac{\log(1 + x)}{x}$

▪ **4.** $\lim_{x \to \infty} \left\{ \dfrac{x^{100}}{\exp[(\log x)^2]} \right\}$

5. Prove by induction that for every natural number n we have $\lim_{x \to \infty} x^n e^{-x} = 0$. Deduce that if α is any real number, then we have $\lim_{x \to \infty} x^\alpha e^{-x} = 0$ and $\lim_{x \to \infty} (\log x)^\alpha / x = 0$.

Evaluate the limits specified in exercises 6–13.

▪ **6.** $\lim_{x \to \infty} \left(\sqrt[4]{x^4 - 5x^3 + 8x^2 - 2x + 1} - x \right)$

▪ **7.** $\lim_{x \to 0} (1 + x)^{1/x}$ ▪ **8.** $\lim_{x \to 0} \dfrac{e - (1 + x)^{1/x}}{x}$

▪ **9.** $\lim_{x \to \infty} x^{1/x}$ ▪ **10.** $\lim_{x \to \infty} x^{[(\log x)/x]}$

▪ **11.** $\lim_{x \to 0} \dfrac{e^x - \log(1 + x) - 1}{x^2(x + 2)}$

▪ **12.** $\lim_{x \to \infty} x \left[\dfrac{(2x + 2)^\alpha - (2x + 1)^\alpha}{(2x + 2)^\alpha} \right]$, where $\alpha > 0$. *Hint:* Put $u = \dfrac{2x + 1}{2x + 2}$.

▪ **13.** $\lim_{x \to \infty} \dfrac{x^{\log x}}{(x + 1)^{\log(x+1)}}$

In exercises 14–16 it might be useful to observe that if f is a differentiable function, then the expression $f(x + 1) - f(x)$ can be written in the form $f'(t)$ for some number t between x and $x + 1$.

▪ **14.** $\lim_{x \to \infty} \{[\log(x + 1)]^\alpha - (\log x)^\alpha\}$

▪ **15.** $\lim_{x \to \infty} \left[\dfrac{\log(x + 1)}{\log x} \right]^x$ ▪ **16.** $\lim_{x \to \infty} \left[\dfrac{\log(x + 1)}{\log x} \right]^{x \log x}$

17. (a) Prove that if $0 < a < b$ and $x \geq 1$, then

$$ xa^{x-1} \leq \dfrac{b^x - a^x}{b - a} \leq xb^{x-1}. $$

(b) Prove that $\lim_{x \to \infty} \{[\log(x + 1)]^{(\log x)} - (\log x)^{(\log x)}\} = \infty$.

(c) Prove that $\lim_{x \to \infty} \{[\log(x + 1)]^{(\log\log x)} - (\log x)^{(\log\log x)}\} = 0$.

Hint: In each case, multiply and divide by $\log(x + 1) - \log x$, use part (a), and use the identity

$$\log(x + 1) - \log x = \frac{1}{x} \log \left[\left(1 + \frac{1}{x} \right)^x \right].$$

18. Evaluate

$$\lim_{x \to \infty} \frac{e^{\sin x}}{x} \quad \text{and} \quad \lim_{x \to \infty} \frac{xe^{\sin x}}{\log x}.$$

One may not use L'Hospital's rule to evaluate either of these limits. Why not?

19. Give a precise statement for L'Hospital's rule for limits from the right and a very short proof of this theorem that makes use of Theorem 7.15.1.

20. Prove the two-sided version of L'Hospital's rule, using Theorems 7.15.1 and 7.15.2.

Chapter 8

The Riemann Integral

8.1 INTRODUCTION

From your studies in elementary calculus you have already become acquainted with the notion of an integral, and you have seen integrals of several different kinds, such as indefinite integrals, definite integrals, double integrals, triple integrals, line integrals, and surface integrals. But even though you may have learned how to evaluate some of these integrals and to use them for a variety of applications, in all probability you have not yet come face to face with the question as to precisely *what* integrals are. In this chapter we are going to explore the very meaning of the word *integral*. After a very brief look at the work of Eudoxus and Archimedes, we shall jump to the nineteenth century and study the integral as it was developed by Cauchy, Darboux, and Riemann. Then we shall end with some remarks that throw light on the inadequacy of this nineteenth-century approach to integration; and if we have done our job well, we shall leave you totally unsatisfied,[1] chafing at the bit, and anxious for your next course in mathematical analysis—the course in which you will see integration according to the twentieth-century theory of Lebesgue.

---

[1] But not dissatisfied, we hope.

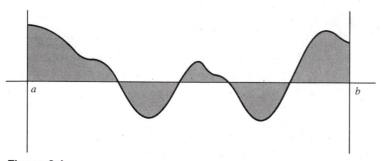

Figure 8.1

Now as you know, the symbol $\int_a^b f(x)\, dx$, written alternatively as $\int_a^b f$, ought to stand for the signed "area" of the region which lies between the vertical lines $x = a$ and $x = b$, the x-axis, and the curve $y = f(x)$ (see Figure 8.1), taking the area of the part of the region that lies above the x-axis as positive and the area of the part below as negative. This would be just fine if we knew what "area" means. But do we? Looking at the Figure 8.1, one might take either of the following two points of view:

(1) Area is *there*! Area has a meaning that is absolute and which has nothing to do with what you or I might think it means. All we have to do in mathematics is to learn how to calculate it. Luckily, integrals do this job, and we know how to work out integrals (some of them).

(2) Area is a figment of our imagination (just like everything else in mathematics); and until we have assigned it a precise meaning by giving it a *definition*, it will have no meaning, and there will be no point in trying to calculate it; for what can be the point of calculating something that does not exist?

Which of these two points of view is yours? Need we ask? Now that you have reached Chapter 8 of this book, you have reached the stage of mathematical maturity at which you would instantly select (2). But in case you have any lingering doubts, we will look at a few examples that give us a graphic reminder of the importance of a proper definition when we wish to speak of a concept like area.

We begin with an easy one: The area of a 3-by-4 rectangle is 12. No one needs to tell you that the area of a 3-by-4 rectangle is its length times its breadth, because you can see this at once by looking at a figure. The sense in which the

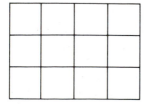

Figure 8.2

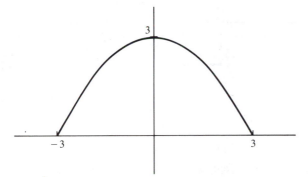

Figure 8.3

area of this rectangle is 12 is that it is possible to "pave" the rectangle with 12 1-by-1 squares. (See Figure 8.2.)

Now let us look at the region that is bounded between the graph $y = 3 - x^2/3$ and the x-axis. (See Figure 8.3.) This area must, of course, be $\int_{-3}^{3} (3 - x^2/3) \, dx$; and so just like the rectangle we looked at a moment ago, this region must have an area of 12. What we are saying is that somehow this curved region must be equivalent to 12 1-by-1 squares, but quite obviously, there is no way of paving the region with squares of any kind. In what way, then, is the region equivalent to 12 1-by-1 squares? What can we possibly *mean* when we say that this area is 12?

In our third example we shall see that even some rectangles can present the kind of difficulties we saw in the curved region of Figure 8.3. Look at the rectangle shown in Figure 8.4. If the area of this rectangle is to mean anything at all, it must be 12. But this rectangle is every bit as bad as the curved region we looked at a moment ago because it is impossible to pave it with 12 1-by-1 squares. Even worse, no matter what positive number p we take, this rectangle cannot be paved by squares of size p by p. To obtain a contradiction, suppose that we could. Suppose that for some natural numbers m and n we could pave the rectangle with m rows of p-by-p squares and that in each row there were n squares. Then we would have

$$\frac{m}{n} = \frac{\sqrt{13} - 1}{\sqrt{13} + 1} = \frac{(\sqrt{13} - 1)(\sqrt{13} - 1)}{(\sqrt{13} + 1)(\sqrt{13} - 1)} = \frac{7 - \sqrt{13}}{6},$$

and this is impossible because it implies that m/n is an irrational number.

Figure 8.4

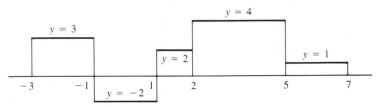

Figure 8.5

In the light of these three examples, we have to accept that whatever we really mean when we say that the area of a region is equal to 12, we don't mean that it must be possible to fill up the region with squares. So what *do* we mean? In the case of rectangles we can get ourselves out of this mess by making the following definition:

The area of a rectangle is its length times its breadth.

Now what about integrals? We are clearly a long way from an understanding of what we mean by the integrals of even the easiest kinds of function. Quadratic functions are out of the question, and even sloping straight lines are still beyond our reach. In fact, the only kind of function that seems to have an "obvious" integral at this stage is a function like the one sketched in Figure 8.5.

If we call this function f, then the obvious meaning of $\int_{-3}^{7} f$ is that

$$\int_{-3}^{7} f = (3)(2) + (-2)(2) + (2)(1) + (4)(3) + (1)(2) = 18.$$

Notice that it shouldn't matter how the function f is defined at the four points where its graph jumps. Functions like this will be called *step functions* (precise definition to follow), and it is with the integration of step functions that our chapter will begin.

8.2 PARTITIONS AND STEP FUNCTIONS

8.2.1 Definition of a Partition.

Suppose a and b are real numbers and that $a \leq b$. A **partition** $\mathcal{P}$ of the interval $[a, b]$ is a finite, strictly increasing sequence $(x_0, x_1, x_2, \ldots, x_n)$, where n is a nonnegative integer and $x_0 = a$ and $x_n = b$:

Given a partition $\mathcal{P}$ as in the definition, the numbers $x_0, x_1, \ldots, x_n$ are called the **points** of the partition $\mathcal{P}$. If $\mathcal{P}$ and $\mathcal{Q}$ are two partitions of $[a, b]$, then we say that $\mathcal{Q}$ is a **refinement** of $\mathcal{P}$ if every point of $\mathcal{P}$ is also a point of $\mathcal{Q}$. Given two partitions $\mathcal{P}$ and $\mathcal{Q}$ of $[a, b]$, the **common refinement** of $\mathcal{P}$ and $\mathcal{Q}$ is the partition of $[a, b]$ whose points are the numbers which are points either of $\mathcal{P}$ or of $\mathcal{Q}$.

The intervals (x_{i-1}, x_i) of a partition $(x_0, x_1, . . ., x_n)$ need not have the same length, but when they do, we call the partition **regular**. Given $a < b$ and a natural number n, the **regular n-partition** of $[a, b]$ means the partition $(x_0, x_1, . . ., x_n)$, where for each $i = 0, 1, 2, . . ., n$ we have

$$x_i = a + \frac{i(b - a)}{n}.$$

In this case the length of each interval (x_{i-1}, x_i) is $(b - a)/n$.

Given a partition $\mathcal{P} = (x_0, x_1, . . ., x_n)$ of an interval $[a, b]$, we shall sometimes have cause to look at the largest of the lengths of the intervals (x_{i-1}, x_i), where $i = 1, 2, . . ., n$. This number will be called the **mesh** of the partition $\mathcal{P}$ and written as $\|\mathcal{P}\|$. Note that if $\mathcal{P}$ is the regular n-partition of $[a, b]$, then $\|\mathcal{P}\| = (b - a)/n$.

8.2.2 Definition of a Step Function. Suppose $\mathcal{P}$ is the partition $(x_0, x_1, x_2, . . ., x_n)$ of $[a, b]$ and that f is a function whose domain includes $[a, b]$. We say that f **steps within** the partition $\mathcal{P}$ if f is constant on each of the open intervals (x_{i-1}, x_i), where $1 \leq i \leq n$.

Notice that a function f which steps within this partition can take any values it likes at the points $x_0, x_1, . . ., x_n$. Note also that if f steps within a partition $\mathcal{P}$, then f will certainly step within every refinement of $\mathcal{P}$.

Given a function f defined on a set S of real numbers, we say that f is a **step function** on S if there exists an interval $[a, b] \subseteq S$ and a partition $\mathcal{P}$ of $[a, b]$ such that f steps within $\mathcal{P}$ and $f(x) = 0$ for all $x \in S \setminus [a, b]$. A step function on $\mathbf{R}$ will be known simply as a *step function*.

8.2.3 Examples. The following four examples illustrate the definitions given in Sections 8.2.1 and 8.2.2.

(1) The regular 3-partition of $[0, 1]$ is not a refinement of the regular 2-partition. The regular 6-partition of $[0, 1]$ is a refinement of both the regular 2-partition and the regular 3-partition of $[0, 1]$, but it is not their common refinement. Their common refinement is the partition $(0, \frac{1}{3}, \frac{1}{2}, \frac{2}{3}, 1)$.

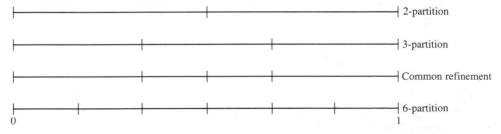

(2) Suppose n is some natural number and $\mathcal{P}$ and $\mathcal{Q}$ are, respectively, the regular 2^n and 2^{n+1} partitions of a given interval $[a, b]$. Then $\mathcal{Q}$ is a refinement of $\mathcal{P}$ and $\|\mathcal{Q}\| = \|\mathcal{P}\|/2$.

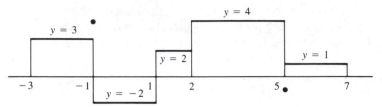

Figure 8.6

(3) Define $f: [-3, \infty) \to \boldsymbol{R}$ as follows (see Figure 8.6):

$f(x) = 3$ if $-3 \le x < -1$, $\quad f(-1) = 4$, $\quad f(x) = -2$ if $-1 < x \le 1$,

$f(x) = 2$ if $1 < x < 2$, $\quad f(x) = 4$ if $2 \le x < 5$, $\quad f(5) = -1$,

$f(x) = 1$ if $5 < x < 7$, $\quad f(x) = 0$ if $x \ge 7$.

Since the interval $[-3, 7]$ is included in $[-3, \infty)$, and since f steps within the partition $(-3, -1, 1, 2, 5, 7)$ of $[-3, 7]$, we see that f is a step function on $[-3, \infty)$. In order to show this, we could just as well have shown that f steps within the partition $(-3, -2, -1, 1, \sqrt{2}, 2, 5, 7, 8)$ of the interval $[-3, 8]$.

(4) In this example we look at the general case of a step function whose domain is an interval $[a, b]$. From the definition we may choose a subinterval $[c, d]$ of $[a, b]$ and a partition $\mathcal{P} = (x_0, x_1, \ldots, x_n)$ of $[c, d]$ such that f steps within $\mathcal{P}$ and $f(x) = 0$ for every point $x \in [a, b] \backslash [c, d]$:

Note that f steps within the partition $(a, x_0, \ldots, x_n, b)$ of the interval $[a, b]$.

8.3 INTEGRATION OF STEP FUNCTIONS

8.3.1 The Sum of a Step Function over a Partition. Suppose that $f: [a, b] \to \boldsymbol{R}$ and that f steps within the partition $\mathcal{P}$ of $[a, b]$, where $\mathcal{P} = (x_0, x_1, x_2, \ldots, x_n)$. The **sum** $\Sigma(\mathcal{P}, f)$ *of f over $\mathcal{P}$* is defined by

$$\Sigma(\mathcal{P}, f) = \sum_{i=1}^{n} \alpha_i (x_i - x_{i-1})$$

where for each i, α_i is the constant value of f on the interval (x_{i-1}, x_i).

Note that in the event that $a = b$, and $\mathcal{P}$ is the partition $(x_0, x_1, x_2, \ldots, x_n)$ of $[a, b]$, then $n = 0$; and since the summation $\Sigma(\mathcal{P}, f)$ has no terms in it, we must have $\Sigma(\mathcal{P}, f) = 0$.

8.3.2 Lemma. Suppose that $f: [a, b] \to R$, that f steps within a partition $\mathscr{P}$ of $[a, b]$, and that a refinement $\mathscr{Q}$ of $\mathscr{P}$ is obtained by inserting one more point t. Then we have $\Sigma(\mathscr{P}, f) = \Sigma(\mathscr{Q}, f)$.

▪ **Proof.** Write $\mathscr{P} = (x_0, x_1, \ldots, x_n)$. For each i, write the constant value of f in (x_{i-1}, x_i) as α_i, and choose j such that $x_{j-1} < t < x_j$:

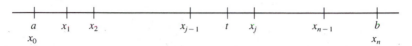

Since all the terms in the sum $\Sigma(\mathscr{Q}, f)$ are precisely the same as the corresponding terms in the sum $\Sigma(\mathscr{P}, f)$, except for the terms that are drawn from the interval (x_{j-1}, x_j), we see at once that

$$\Sigma(\mathscr{Q}, f) - \Sigma(\mathscr{P}, f) = \alpha_j(t - x_{j-1}) + \alpha_j(x_j - t) - \alpha_j(x_j - x_{j-1}) = 0. \qquad ▪$$

8.3.3 Lemma. Suppose that $f: [a, b] \to R$, that f steps within a partition $\mathscr{P}$ of $[a, b]$, and that $\mathscr{Q}$ is a refinement of $\mathscr{P}$. Then we have $\Sigma(\mathscr{P}, f) = \Sigma(\mathscr{Q}, f)$.

▪ **Proof.** Since $\mathscr{Q}$ may be obtained from $\mathscr{P}$ by repeatedly inserting one new point finitely many times, the lemma follows at once from Lemma 8.3.2. ▪

8.3.4 Theorem. Suppose that $f: [a, b] \to R$, that $\mathscr{P}$ and $\mathscr{Q}$ are partitions of $[a, b]$, and that f steps within both of these two partitions. Then we have $\Sigma(\mathscr{P}, f) = \Sigma(\mathscr{Q}, f)$.

▪ **Proof.** Denote the common refinement of $\mathscr{P}$ and $\mathscr{Q}$ as $\mathscr{R}$. Then by Lemma 8.3.3 it follows that $\Sigma(\mathscr{P}, f) = \Sigma(\mathscr{R}, f) = \Sigma(\mathscr{Q}, f)$. ▪

In the light of this theorem, we can define the integral of a step function as shown in the next subsection.

8.3.5 Definition of the Integral of a Step Function. Suppose f is a step function on the interval $[a, b]$. Then the **integral** $\int_a^b f$ of f on the interval $[a, b]$ is defined to be the sum $\Sigma(\mathscr{P}, f)$, where $\mathscr{P}$ is any partition of $[a, b]$ such that f steps within $\mathscr{P}$. An alternative notation for the integral $\int_a^b f$ is $\int_a^b f(x)\, dx$.

8.3.6 Theorem: Linearity of the Integral of Step Functions. Suppose that f and g are step functions on the interval $[a, b]$ and that c is any real number.

(1) $f + g$ is a step function on $[a, b]$, and $\int_a^b (f + g) = \int_a^b f + \int_a^b g$.

(2) cf is a step function on $[a, b]$, and $\int_a^b cf = c \int_a^b f$.

■ **Proof.** Choose partitions $\mathcal{P}$ and $\mathcal{Q}$ of $[a, b]$ such that f steps within $\mathcal{P}$ and g steps within $\mathcal{Q}$, and define $\mathcal{R}$ to be the common refinement of $\mathcal{P}$ and $\mathcal{Q}$. Then both f and g step within $\mathcal{R}$. Now write $\mathcal{R}$ in the form $(x_0, x_1, \ldots, x_n)$; and for each i, denote the constant values of f and g in the interval (x_{i-1}, x_i) as α_i and β_i, respectively. Then since $f + g$ takes the constant value $\alpha_i + \beta_i$ in each interval (x_{i-1}, x_i), we see that $f + g$ steps within $\mathcal{R}$ and that

$$\int_a^b (f + g) = \Sigma(\mathcal{R}, f + g) = \sum_{i=1}^n (\alpha_i + \beta_i)(x_i - x_{i-1})$$

$$= \sum_{i=1}^n \alpha_i(x_i - x_{i-1}) + \sum_{i=1}^n \beta_i(x_i - x_{i-1}) = \int_a^b f + \int_a^b g.$$

This proves part (1). The proof of part (2) is similar (and easier). ■

8.3.7 Theorem: Nonnegativity of the Integral of Step Functions. Note first that a function f is said to be *nonnegative* if $f(x) \geq 0$ for every point x in its domain. We now state the theorem:

Suppose f is a nonnegative step function on $[a, b]$. Then $\int_a^b f \geq 0$.

This theorem can be stated a little more generally. Given two functions f and g with the same domain, we write $f \leq g$ if for every point x in this domain, we have $f(x) \leq g(x)$. The more general form of the theorem now says:

If f and g are step functions on $[a, b]$, and if $f \leq g$, then we have

$$\int_a^b f \leq \int_a^b g.$$

■ **Proof.** (Of the more general version.) Let f and g be step functions on $[a, b]$, and suppose that $f \leq g$. As in the proof of Lemma 8.3.6, choose a partition $\mathcal{P} = (x_0, \ldots, x_n)$ of $[a, b]$ such that both f and g step within $\mathcal{P}$; and for each i, denote the constant values of f and g in (x_{i-1}, x_i) by α_i and β_i, respectively. Then for each i we have $\alpha_i \leq \beta_i$, and it follows at once that $\Sigma(\mathcal{P}, f) \leq \Sigma(\mathcal{P}, g)$. ■

8.3.8 Theorem: Additivity of the Integral of Step Functions. Suppose that $a < b < c$ and that f is a step function on $[a, c]$. Then the restrictions of f to $[a, b]$ and $[b, c]$ are step functions on $[a, b]$ and $[b, c]$, respectively, and we have

$$\int_a^c f = \int_a^b f + \int_b^c f.$$

■ **Proof.** See the figure.

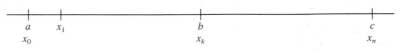

Choose a partition $\mathcal{P}$ of $[a, c]$ such that f steps within $\mathcal{P}$, and refine $\mathcal{P}$ (if necessary) to ensure that b is one of its points. We shall suppose that $\mathcal{P}$ is the partition $(x_0, \ldots, x_k, \ldots, x_n)$ and that $b = x_k$. Define $\mathcal{P}_1 = (x_0, \ldots, x_k)$ and $\mathcal{P}_2 = (x_k, \ldots, x_n)$. Then $\mathcal{P}_1$ and $\mathcal{P}_2$ are partitions of $[a, b]$ and $[b, c]$, respectively, and the restrictions of f to $[a, b]$ and $[b, c]$ obviously step within $\mathcal{P}_1$ and $\mathcal{P}_2$. We now complete the proof by observing that

$$\Sigma(\mathcal{P}, f) = \Sigma(\mathcal{P}_1, f) + \Sigma(\mathcal{P}_2, f). \qquad \blacksquare$$

8.3.9   Some Extensions of Our Notation

(1) Suppose f is a step function on $[a, b]$. In the light of the additivity theorem 8.3.8, it would be nice to know that

$$\int_a^b f = \int_a^a f + \int_a^b f$$

and as a matter of fact, this is true because it follows at once from the last remark of Section 8.3.1 that $\int_a^a f = 0$.

(2) Suppose f is a step function on $[a, b]$, where $a < b$. We would like to assign a meaning to $\int_b^a f$. Now looking at the additivity theorem, we would like to know that

$$\int_a^b f + \int_b^a f = \int_a^a f = 0;$$

and accordingly, we define $\int_b^a f = - \int_a^b f$. With this definition it is easy to see that if a, b, and c are any real numbers, and if f is a step function on the interval that runs from the smallest of these numbers to the largest, then

$$\int_a^c f = \int_a^b f + \int_b^c f.$$

(3) Suppose f takes the constant value α on $[a, b]$. Then obviously, $\int_a^b f = \alpha(b - a)$. Now although there is an important distinction between the number α and the function f which takes the constant value α on $[a, b]$, we shall allow ourselves a little license and write $\int_a^b \alpha = \alpha(b - a)$.

(4) Suppose f is a step function. Recall that this means that there exists an interval $[a, b]$ such that f steps within some partition of $[a, b]$ and $f(x) = 0$ whenever $x \in \mathbf{R} \backslash [a, b]$. Now suppose that $[c, d]$ is another interval outside of which f is identically zero. It is easy to see that $\int_a^b f = \int_c^d f$, and this suggests a definition of the integral $\int_{-\infty}^{\infty} f$; we simply define $\int_{-\infty}^{\infty} f = \int_a^b f$. Although this integral may look "improper," it isn't, of course.

8.4 EXERCISES

These exercises, though not difficult, are very important. Do them carefully.

1. Prove that if f is a step function on a set S, then so is $|f|$.

2. Prove that if f is a step function on $[a, b]$, then

$$\left| \int_a^b f \right| \leq \int_a^b |f|.$$

Hint: Observe that $-|f| \leq f \leq |f|$, and use nonnegativity.

3. Prove that if f and g are step functions on $[a, b]$, then so is their product fg.

4. Given two functions f and g from a set S to $\mathbf{R}$, the *minimum* $f \wedge g$ of f and g is defined on S by

$$f \wedge g(x) = f(x) \qquad \text{whenever} \qquad f(x) \leq g(x),$$

and

$$f \wedge g(x) = g(x) \qquad \text{whenever} \qquad f(x) > g(x).$$

Similarly, the *maximum* $f \vee g$ of f and g is defined by

$$f \vee g(x) = f(x) \qquad \text{whenever} \qquad f(x) \geq g(x),$$

and

$$f \vee g(x) = g(x) \qquad \text{whenever} \qquad f(x) < g(x).$$

Prove that

$$f \wedge g = \frac{f + g - |f - g|}{2} \qquad \text{and} \qquad f \vee g = \frac{f + g + |f - g|}{2}.$$

5. Given two functions f and g from a set S to $\mathbf{R}$, prove that $f \wedge g \leq f \leq f \vee g$ and $f \wedge g \leq g \leq f \vee g$.

6. Prove that if f and g are step functions on an interval $[a, b]$, then so are $f \vee g$ and $f \wedge g$.

▪ 7. Given that f is a step function and that the set $\{x \in \mathbf{R} \mid f(x) \neq 0\}$ is finite, prove that $\int_{-\infty}^{\infty} f = 0$.

8. Give an example of a step function f such that $\int_{-\infty}^{\infty} f = 0$ but the set $\{x \in \mathbf{R} \mid f(x) \neq 0\}$ is infinite.

9. Prove that if f is a nonnegative step function and if $\int_{-\infty}^{\infty} f = 0$, then the set $\{x \in \mathbf{R} \mid f(x) \neq 0\}$ is finite.

10. Suppose that f is a step function. Justify the assertion that was made in Section 8.3.9(4) that if $[a, b]$ and $[c, d]$ are any two intervals outside of which f is identically zero, then $\int_a^b f = \int_c^d f$.

■ **11.** Given two step functions f and g (on $\mathbf{R}$) and a real number α, prove that $\int_{-\infty}^{\infty} \alpha f = \alpha \int_{-\infty}^{\infty} f$ and $\int_{-\infty}^{\infty} (f + g) = \int_{-\infty}^{\infty} f + \int_{-\infty}^{\infty} g$.

12. Given an interval $[a, b]$ and natural numbers m and n, prove that the regular n-partition of $[a, b]$ is a refinement of the regular m-partition if and only if n is a multiple of m.

8.5 ELEMENTARY SETS

8.5.1 Introduction. Thus far we have defined the integral $\int_a^b f$, where f is a step function whose domain includes the interval $[a, b]$. Now if you think back to the part of your elementary calculus that dealt with *double* and *triple integrals,* you will remember that there is another way in which integrals are traditionally denoted. In elementary calculus you saw that if A is a given "region" of points in the plane, and if f is a function defined on A, then it is sometimes possible to speak of the *integral of f over the region A,* which is denoted as $\iint_A f(x, y) \, dx \, dy$. In this book, where most of our attention is confined to the theory of functions of one variable, we shall not, of course, be looking at integrals of this type; but a glance at this integral might motivate us to ask whether, if f is a function defined on a certain set A of real numbers, there should not perhaps be a meaning to the integral $\int_A f$. In the event that A is an interval $[a, b]$, then the obvious meaning of $\int_A f$ is $\int_a^b f$, and the most natural extension of this idea in this theory of integration of step functions with which we are presently engaged is an extension to the sets A which can be written as finite unions of intervals. It is for this purpose that we introduce the notion of an elementary set.

8.5.2 Definition of a Characteristic Function. Given $A \subseteq \mathbf{R}$, the **characteristic function** of A is the function $\chi_A : \mathbf{R} \to \mathbf{R}$ defined by $\chi_A(x) = 1$ when $x \in A$, and $\chi_A(x) = 0$ when $x \in \mathbf{R} \backslash A$.

8.5.3 Definition of an Elementary Set. Given a bounded set E of real numbers, we say that E is an **elementary set** when χ_E is a step function.

Note that $\chi_\emptyset$ is the constant function zero, and therefore, $\emptyset$ is an elementary set. Given a bounded set E and given a lower bound a and an upper bound b of E, the condition that E be an elementary set is simply the condition that the restriction of χ_E to $[a, b]$ be a step function on $[a, b]$.

8.5.4 Some Examples. As we shall soon see, an elementary set is a set which is the union of finitely many bounded intervals. The following examples help illustrate this idea.

(1) Suppose $E = [0, 1) \cup (1, 2) \cup [3, 4] \cup \{5\} \cup (6, 7] \cup [8, 9)$. If we define $\mathscr{P}$ to be the partition $(0, 1, 2, 3, 4, 5, 6, 7, 8, 9)$ of $[0, 9]$, then obviously χ_E steps within $\mathscr{P}$; and it follows that E is elementary:

```
———  ———   ———   •    ———    ———
 0    1     2     3    4     5     6     7     8     9
```

(2) Suppose $E = [0, 1]\setminus\{1/m \mid m \in \mathbf{Z}^+\}$. Then given any partition $(x_0, \ldots, x_n)$ of $[0, 1]$, it is clear that χ_E can't be constant on the interval (x_0, x_1); and it follows that E is not an elementary set.

(3) Suppose $E = [0, 1] \cap \mathbf{Q}$. This set is even worse than the set we looked at in (2), and it is certainly not elementary.

(4) Suppose $E = [0, 1] \cup [2, \infty)$. The set E is not elementary because it is not bounded. Note that there is no bounded interval outside of which χ_E is identically zero.

8.5.5 Theorem.

(1) Every bounded interval is an elementary set.

(2) The intersection of two elementary sets is an elementary set.

(3) The union of two elementary sets is an elementary set.

(4) The difference $A \setminus B$ of two elementary sets A and B is an elementary set.

(5) A set E is elementary if and only if it is the union of finitely many bounded intervals.

■ **Proof.**   Part (1) is obvious. Now suppose that A and B are elementary sets. By definition, χ_A and χ_B are step functions. Since

$$\chi_{A \cap B} = (\chi_A)(\chi_B),$$

and since the product of two step functions must be a step function, the set $A \cap B$ must be elementary. Since

$$\chi_{A \cup B} = \chi_A + \chi_B - (\chi_A)(\chi_B),$$

we see that $A \cup B$ must be elementary. And finally, since

$$\chi_{A \setminus B} = \chi_A - (\chi_A)(\chi_B),$$

we see that $A \setminus B$ must be elementary. This proves parts (2), (3), and (4). It is also clear at this stage that the union of finitely many bounded intervals, being a finite union of elementary sets, must be elementary; and therefore, to complete the proof, we need only show that every elementary set must be the union of finitely many bounded intervals. Let E be an elementary set. Choose a lower bound a and an upper bound b of E. Using the fact that χ_E is a step function, choose a partition $\mathcal{P} = (x_0, \ldots, x_n)$ of $[a, b]$ such that χ_E steps within $\mathcal{P}$. Then E is the union of some (possibly none) of the intervals (x_{i-1}, x_i) and some (possibly none) of the points of $\mathcal{P}$.   ■

8.5.6   Integral of a Step Function over an Elementary Set.   Suppose f is a step function and E is an elementary set. The **integral** of f **over** E, which is written as $\int_E f$, is defined to be $\int_{-\infty}^{\infty} f\chi_E$.

To motivate this definition, observe that the function $f\chi_E$ takes the value $f(x)$ whenever $x \in E$ and 0 whenever $x \in R\backslash E$.

8.5.7   Theorem: An Improvement on the Additivity Theorem.   Suppose that f is a step function and that E and F are mutually disjoint elementary sets. Then we have

$$\int_{E\cup F} f = \int_E f + \int_F f.$$

■ ***Proof.***   The result follows at once from the fact that $f\chi_{E\cup F} = f\chi_E + f\chi_F$.   ■

8.6   EXERCISES

1. Given that f is a step function and that $E = [a, b]$, where $a \le b$, prove that $\int_E f = \int_a^b f$.

2. Given that f is a step function, that E is an elementary set, and that $f(x) = 0$ whenever $x \in R\backslash E$, prove that $\int_E f = \int_{-\infty}^{\infty} f$.

3. Given two step functions f and g and an elementary set E, prove that $\int_E (f + g) = \int_E f + \int_E g$.

4. Given two step functions f and g and an elementary set E, and given that $f(x) \le g(x)$ for every $x \in E$, prove that $\int_E f \le \int_E g$.

■ 5. Given a step function f and an elementary set E, prove that $|\int_E f| \le \int_E |f|$. See Theorem 8.7.6 for a solution to this exercise.

6. Given two elementary sets E and F, prove that $\int_E \chi_F = \int_F \chi_E$.

*■ 7. Prove that every elementary set can be written as the union of finitely many intervals, no two of which intersect.

8.7   THE LEBESGUE MEASURE OF AN ELEMENTARY SET

8.7.1   Introduction and Definition.   In its most general form the concept of Lebesgue measure plays a fundamental role in some of the more advanced theories of integration, where it provides a precise definition of the "length" of a set of numbers. In this book we shall not be looking at the most general case of Lebesgue measure. Instead, we shall confine our attention to elementary sets, but even in this restricted sense Lebesgue measure will be very useful to us. To motivate

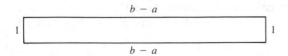

the idea of Lebesgue measure, note that in the simplest case, when an elementary set is an interval of the form (a, b), $(a, b]$, $[a, b)$, or $[a, b]$, the natural meaning for its length is the number $b - a$. In general, an elementary set can be written as the union of finitely many nonoverlapping intervals [see Exercise 8.6(7)], and its Lebesgue measure ought to be the sum of the lengths of these intervals. Now in order to give a precise definition, we shall look at the length of an interval in another way (see Figure 8.7). We observe that the length of an interval is the same as the area of a rectangle of which the interval forms one side, and the length of the other side is 1.

Since the area of this rectangle is the same as $\int_a^b 1$ (see Figure 8.8), we see that if E is any one of the previously mentioned four intervals, then the length of E is $\int_{-\infty}^{\infty} \chi_E$.

With this thought in mind we can say quite simply what we mean by the **length** or **Lebesgue measure** $m(E)$ of an elementary set E. We define

$$m(E) = \int_{-\infty}^{\infty} \chi_E.$$

8.7.2 Example. Suppose $E = (-1, 0) \cup [1, 3) \cup (4, 9] \cup \{10\}$. This set E is elementary because it is the union of four bounded intervals:

$$\begin{array}{cccccc} \underline{} & \underline{} & \underline{} & \underline{} & \bullet \\ -1 \quad 0 & 1 \quad\quad 3 & 4 & \quad\quad 9 & 10 \end{array}$$

What should the length of this set be? To answer this question, we need to take a look at χ_E. We see that

$$\chi_E(x) = 0 \quad \text{for} \quad x \leq -1, \qquad \chi_E(x) = 1 \quad \text{for} \quad -1 < x < 0,$$

$$\chi_E(x) = 0 \quad \text{for} \quad 0 \leq x < 1, \qquad \chi_E(x) = 1 \quad \text{for} \quad 1 \leq x < 3,$$

$$\chi_E(x) = 0 \quad \text{for} \quad 3 \leq x \leq 4, \qquad \chi_E(x) = 1 \quad \text{for} \quad 4 < x \leq 9,$$

$$\chi_E(x) = 0 \quad \text{for} \quad 9 < x < 10, \qquad \chi_E(10) = 1,$$

$$\chi_E(x) = 0 \quad \text{for} \quad x > 10.$$

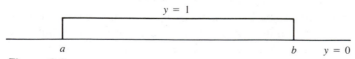

Figure 8.8

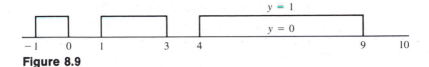

Figure 8.9

See Figure 8.9. Therefore,

$$\int_{-\infty}^{\infty} \chi_E = \int_{-1}^{0} 1 + \int_{1}^{3} 1 + \int_{4}^{9} 1 + \int_{10}^{10} 1 = 1 + 2 + 5 + 0 = 8,$$

and we see that $m(E)$ is the sum of the lengths of the four intervals $(-1, 0)$, $[1, 3)$, $(4, 9]$, and $\{10\}$.

8.7.3 Remark. The equation

$$m(E) = \int_{-\infty}^{\infty} \chi_E$$

applies only to elementary sets E in this simple theory of the integration of step functions which we are presently studying. However, the same equation can be used in more sophisticated theories of integration to relate the integral with Lebesgue measure in a much wider class of sets. The ultimate theory of integration of functions of a real variable is the one which was developed by Lebesgue and published by him in 1902 in his classic memoir *Integrale longueur, aire*, and in this theory the family of sets E for which the equation makes sense is very wide indeed. In fact, one can show that it is impossible to go any further. Although the Lebesgue theory is too advanced for us (alas, for it is the nicest one to work with, in addition to being the most powerful), we still use Lebesgue's name and call the length $m(E)$ of an elementary set E the *Lebesgue measure* of E. For some further information about the history of the measure concept, see Kline [8], pages 1040–1051; and you might possibly want to look at the work of Hawkins referred to on page 1051 of [8].

8.7.4 Theorem

(1) For every elementary set A we have $m(A) \geq 0$; and if A is empty, then we have $m(A) = 0$.

(2) If A and B are any two elementary sets, then we have

$$m(A \cup B) \leq m(A) + m(B).$$

(3) If A and B are any two mutually disjoint elementary sets, then

$$m(A \cup B) = m(A) + m(B).$$

(4) If A and B are two elementary sets and $B \subseteq A$, then we have

$$m(A \setminus B) = m(A) - m(B).$$

(5) If A and B are two elementary sets and $B \subseteq A$, then we have

$$m(B) \le m(A).$$

■ **Proof.** Part (1) is clear. Now since $\chi_{A \cup B} \le \chi_A + \chi_B$, it follows from nonnegativity and linearity that

$$\int_{-\infty}^{\infty} \chi_{A \cup B} \le \int_{-\infty}^{\infty} (\chi_A + \chi_B) = \int_{-\infty}^{\infty} \chi_A + \int_{-\infty}^{\infty} \chi_B.$$

This proves part (2). Part (3) follows similarly because if A and B are mutually disjoint, we have $\chi_{A \cup B} = \chi_A + \chi_B$. Part (4) follows from the fact that if $B \subseteq A$, then $\chi_A - \chi_B = \chi_{A \setminus B}$; and part (5) follows at once from part (4). ■

To motivate our next theorem, let us look at the simplest kind of elementary set—in other words, at a bounded interval I. Let the left and right endpoints of I be a and b, respectively:

We see that for every sufficiently large natural number n the set $H_n = [a + 1/n, b - 1/n]$ is a *closed* interval which is a subset of A and that $m(H_n) \to m(A)$ as $n \to \infty$. Similarly, for every n the set $U_n = (a - 1/n, b + 1/n)$ is an *open* interval which includes A, and $m(U_n) \to m(A)$ as $n \to \infty$. In the next theorem we shall extend this idea to arbitrary elementary sets.

8.7.5 Theorem. Given any elementary set A.

(1) $m(A) = \sup\{m(H) \mid H$ is closed and elementary and $H \subseteq A\}$.

(2) $m(A) = \inf\{m(U) \mid U$ is open and elementary and $A \subseteq U\}$.

Note that from part (5) of Theorem 8.7.4 it follows that if H, A, and U are elementary sets and $H \subseteq A \subseteq U$, then $m(H) \le m(A) \le m(U)$. Therefore, what we are saying in Theorem 8.7.5 is if A is an elementary set, then we can approximate $m(A)$ as closely as we please by the measures of a closed elementary set $H \subseteq A$ and an open elementary set $U \supseteq A$.

■ **Proof.** For the purpose of this proof let us call an elementary set A *admissible* if the conditions (1) and (2) hold. What we saw in the previous remarks is that every bounded interval is admissible. We shall show that the union of any two admissible sets is admissible. From this it will follow that the union of finitely many admissible sets is admissible; and since every elementary set is a union of finitely many intervals, it will then follow that all elementary sets are admissible.

Suppose that A and B are admissible. To show that $A \cup B$ satisfies (1), let $\varepsilon > 0$. Choose closed elementary subsets H and K of A and B, respectively, such that $m(H) > m(A) - \varepsilon/2$ and $m(K) > m(B) - \varepsilon/2$. Then $H \cup K$ is a closed elementary subset of $A \cup B$; and since

$$(A \cup B) \backslash (H \cup K) \subseteq (A \backslash H) \cup (B \backslash K)$$

it follows from Theorem 8.7.4 that

$$m(A \cup B) - m(H \cup K) \leq m[(A \backslash H) \cup (B \backslash K)] \leq m(A \backslash H) + m(B \backslash K)$$

$$= m(A) - m(H) + m(B) - m(K) < \frac{\varepsilon}{2} + \frac{\varepsilon}{2} = \varepsilon;$$

and therefore, $m(H \cup K) > m(A \cup B) - \varepsilon$.

Finally, to show that $A \cup B$ satisfies (2), let $\varepsilon > 0$. Choose open elementary sets U and V such that $A \subseteq U$ and $B \subseteq V$, and $m(U) < m(A) + \varepsilon/2$ and $m(V) < m(B) + \varepsilon/2$. Then $U \cup V$ is an open elementary set, and $A \cup B \subseteq U \cup V$; and since

$$(U \cup V) \backslash (A \cup B) \subseteq (U \backslash A) \cup (V \backslash B),$$

one can use an argument similar to the one we used previously to show that $m(U \cup V) < m(A \cup B) + \varepsilon$. Therefore, $A \cup B$ is admissible, and the proof is complete. ∎

The next theorem provides a solution to Exercise 8.6(5).

8.7.6 Theorem. Given an elementary set A and a step function f, we have

$$\left| \int_A f \right| \leq \int_A |f|.$$

■ **Proof.** For every number x we have

$$-|f(x)|\chi_A(x) \leq f(x)\chi_A(x) \leq |f(x)|\chi_A(x),$$

and the result now follows at once from nonnegativity (Theorem 8.3.7). ∎

8.7.7 Theorem. Suppose A is an elementary set, f is a step function, and $k \in \mathbf{R}$; and suppose that for every $x \in A$ we have $|f(x)| \leq k$. Then

$$\left| \int_A f \right| \leq k \cdot m(A).$$

■ **Proof.** Since $|f|\chi_A \leq k\chi_A$, we have

$$\left| \int_A f \right| \leq \int_A |f| = \int_{-\infty}^{\infty} |f|\chi_A \leq \int_{-\infty}^{\infty} k\chi_A = k \int_{-\infty}^{\infty} \chi_A = k \cdot m(A).$$

∎

8.8 EXERCISES

■ **1.** Prove that if A and B are elementary sets, then

$$m(A \cup B) = m(A) + m(B) - m(A \cap B).$$

2. Prove that if A is a finite set, then A must be elementary and $m(A) = 0$. Compare this with Exercise 8.4(7).

3. Prove that if A is an elementary set and $m(A) = 0$, then A is finite. Compare this with Exercise 8.4(9).

■ 4. Given that A is an elementary set which is not closed and that H is a closed elementary subset of A, prove that $m(H) < m(A)$.

5. Given that A is an elementary set which is not open and that U is an open elementary set which includes A, prove that $m(A) < m(U)$.

■ 6. Prove that if $A = \boldsymbol{Q} \cap [0, 1]$, then whenever H is an elementary set and $H \subseteq A$, we have $m(H) = 0$; and whenever U is an elementary set and $A \subseteq U$, we have $m(U) = 1$.

■ 7. Suppose that (x_n) is a strictly decreasing sequence in $[0, 1]$, that $x_n \to 0$, and that $A = \cup_{n=1}^{\infty} [x_{2n+1}, x_{2n}]$. Prove that although A is not elementary, we have

$$\sup\{m(H) \mid H \text{ is elementary and } H \subseteq A\}$$
$$= \inf\{m(U) \mid U \text{ is elementary and } A \subseteq U\}.$$

*8. If A is a bounded set of real numbers and if

$$\sup\{m(H) \mid H \text{ is elementary and } H \subseteq A\}$$
$$= \inf\{m(U) \mid U \text{ is elementary and } A \subseteq U\},$$

can you suggest a definition for the Lebesgue measure $m(A)$ of this set A? Using this definition, prove that if A and B are two disjoint sets of this type, then $m(A \cup B) = m(A) + m(B)$. What else can you prove?

8.9 THE ROAD TO A MORE GENERAL THEORY OF INTEGRATION

Although the theory of integration of step functions which we have developed up to this point has much of the character and the flavor of integral calculus, most of what we have been doing is *pure algebra*![2] In order to define the integral of a step function, we need only the operations of arithmetic: addition, subtraction, multiplication, and division. We do not require such notions as limit, supremum, or infimum, which constitute the bread and butter of calculus. These limit notions begin to appear, however, the moment we extend the theory to a wider class of functions. There are several different ways of extending the theory to a wider class of functions, and each yields its own theory of integration. Roughly speaking, each theory begins with the integration of step functions as we have described it and then approaches the integration of more general functions by some sort of limit or supremum or infimum as a function is approximated ever more closely by

---

[2] The notable exception is Theorem 8.7.5, which involves open and closed sets and suprema and infima.

step functions. The way these theories differ from one another is in precisely how the approximation of a given function by step functions takes place.

At the heart of every one of these theories, however, is a technique which was discovered by Eudoxus and then developed into a sophisticated theory by Archimedes somewhere around the year 250 B.C.E. Archimedes calculated the areas and volumes of a number of geometric figures by approximating them both from inside and from outside by polygonal figures (see Figure 8.10). He reasoned that if A is a plane region and if s and S are two polygons such that $s \subseteq A \subseteq S$, then

$$\text{area } (s) \leq \text{area } (A) \leq \text{area } (S).$$

Therefore, given any number $\varepsilon > 0$, one can guarantee that area (s) and area (S) must both approximate area (A) to within ε simply by making

$$\text{area } (S) - \text{area } (s) < \varepsilon.$$

The analogue of this idea for functions is as follows: Suppose that f is a bounded function defined on an interval $[a, b]$ and that s and S are step functions on $[a, b]$ such that $s \leq f \leq S$. Then whatever $\int_a^b f$ ought to mean, we should have

$$\int_a^b s \leq \int_a^b f \leq \int_a^b S;$$

and therefore (see Figure 8.11), given any number $\varepsilon > 0$, we can guarantee that both $\int_a^b s$ and $\int_a^b S$ approximate $\int_a^b f$ to within ε simply by making

$$\int_a^b S - \int_a^b s < \varepsilon.$$

This simple method of approximating a given function by step functions is the one that is used in the theory of integration which was developed by Cauchy, Darboux, Riemann, and others during the nineteenth century, and which has come to be called *Riemann integration*. This is the theory which we shall be studying; but perhaps after you have read this book, you will feel ready to take a look at some of the more sophisticated and more satisfactory theories of integration that form the pillar of mathematical analysis today.

Finally, before we begin our study of the Riemann integral, let us say what one expects of any theory of integration of functions of one variable. An integral

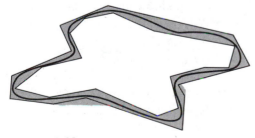

Figure 8.10

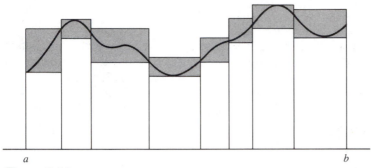

Figure 8.11

of the form $\int_a^b f$ or $\int_A f$ will be defined for certain functions f, and these functions will be said to be *integrable* in the given theory. For example, in the theory of integration of step functions that appears earlier, only the step functions were integrable. In the Riemann theory the functions we integrate are called *Riemann-integrable*; in other theories such as the Lebesgue theory, the Perron theory, or the Kurzweil-Henstock theory, one would speak of functions which are *Lebesgue-integrable*, *Perron-integrable*, or *Kurzweil-integrable*. Whatever the theory, the sum and difference of two integrable functions must be integrable, and a constant multiple of an integrable function must be integrable. The integral must have the property of linearity that we saw in Theorem 8.3.6, and it must have the properties of nonnegativity and additivity that we saw in Theorems 8.3.7 and 8.3.8. Finally, a theory of integration must provide us with the two important theorems which we call the *fundamental theorem of calculus* and the *bounded convergence theorem*. We shall see these two theorems for the theory of Riemann integration in Sections 8.17 and 8.20 respectively.

8.10 RIEMANN INTEGRABILITY AND THE RIEMANN INTEGRAL

8.10.1 Lower Integrals. Suppose f is a bounded function defined on an interval $[a, b]$. The **lower integral** $\underline{\int_a^b} f$ of the function f is defined to be the supremum of the set of all integrals $\int_a^b s$ that can be obtained by taking s to be a step function such that $s \leq f$ (see Figure 8.12). More precisely,

$$\underline{\int_a^b} f = \sup\left\{ \int_a^b s \;\middle|\; s \text{ is a step function on } [a, b] \text{ and } s \leq f \right\}.$$

Figure 8.12 shows a step function $s \leq f$. In this figure we have drawn the graphs in such a way that the functions appear to be nonnegative, but in general, they don't have to be.

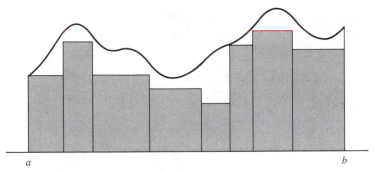

Figure 8.12

8.10.2 Upper Integrals. Upper integrals are defined in much the same way that we defined lower integrals.

Suppose f is a bounded function defined on an interval $[a, b]$. The **upper integral** $\overline{\int_a^b} f$ of the function f is defined to be the infimum of the set of all integrals $\int_a^b S$ that can be obtained by taking S to be a step function such that $S \geq f$ (see Figure 8.13). More precisely,

$$\overline{\int_a^b} f = \inf\left\{\int_a^b S \ \middle| \ S \text{ is a step function on } [a, b] \text{ and } S \geq f\right\}.$$

8.10.3 Theorem. If f is a bounded function defined on an interval $[a, b]$, then

$$\underline{\int_a^b} f \leq \overline{\int_a^b} f.$$

■ **Proof.** Given step functions s and S on $[a, b]$ such that $s \leq f \leq S$, it follows from the nonnegativity that $\int_a^b s \leq \int_a^b S$. The theorem follows at once from this fact. ■

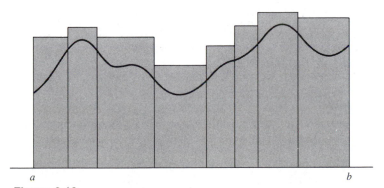

Figure 8.13

8.10.4 Riemann Integrability and the Riemann Integral. Suppose f is a step function on a given interval $[a, b]$. Then since $\int_a^b f$ is both the largest member of the set

$$\left\{ \int_a^b s \;\middle|\; s \text{ is a step function on } [a, b] \text{ and } s \leq f \right\}$$

and the smallest member of

$$\left\{ \int_a^b S \;\middle|\; S \text{ is a step function on } [a, b] \text{ and } S \geq f \right\},$$

we have

$$\underline{\int_a^b} f = \int_a^b f = \overline{\int_a^b} f.$$

This suggests the following definition of Riemann integrability:

A bounded function f on an interval $[a, b]$ is said to be **Riemann-integrable** on $[a, b]$ when $\underline{\int_a^b} f = \overline{\int_a^b} f$; and in the event that f is Riemann-integrable on $[a, b]$, we define

$$\int_a^b f = \underline{\int_a^b} f = \overline{\int_a^b} f.$$

Another way of saying this is to say that $\int_a^b f$ is the unique number I which satisfies the inequality $\int_a^b s \leq I \leq \int_a^b S$ for every choice of step functions s and S such that $s \leq f \leq S$. We mention finally that an alternative notation for $\int_a^b f$ is $\int_a^b f(x)\,dx$.

8.11 SQUEEZING A FUNCTION: A NECESSARY AND SUFFICIENT CONDITION FOR RIEMANN INTEGRABILITY

8.11.1 Theorem. Suppose f is a bounded function on an interval $[a, b]$.

(1) There exists a sequence (s_n) of step functions on $[a, b]$ such that $s_n \leq f$ for every $n \in \mathbf{Z}^+$ and such that

$$\int_a^b s_n \to \underline{\int_a^b} f \quad \text{as} \quad n \to \infty.$$

(2) There exists a sequence (S_n) of step functions on $[a, b]$ such that $S_n \geq f$ for every $n \in \mathbf{Z}^+$ and such that

$$\int_a^b S_n \to \overline{\int_a^b} f \quad \text{as} \quad n \to \infty.$$

■ **Proof.**   Define $E = \{\int_a^b s \mid s$ is a step function on $[a, b]$ and $s \leq f\}$. Then $\underline{\int_a^b} f = \sup E$. For every natural number n, using the fact that $\underline{\int_a^b} f - 1/n$ is not an upper bound of E, choose a step function $s_n \leq f$ such that $\int_a^b s_n > \underline{\int_a^b} f - 1/n$. Part (1) is now clear, and part (2) follows similarly.   ■

8.11.2   Definition.   Suppose f is a bounded function on an interval $[a, b]$. We say that a pair of sequences (s_n) and (S_n) of step functions on $[a, b]$ **squeezes** the function f on $[a, b]$ if $s_n \leq f \leq S_n$ for every $n \in \mathbf{Z}^+$ and

$$\int_a^b (S_n - s_n) \to 0 \qquad \text{as} \qquad n \to \infty.$$

8.11.3   Theorem.   Suppose f is a bounded function on an interval $[a, b]$. Then the following two conditions are equivalent:

(1) f is Riemann-integrable on $[a, b]$.

(2) There exists a pair of sequences (s_n) and (S_n) of step functions which squeezes f on $[a, b]$.

■ **Proof.**   To prove that $(1) \Rightarrow (2)$, assume that (1) holds. Using Theorem 8.11.1, choose a pair (s_n) and (S_n) of sequences of step functions on $[a, b]$ such that $s_n \leq f \leq S_n$ for all $n \in \mathbf{Z}^+$, and $\int_a^b s_n \to \underline{\int_a^b} f$ and $\int_a^b S_n \to \overline{\int_a^b} f$. Then clearly,

$$\int_a^b (S_n - s_n) \to \int_a^b f - \int_a^b f = 0,$$

and so the pair (s_n) and (S_n) squeezes f.

To prove that $(2) \Rightarrow (1)$, assume that (2) holds; and choose a pair (s_n) and (S_n) of step functions which squeezes f on $[a, b]$. For each n we have

$$0 \leq \overline{\int_a^b} f - \underline{\int_a^b} f \leq \int_a^b (S_n - s_n),$$

and since the latter expression tends to zero as $n \to \infty$, it follows that $\overline{\int_a^b} f - \underline{\int_a^b} f = 0$.   ■

8.11.4   Theorem.   Suppose f is Riemann-integrable on $[a, b]$ and that the pair of sequences (s_n) and (S_n) squeezes f on $[a, b]$. Then as $n \to \infty$, we have $\int_a^b s_n \to \int_a^b f$ and $\int_a^b S_n \to \int_a^b f$.

■ **Proof.**   This theorem is really a simple fact about sequences. For each n we have

$$\int_a^b s_n \leq \int_a^b f \leq \int_a^b S_n;$$

and therefore,

$$0 \le \int_a^b f - \int_a^b s_n \le \int_a^b S_n - \int_a^b s_n = \int_a^b (S_n - s_n).$$

It follows from the sandwich theorem (Theorem 4.3.11) that $\int_a^b f - \int_a^b s_n \to 0$, in other words, $\int_a^b s_n \to \int_a^b f$. A similar argument shows that $\int_a^b S_n \to \int_a^b f$. ∎

8.11.5 Example. We can at last find the area of a triangle (see Figure 8.14). Define $f(x) = x$ for $0 \le x \le 1$. For each natural number n, define $\mathcal{P}_n$ to be the partition $(0/n, 1/n, \ldots, n/n)$ of $[0, 1]$ (in other words, $\mathcal{P}_n$ is the regular n-partition of $[0, 1]$), and define two step functions s_n and S_n as follows: $s_n(x) = S_n(x) = x$ whenever x is a point of $\mathcal{P}_n$, and s_n and S_n take the constant values $(i - 1)/n$ and i/n, respectively, in each interval $((i - 1)/n, i/n)$. It is clear that $s_n \le f \le S_n$ for every n; and since

$$\int_0^1 (S_n - s_n) = \sum_{i=1}^n \left(\frac{i}{n} - \frac{i - 1}{n}\right) \frac{1}{n} = \frac{1}{n} \to 0,$$

it follows that the pair of sequences (s_n) and (S_n) squeezes f.

This shows that f is Riemann-integrable on $[0, 1]$. To find the value of $\int_0^1 f$, we note that for each n

$$\int_0^1 S_n = \sum_{i=1}^n \frac{i}{n} \cdot \frac{1}{n} = \frac{1}{n^2}\left[\frac{n(n + 1)}{2}\right] = \frac{1}{2}\left(1 + \frac{1}{n}\right) \to \frac{1}{2};$$

and therefore, $\int_0^1 f = \frac{1}{2}$, or alternatively, $\int_0^1 x \, dx = \frac{1}{2}$.

8.11.6 Example. The previous example showed a very simple case of an integrable function. Now we go to the opposite extreme and give an example of a

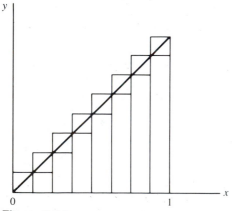

Figure 8.14

function that is too discontinuous to be Riemann-integrable. We define

$$f(x) = 1 \quad \text{whenever} \quad x \in [0, 1] \cap \mathbf{Q},$$

and

$$f(x) = 0 \quad \text{whenever} \quad x \in [0, 1] \backslash \mathbf{Q}.$$

It is easy to see that this function is discontinuous at every point of the interval $[0, 1]$. Now suppose that $\mathscr{P}$ is any partition $(x_0, x_1, \ldots, x_n)$ of $[0, 1]$, that s and S step within $\mathscr{P}$, and that $s \leq f \leq S$. Then for each i, if α_i and β_i are the constant values of s and S in the interval (x_{i-1}, x_i), since the interval (x_{i-1}, x_i) must contain both rationals and irrationals, we see that $\alpha_i \leq 0$ and $\beta_i \geq 1$. Therefore, $\underline{\int_0^1} f \leq 0$ and $\overline{\int_0^1} f \geq 1$; and we deduce that f cannot be Riemann-integrable on $[0, 1]$.

Now let's pause for a little philosophy. Here we have a function f that is not integrable. Why isn't it? Considering that f is discontinuous at every point of the interval $[0, 1]$, one might be inclined to feel that f is just too wild to have a role in the theory of integration. But one would be wrong, for even though f is too discontinuous to be integrable in *Riemann* integration theory, it is quite easy to integrate in some of the more sophisticated theories. The fact that f is not Riemann-integrable should therefore be seen as a defect of Riemann integration theory. By the time you reach the end of this chapter, you will see that from some points of view f is not wild at all. On the contrary, f seems to behave just like the constant function zero, and in a satisfactory theory of integration we should have $\int_0^1 f = 0$.

8.12 EXERCISES

1. Suppose that $a < b$, that $f(x) = x$ for all $x \in [a, b]$, and that for each natural number n, $\mathscr{P}_n = (x_0, \ldots, x_n)$ is the regular n-partition of $[a, b]$. For each n we define two step functions s_n and S_n as follows: $s_n(x_i) = S_n(x_i) = f(x_i)$ for every $i = 0, 1, \ldots, n$, and in each interval (x_{i-1}, x_i), s_n and S_n take the constant values $f(x_{i-1})$ and $f(x_i)$, respectively. Prove that the pair of sequences (s_n) and (S_n) squeezes f on $[a, b]$, and use this to show that $\int_a^b x\, dx = b^2/2 - a^2/2$.

2. Repeat the method of exercise 1, taking $f(x) = x^2$ for each $x \in [a, b]$, and deduce that $\int_a^b x^2\, dx = b^3/3 - a^3/3$.

■ 3. Suppose that $a < b$, that f is any increasing function on the interval $[a, b]$, and that $\mathscr{P}_n$, s_n, and S_n are defined for every natural number n as in exercise 1. Prove that

$$\int_a^b (S_n - s_n) = \frac{[f(b) - f(a)](b - a)}{n}$$

and deduce that the pair of sequences (s_n) and (S_n) squeezes the function f on $[a, b]$. A solution to this exercise will be provided in Theorem 8.16.1.

4. Suppose $f(x) = \sqrt{x}$ for every $x \in [0, 1]$. For each natural number n suppose that $\mathcal{P}_n$ is the partition $(x_0, x_1, \ldots, x_{n^2})$, where for each i we have $x_i = i^2/n^2$; and suppose that s_n and S_n are defined for each n, as in exercise 1. Prove that the pair of sequences (s_n) and (S_n) squeezes f on $[0, 1]$, and use this to show that $\int_0^1 \sqrt{x}\, dx = \frac{2}{3}$.

5. Use a method similar to that of exercise 4 to show that $\int_0^1 x^{1/3}\, dx = \frac{3}{4}$.

8.13 SIMPLE PROPERTIES OF THE RIEMANN INTEGRAL

In this section we shall show that the Riemann integral has the properties of linearity, nonnegativity, and additivity that we saw previously for step functions.

8.13.1 Theorem: Linearity of the Riemann Integral. Suppose f and g are Riemann-integrable on $[a, b]$. Then $f + g$ is Riemann-integrable on $[a, b]$, and we have

$$\int_a^b (f + g) = \int_a^b f + \int_a^b g.$$

■ **Proof.** Choose a pair of sequences (s_n) and (S_n) of step functions which squeezes the function f and a pair of sequences (s_n^*) and (S_n^*) which squeezes the function g. For every n we have

$$s_n + s_n^* \leq f + g \leq S_n + S_n^*.$$

Therefore, since

$$\int_a^b [(S_n + S_n^*) - (s_n + s_n^*)] = \int_a^b (S_n - s_n) + \int_a^b (S_n^* - s_n^*) \to 0,$$

we deduce that the pair of sequences $(s_n + s_n^*)$ and $(S_n + S_n^*)$ must squeeze $f + g$ on $[a, b]$. This shows that $f + g$ must be Riemann-integrable on $[a, b]$ and that

$$\int_a^b (f + g) = \lim_{n \to \infty} \int_a^b (s_n + s_n^*) = \lim_{n \to \infty} \left(\int_a^b s_n + \int_a^b s_n^* \right) = \int_a^b f + \int_a^b g. \qquad ■$$

8.13.2 Theorem: Linearity of the Riemann Integral. Suppose that f is Riemann-integrable on $[a, b]$ and that α is any real number. Then the function αf is Riemann-integrable on $[a, b]$, and

$$\int_a^b \alpha f = \alpha \int_a^b f.$$

■ **Proof.** Choose a pair of sequences (s_n) and (S_n) that squeezes f on $[a, b]$. Now for every n, since $s_n \leq f \leq S_n$, we see that if $\alpha > 0$, then we have $\alpha s_n \leq \alpha f \leq \alpha S_n$;

and if $\alpha \leq 0$, then we have $\alpha S_n \leq \alpha f \leq \alpha s_n$. Furthermore,

$$\int_a^b (\alpha S_n - \alpha s_n) = \alpha \int_a^b (S_n - s_n) \to 0.$$

Therefore, in both of these cases the pair of sequences (αs_n) and (αS_n) squeezes αf; and it follows that αf is Riemann-integrable on $[a, b]$ and that

$$\int_a^b \alpha f = \lim_{n \to \infty} \int_a^b \alpha s_n = \lim_{n \to \infty} \alpha \int_a^b s_n = \alpha \int_a^b f. \quad \blacksquare$$

8.13.3 Theorem: Nonnegativity of the Riemann Integral. Suppose that f and g are Riemann-integrable on $[a, b]$ and that $f \leq g$. Then

$$\int_a^b f \leq \int_a^b g.$$

■ **Proof.** Given any step function s on $[a, b]$ such that $s \leq f$, we certainly have $s \leq g$ and, therefore, $\int_a^b s \leq \underline{\int_a^b} g$. Since $\underline{\int_a^b} g$ is an upper bound of the set $\{\int_a^b s \mid s$ is a step function and $s \leq f\}$, it therefore follows at once from the definition of a lower integral that $\underline{\int_a^b} f \leq \underline{\int_a^b} g$. ■

8.13.4 Theorem: Additivity of the Riemann Integral. Suppose that $a \leq b \leq c$ and that f is Riemann-integrable on $[a, c]$. Then f must be Riemann-integrable on each of the intervals $[a, b]$ and $[b, c]$, and we have

$$\int_a^c f = \int_a^b f + \int_b^c f.$$

■ **Proof.** Choose a pair of sequences (s_n) and (S_n) of step functions which squeezes f on $[a, c]$. Now for each n we have

$$\int_a^b (S_n - s_n) \leq \int_a^c (S_n - s_n) \quad \text{and} \quad \int_b^c (S_n - s_n) \leq \int_a^c (S_n - s_n).$$

Therefore, the pair of sequences (s_n) and (S_n) squeezes f on each of the intervals $[a, b]$ and $[b, c]$, and it follows that

$$\int_a^c f = \lim_{n \to \infty} \int_a^c s_n = \lim_{n \to \infty} \left(\int_a^b s_n + \int_b^c s_n \right) = \int_a^b f + \int_b^c f. \quad \blacksquare$$

8.13.5 Definition. If $a < b$ and f is Riemann-integrable on $[a, b]$, we define

$$\int_b^a f = -\int_a^b f.$$

With this definition we can now state the additivity theorem a little more generally.

8.13.6   Theorem: A Better Additivity Theorem.   Suppose that a, b, and c are real numbers and that f is Riemann-integrable on the interval that runs from the smallest of these numbers to the largest. Then

$$\int_a^c f = \int_a^b f + \int_b^c f.$$

8.14   SOME NECESSARY AND SUFFICIENT CONDITIONS FOR A FUNCTION TO BE INTEGRABLE

8.14.1   Theorem.   Suppose f is a bounded function on an interval $[a, b]$. Then the following two conditions are equivalent:

(1) f is Riemann-integrable on $[a, b]$.

(2) For every number $\varepsilon > 0$ there exist two step functions s and S on $[a, b]$ such that $s \leq f \leq S$, and $\int_a^b (S - s) < \varepsilon$.

■ **Proof.**   To prove that $(1) \Rightarrow (2)$, assume that (1) holds; and choose a pair of sequences (s_n) and (S_n) of step functions which squeezes f on $[a, b]$. Then given $\varepsilon > 0$, we must have $\int_a^b (S_n - s_n) < \varepsilon$ for n large enough, and it follows that condition (2) must hold.

Now to prove that $(2) \Rightarrow (1)$, assume that (2) holds. For every natural number n, choose two step functions s_n and S_n on $[a, b]$ such that $s_n \leq f \leq S_n$ and $\int_a^b (S_n - s_n) < 1/n$. Then obviously, the pair of sequences (s_n) and (S_n) must squeeze f on $[a, b]$, and so f must be integrable on $[a, b]$.   ■

The next theorem describes the integrability of a given function f in a slightly different way. Instead of looking for two step functions s and S on either side of f such that $S - s$ has a small integral, we shall now look for two step functions which are close to each other in "most" of the interval $[a, b]$.

8.14.2   Theorem.   Suppose f is a bounded function on an interval $[a, b]$. Then the following two conditions are equivalent:

(1) f is Riemann-integrable on $[a, b]$.

(2) For every number $\varepsilon > 0$ there exist two step functions s and S on $[a, b]$ such that $s \leq f \leq S$; and if E is the set

$$E = \{x \in [a, b] \mid S(x) - s(x) \geq \varepsilon\},$$

then we have $m(E) < \varepsilon$.

■ **Proof.**   To prove that $(1) \Rightarrow (2)$, assume that (1) holds. Let $\varepsilon > 0$, and choose two step functions s and S such that $s \leq f \leq S$ and $\int_a^b (S - s) < \varepsilon^2$. Define $E =$

$\{x \in [a, b] \mid S(x) - s(x) \geq \varepsilon\}$. Since s and S are step functions, E must be an elementary set; and we deduce that

$$\varepsilon^2 > \int_a^b (S - s) \geq \int_E (S - s) \geq \int_E \varepsilon = \varepsilon m(E),$$

and from this it follows that $m(E) < \varepsilon$.

Now to prove that $(2) \Rightarrow (1)$, assume that (2) holds. Using the fact that f is bounded, choose numbers α and β such that $\alpha \leq f(x) \leq \beta$ for all $x \in [a, b]$. Now we use condition (2). For every natural number n, choose two step functions s_n and S_n such that $s_n \leq f \leq S_n$ and such that, defining $E_n = \{x \in [a, b] \mid S_n(x) - s_n(x) \geq 1/n\}$, we have $m(E_n) < 1/n$.

It would have been nice if we could now have said that the pair of sequences (s_n) and (S_n) squeezes f, but this needn't be true. The trouble is that although $S_n - s_n$ must be small except in E_n, inside the set E_n this function might be enormous. So we need to perform a little surgery on s_n and S_n. For each n we define $s_n^* = s_n \vee \alpha$ and $S_n^* = S_n \wedge \beta$. [See Exercise 8.4(4) for the meaning of $\wedge$ and $\vee$.] Then for each n, s_n^* and S_n^* are step functions, and $\alpha \leq s_n^* \leq f \leq S_n^* \leq \beta$; and for every point $x \notin E_n$ we have $S_n^*(x) - s_n^*(x) < 1/n$. Now for each n, writing $F_n = [a, b] \backslash E_n$, we see that both E_n and F_n are elementary sets; and so by Theorem 8.5.7 we have

$$\int_a^b (S_n^* - s_n^*) = \int_{E_n} (S_n^* - s_n^*) + \int_{F_n} (S_n^* - s_n^*)$$

$$\leq (\beta - \alpha)m(E_n) + \frac{1}{n} m(F_n) < \frac{(\beta - \alpha)}{n} + \frac{(b - a)}{n}.$$

It follows that the pair of sequences (s_n^*) and (S_n^*) must squeeze f on $[a, b]$. ∎

8.14.3   Lemma: A Simple Fact About Suprema and Infima of Functions.[3]

Suppose that $f: S \to R$, where S is a nonempty set; and suppose that the function f is bounded. Put

$$\alpha = \inf\{f(x) \mid x \in S\}, \qquad \beta = \sup\{f(x) \mid x \in S\},$$

and

$$\delta = \sup\{f(x) - f(t) \mid x \text{ and } t \in S\} = \sup\{|f(x) - f(t)| \mid x \text{ and } t \in S\}.$$

Then we have $\delta = \beta - \alpha$.

■ **Proof.**   Given any points x and t in S, $|f(x) - f(t)|$ is either $f(x) - f(t)$ or $f(t) - f(x)$. Therefore, the two expressions in the definition of δ are the same. Now whenever x and t are points of S, since $f(x) \leq \beta$ and $f(t) \geq \alpha$, we have $f(x) - f(t) \leq \beta - \alpha$; and it follows that $\delta \leq \beta - \alpha$. Now to obtain a contradiction, assume $\delta < \beta - \alpha$. Using the fact that $\delta + \alpha < \beta$, choose a point $x \in S$ such

[3] This lemma is really a carbon copy of Exercise 2.5(13).

that $\delta + \alpha < f(x)$. Now using the fact that $\alpha < f(x) - \delta$, choose a point $t \in S$ such that $f(t) < f(x) - \delta$. But this gives $\delta < f(x) - f(t)$, which is impossible. ∎

8.14.4 The Upper, Lower, and Oscillation Functions of a Given Function on a Partition.

Suppose f is a bounded function defined on the interval $[a, b]$, and that $\mathscr{P}$ is the partition $(x_0, x_1, \ldots, x_n)$ of $[a, b]$. The **upper** and **lower functions** $u(\mathscr{P}, f)$ and $\ell(\mathscr{P}, f)$ of f on $\mathscr{P}$ are defined as follows: For every $i = 0, 1, \ldots,$ n we define $u(\mathscr{P}, f)(x_i) = \ell(\mathscr{P}, f)(x_i) = f(x_i)$; and for every $i = 1, 2, \ldots, n$ we let $u(\mathscr{P}, f)$ and $\ell(\mathscr{P}, f)$ take the constant values $\sup\{f(x) \mid x \in (x_{i-1}, x_i)\}$ and $\inf\{f(x) \mid x \in (x_{i-1}, x_i)\}$, respectively, in the interval (x_{i-1}, x_i).

Notice that $u(\mathscr{P}, f)$ is the smallest of all the step functions S which step within the partition $\mathscr{P}$ and satisfy $f \leq S$, and that $\ell(\mathscr{P}, f)$ is the largest of all the step functions s which step within $\mathscr{P}$ and satisfy $s \leq f$.

We define the **oscillation function** $\omega(\mathscr{P}, f)$ of f on $\mathscr{P}$ by

$$\omega(\mathscr{P}, f) = u(\mathscr{P}, f) - \ell(\mathscr{P}, f).$$

The function $\omega(\mathscr{P}, f)$ is zero at each point x_i; and in view of Lemma 8.14.3, we see that in each of the intervals (x_{i-1}, x_i), $\omega(\mathscr{P}, f)$ takes the constant value

$$\sup\{f(x) \mid x \in (x_{i-1}, x_i)\} - \inf\{f(x) \mid x \in (x_{i-1}, x_i)\}$$

$$= \sup\{f(x) - f(t) \mid x \text{ and } t \in (x_{i-1}, x_i)\}$$

$$= \sup\{|f(x) - f(t)| \mid x \text{ and } t \in (x_{i-1}, x_i)\}.$$

Now using the upper, lower, and oscillation functions, we can combine Theorems 8.14.1 and 8.14.2 and restate them.

8.14.5 Theorem.

Suppose f is a bounded function defined on $[a, b]$. Then the following three conditions are equivalent:

(1) f is Riemann-integrable on $[a, b]$.

(2) For every number $\varepsilon > 0$ there exists a partition $\mathscr{P}$ of $[a, b]$ such that $\int_a^b \omega(\mathscr{P}, f) < \varepsilon$.

(3) For every number $\varepsilon > 0$ there exists a partition $\mathscr{P} = (x_0, x_1, \ldots, x_n)$ of $[a, b]$ such that if we define E to be the set $\{x \in [a, b] \mid \omega(\mathscr{P}, f)(x) \geq \varepsilon\}$, then we have $m(E) < \varepsilon$.

∎ **Proof.** In the light of Theorems 8.14.1 and 8.14.2, we see at once that each of the conditions (2) and (3) implies the integrability of f. Now to prove that (1) $\Rightarrow$ (2), assume that (1) holds. Let $\varepsilon > 0$. Using Theorem 8.14.1, choose two step functions s and S such that $s \leq f \leq S$ and $\int_a^b (S - s) < \varepsilon$. Choose a partition $\mathscr{P}$ of $[a, b]$ such that both s and S step within $\mathscr{P}$. Then since $u(\mathscr{P}, f) - \ell(\mathscr{P}, f) \leq S - s$, we see that $\int_a^b [u(\mathscr{P}, f) - \ell(\mathscr{P}, f)] < \varepsilon$, as required. We can use almost exactly the same argument to prove that (1) $\Rightarrow$ (3), but this time, of course, we use Theorem 8.14.2 when we choose the two functions s and S. ∎

8.15 EXERCISES

1. Given that f is the function defined on $[0, 1]$ by

$$f(x) = 1 \quad \text{if} \quad x = \frac{1}{n} \quad \text{for some natural number } n,$$

and

$$f(x) = 0 \quad \text{otherwise},$$

prove that f is Riemann-integrable on $[0, 1]$ and that $\int_0^1 f = 0$.

2. The function f is defined on $[0, 1]$ as follows:

$$f(x) = 1 \quad \text{whenever} \quad \frac{1}{2n + 1} \le x \le \frac{1}{2n} \quad \text{for some } n \in \mathbf{Z}^+,$$

and

$$f(x) = 0 \quad \text{otherwise}.$$

Prove that f is Riemann-integrable on $[0, 1]$. Draw yourself a rough sketch of the graph of f. You will see in Exercise 10.7(13) that $\int_0^1 f = 1 - \log 2$.

3. Prove Theorem 8.13.6.

■ **4.** Suppose that f is a nonnegative function defined on $[a, b]$, and that for every number $\varepsilon > 0$ the set $\{x \in [a, b] \mid f(x) \ge \varepsilon\}$ is finite. Prove that f must be Riemann-integrable on $[a, b]$ and that $\int_a^b f = 0$. *Hint:* For a given ε, take $\mathcal{P}$ to be the partition whose points are the members of the given finite set, and take a look at $\int_a^b u(\mathcal{P}, f)$.

■ **5.** Prove that if f is the ruler function of Section 5.8, then f is Riemann-integrable on $[0, 1]$ and $\int_0^1 f = 0$.

6. Given a bounded nonnegative function f defined on an interval $[a, b]$, prove that the following two conditions are equivalent:
   (a) f is Riemann-integrable on $[a, b]$ and $\int_a^b f = 0$.
   (b) For every number $\varepsilon > 0$ there exists an elementary set E such that $m(E) < \varepsilon$ and $\{x \mid f(x) \ge \varepsilon\} \subseteq E$.

■ **7.** Prove the following converse to Theorem 8.13.4: If $a \le b \le c$ and f is a function defined on the interval $[a, c]$ which is Riemann-integrable on both of the intervals $[a, b]$ and $[b, c]$, then f is Riemann-integrable on $[a, c]$.

8. A bounded set A is said to be *Riemann-measurable* if there exists an interval $[a, b]$ such that $A \subseteq [a, b]$ and χ_A is Riemann-integrable on $[a, b]$.
   (a) Prove that if A is a Riemann-measurable set, and the interval $[a, b]$ is chosen as above, and $[c, d]$ is any other interval including A, then χ_A is also Riemann-integrable on $[c, d]$ and $\int_a^b \chi_A = \int_c^d \chi_A$.
   If A is a Riemann-measurable set, then the *Lebesgue measure* $m(A)$ of A is defined to be $\int_a^b \chi_A$, where $[a, b]$ is any interval that includes A.

(b) Prove that if A and B are Riemann-measurable and $A \subseteq B$, then $m(A) \leq m(B)$.

(c) Prove that if A and B are Riemann-measurable, then $m(A \cup B) \leq m(A) + m(B)$.

(d) Prove that if A and B are Riemann-measurable and $A \cap B$ is empty, then $m(A \cup B) = m(A) + m(B)$.

(e) Given any two Riemann-measurable sets A and B, prove that $m(A \cup B) = m(A) + m(B) - m(A \cap B)$.

8.16 SOME IMPORTANT KINDS OF INTEGRABLE FUNCTIONS

8.16.1 Theorem: Monotone Functions Are Integrable. If f is monotone on $[a, b]$, then f is Riemann-integrable on $[a, b]$.

■ *Proof.* We shall assume that f is increasing and leave the similar result for decreasing functions as an exercise.[4] For each natural number n, define $\mathcal{P}_n$ to be the regular n-partition of $[a, b]$; in other words, $\mathcal{P}_n$ is the partition $(x_0, x_1, \ldots, x_n)$, where for each i we have $x_i = a + i(b - a)/n$. The rest of this proof is the solution of Exercise 8.12(3). Define s_n and S_n to be the step functions which take the value $f(x_i)$ at each point x_i and which take the constant values $f(x_{i-1})$ and $f(x_i)$, respectively, in each interval (x_{i-1}, x_i). We deduce that

$$\int_a^b (S_n - s_n) = \sum_{i=1}^n [f(x_i) - f(x_{i-1})] \frac{(b - a)}{n}$$

$$= \frac{(b - a)}{n} \sum_{i=1}^n [f(x_i) - f(x_{i-1})]$$

$$= \frac{[f(b) - f(a)](b - a)}{n} \to 0 \quad \text{as} \quad n \to \infty.$$

Therefore, since the pair of sequences (s_n) and (S_n) squeezes f on $[a, b]$, f must be integrable. ■

8.16.2 Theorem: Continuous Functions Are Integrable. Suppose f is continuous on the interval $[a, b]$. Then f is Riemann-integrable on $[a, b]$.

■ *Proof.* Our method of showing that f is Riemann-integrable will be to show that f satisfies condition (3) of Theorem 8.14.5. We need to show that for every number $\varepsilon > 0$ there exists a partition $\mathcal{P}$ of $[a, b]$ such that if $E = \{x \mid \omega(\mathcal{P}, f)(x) \geq$

---

[4] It is worth mentioning that if f is decreasing, then $-f$ is increasing; and so the result for decreasing functions doesn't have to be proved at all.

$\varepsilon\}$, then $m(E) < \varepsilon$. We shall actually show more than this, for we shall see that if $\varepsilon > 0$, then the condition holds for *every* partition $\mathscr{P}$ which has sufficiently small mesh. Recall the definition given in Section 8.2.1 of the mesh $\|\mathscr{P}\|$ of a partition $\mathscr{P}$.

Let $\varepsilon > 0$. Using the fact that f is uniformly continuous on $[a, b]$, choose a number $\delta > 0$ such that whenever x and t are points in $[a, b]$ and $|x - t| < \delta$, we have $|f(x) - f(t)| < \varepsilon/2$. Now choose a partition $\mathscr{P} = (x_0, x_1, \ldots, x_n)$ of $[a, b]$ such that $\|\mathscr{P}\| < \delta$. We could, for example, choose $\mathscr{P}$ to be the regular n-partition of $[a, b]$ with n chosen large enough to make $(b - a)/n < \delta$. We see that whenever two points x and t belong to a given interval (x_{i-1}, x_i), we have $|x - t| < \delta$ and, consequently, $|f(x) - f(t)| < \varepsilon/2$. Therefore, $\omega(\mathscr{P}, f)(x) \leq \varepsilon/2 < \varepsilon$ for every $x \in [a, b]$, and the set $E = \{x \mid \omega(\mathscr{P}, f)(x) \geq \varepsilon\}$, being empty, must satisfy $m(E) < \varepsilon$.   ∎

8.16.3* Junior Version of the Lebesgue Criterion for Riemann Integrability.

(Section 8.16.3 can be omitted without loss of continuity.) As one might suspect from an examination of the proof of Theorem 8.16.2, this theorem can be greatly improved. Theorem 8.16.2 is actually the simplest of a group of theorems that say that a bounded function which is continuous in "most" of an interval $[a, b]$ will be Riemann-integrable on $[a, b]$. The sharpest of all these results is Theorem B.6.3, which is known as the **Lebesgue criterion for Riemann integrability** and which makes use of the concept *almost every*, which is defined in Appendix B. Take a glance at the statement of Theorem B.6.3. It might tempt you to read Appendix B after you have completed this chapter. In the meantime, we shall prove the following simpler theorem which we call the **junior version of the Lebesgue criterion**. This result is already a much sharper theorem than Theorem 8.16.2.

Theorem*. Suppose f is a bounded function defined on the interval $[a, b]$. Then a *sufficient* condition for f to be Riemann-integrable on $[a, b]$ is that for every number $\varepsilon > 0$ there exists an elementary subset E of $[a, b]$ such that $m(E) < \varepsilon$ and f is continuous in the set $[a, b] \backslash E$.

■ *Proof.* Just as in the proof of Theorem 8.16.2, our method of showing that f is Riemann-integrable will be to show that f satisfies condition (3) of Theorem 8.14.5.

Let $\varepsilon > 0$. Choose an elementary subset E of $[a, b]$ such that $m(E) < \varepsilon$ and f is continuous in $[a, b] \backslash E$. Now using Theorem 8.7.5, choose an open elementary set $U \supseteq E$ such that $m(U) < \varepsilon$. Since f is continuous on the closed, bounded set $[a, b] \backslash U$, f must be uniformly continuous on this set. Choose a number $\delta > 0$ such that whenever x and t are points of $[a, b] \backslash U$ and $|x - t| < \delta$, we have $|f(x) - f(t)| < \varepsilon/2$.

Choose a partition $\mathscr{P}_1$ of $[a, b]$ such that $\|\mathscr{P}_1\| < \delta$, and choose a partition $\mathscr{P}_2$ of $[a, b]$ such that the step function χ_U steps within $\mathscr{P}_2$. Now define $\mathscr{P}$ to be the common refinement of $\mathscr{P}_1$ and $\mathscr{P}_2$, and write $\mathscr{P}$ in the form $(x_0, x_1, \ldots, x_n)$. What is important about this partition $\mathscr{P}$ is that for each $i = 1, 2, \ldots, n$ the interval

(x_{i-1}, x_i) is either included in U or is disjoint from U, depending upon whether the constant value of χ_U in (x_{i-1}, x_i) is 1 or 0. And in the event that (x_{i-1}, x_i) is disjoint from U, the fact that $x_i - x_{i-1} < \delta$ guarantees that $|f(x) - f(t)| < \varepsilon/2$ whenever x and t belong to (x_{i-1}, x_i); and so the constant value of $\omega(\mathcal{P}, f)$ in (x_{i-1}, x_i) does not exceed $\varepsilon/2$. Therefore, if $A = \{x \in [a, b] \mid \omega(\mathcal{P}, f)(x) \geq \varepsilon\}$, then we have $A \subseteq U$; and it follows that $m(A) < \varepsilon$. ∎

To see that the converse of this theorem is false, look at the ruler function again. In Exercise 8.15(5) we saw that the ruler function is Riemann-integrable on $[0, 1]$, but we also know from Exercise 6.5(15) that the ruler function is continuous at every irrational point of $[0, 1]$ and discontinuous at every rational. Therefore, whenever E is an elementary set and the ruler function is continuous on $[0, 1]\backslash E$, since $E \supseteq \mathbf{Q} \cap [0, 1]$, we see that E must contain all but finitely many points of the interval $[0, 1]$; and consequently, $m(E) \geq 1$. This tells us that although the ruler function is Riemann-integrable on $[0, 1]$, it does not satisfy the junior version of the Lebesgue criterion for integrability.

On the other hand, some sort of continuity seems to be needed for a function to be Riemann-integrable. As we saw in Section 8.11.6, a function which is really violently discontinuous might not be Riemann-integrable.

8.16.4 Theorem. Suppose f is Riemann-integrable on $[a, b]$. Then $|f|$ must be Riemann-integrable on $[a, b]$.

■ **Proof.** Once again, our method of showing that a function is Riemann-integrable will be to show that it satisfies condition (3) of Theorem 8.14.5. Let $\varepsilon > 0$, and choose a partition $\mathcal{P}$ of $[a, b]$ such that if $E = \{x \mid \omega(\mathcal{P}, f)(x) \geq \varepsilon\}$, then $m(E) < \varepsilon$. But given any numbers x and t in $[a, b]$, we have $\|f(x)| - |f(t)\| \leq |f(x) - f(t)|$; and so if $E^* = \{x \mid \omega(\mathcal{P}, |f|)(x) \geq \varepsilon\}$, then $E^* \subseteq E$; and so $m(E^*) < \varepsilon$. ∎

8.16.5 Theorem. The product of two Riemann-integrable functions is Riemann-integrable.

■ **Proof.** We shall show first that if f is Riemann-integrable on $[a, b]$, then so is f^2. Suppose that f is integrable on $[a, b]$. Using the fact that f is bounded, choose a positive number K such that $|f(x)| \leq K$ for all $x \in [a, b]$. Now clearly, if x and t are any points in $[a, b]$, we have

$$|f^2(x) - f^2(t)| = |f(x) - f(t)| \cdot |f(x) + f(t)| \leq 2K|f(x) - f(t)|$$

and it follows that if $\mathcal{P}$ is any partition of $[a, b]$, we have $\omega(\mathcal{P}, f^2) \leq 2K\omega(\mathcal{P}, f)$. To show that f^2 satisfies condition (3) of Theorem 8.14.5, let $\varepsilon > 0$. Define α to be the smaller of the two numbers ε and $\varepsilon/2k$, and using Theorem 8.14.5, choose a partition $\mathcal{P}$ of $[a, b]$ such that if $E = \{x \mid \omega(\mathcal{P}, f)(x) \geq \alpha\}$, then $m(E) < \alpha$. We see at once that if $E^* = \{x \mid \omega(\mathcal{P}, f^2)(x) \geq \varepsilon\}$, then $E^* \subseteq E$; and therefore, $m(E^*) < \alpha \leq \varepsilon$.

This shows that the square of an integrable function must be integrable. Suppose now that f and g are two integrable functions on $[a, b]$. Since

$$fg = \tfrac{1}{4}(f + g)^2 - \tfrac{1}{4}(f - g)^2$$

it follows at once that fg is integrable. ∎

8.16.6* The Composition Theorem. Our last theorem in this section extends the previous two results. It can be omitted without loss of continuity.

Given any function f, we can look at the function $|f|$ as being the composition $\phi \circ f$, where $\phi(u) = |u|$ for all u; and in the same way, we can look at the function f^2 as being $\phi \circ f$, where $\phi(u) = u^2$ for all u. Notice that in both of these cases if the function f is bounded, then ϕ will be uniformly continuous on the range of f; consequently, it would be interesting to know that whenever a function f is Riemann-integrable on an interval $[a, b]$ and ϕ is uniformly continuous on the range of f, then the function $\phi \circ f$ must be Riemann-integrable on $[a, b]$. As a matter of fact, the condition of uniform continuity of ϕ can even be replaced by plain continuity, but this makes the theorem too hard for us to prove at present. The stronger theorem will be proved in Appendix B.

Theorem*. Suppose that f is Riemann-integrable on an interval $[a, b]$ and that ϕ is uniformly continuous on the range of f. Then the composition $\phi \circ f$ of ϕ with f is Riemann-integrable on $[a, b]$.

■ *Proof.* Put $g = \phi \circ f$; in other words, $g(x) = \phi(f(x)) \; \forall \, x \in [a, b]$. As you may have guessed, we shall show that g is integrable by showing that g satisfies condition (3) of Theorem 8.14.5.

Let $\varepsilon > 0$; and using the fact that ϕ is uniformly continuous on the range of f, choose a number $\delta > 0$ such that whenever y and z are points in the range of f and $|y - z| < \delta$, we have $|\phi(y) - \phi(z)| < \varepsilon/2$. Define α to be the smaller of ε and δ. Now using the integrability of f, choose a partition $\mathcal{P} = (x_0, \ldots, x_n)$ of $[a, b]$ such that if $E = \{x \mid \omega(\mathcal{P}, f)(x) \geq \alpha\}$, then $m(E) < \alpha$. Now for any $i = 1, 2, \ldots, n$, unless $(x_{i-1}, x_i) \subseteq E$, we know that whenever x and t are points of (x_{i-1}, x_i), we have $|f(x) - f(t)| < \alpha \leq \delta$; and consequently, $|\phi(f(x)) - \phi(f(t))| < \varepsilon/2$. This tells us that in any interval (x_{i-1}, x_i) which is not included in E, the constant value of $\omega(\mathcal{P}, g)$ can't exceed $\varepsilon/2$. It follows that if $A = \{x \mid \omega(\mathcal{P}, g)(x) \geq \varepsilon\}$, then $A \subseteq E$; and so $m(A) \leq m(E) < \alpha \leq \varepsilon$. ∎

8.17 THE FUNDAMENTAL THEOREM OF CALCULUS

The theorems of this section provide the important link between the concepts of differentiation and integration. They are known collectively as the **fundamental theorem of calculus**.

8.17.1 Theorem. Suppose that f is differentiable on $[a, b]$, and suppose further that f' is Riemann-integrable on $[a, b]$. Then

$$\int_a^b f' = f(b) - f(a).$$

■ **Proof.** We shall prove the theorem by showing that if s and S are any two step functions such that $s \leq f' \leq S$, then

$$\int_a^b s \leq f(b) - f(a) \leq \int_a^b S.$$

Assume then that s and S are step functions and $s \leq f' \leq S$. Choose a partition $\mathcal{P} = (x_0, \ldots, x_n)$ of $[a, b]$ such that both s and S step within $\mathcal{P}$, and denote the constant values of s and S in each interval (x_{i-1}, x_i) as α_i and β_i, respectively. We shall now apply the first mean value theorem (Theorem 7.12.6) to f in each of the intervals $[x_{i-1}, x_i]$. For each $i = 1, 2, \ldots, n$, choose a point t_i in (x_{i-1}, x_i) such that

$$f(x_i) - f(x_{i-1}) = (x_i - x_{i-1})f'(t_i).$$

Now since $\alpha_i \leq f'(t_i) \leq \beta_i$ for every i, we see that

$$\int_a^b s = \sum_{i=1}^n \alpha_i(x_i - x_{i-1}) \leq \sum_{i=1}^n f'(t_i)(x_i - x_{i-1})$$

$$\leq \sum_{i=1}^n \beta_i(x_i - x_{i-1}) = \int_a^b S.$$

Therefore,

$$\int_a^b s \leq \sum_{i=1}^n [f(x_i) - f(x_{i-1})] \leq \int_a^b S,$$

and it follows that $\int_a^b s \leq f(b) - f(a) \leq \int_a^b S$, as required. ■

8.17.2 Theorem. Suppose f is Riemann-integrable on $[a, b]$; and for every $x \in [a, b]$, define

$$F(x) = \int_a^x f.$$

Then F is lipschitzian (and therefore uniformly continuous) on $[a, b]$.

■ **Proof.** Choose a number K such that $|f(x)| \leq K$ for all points $x \in [a, b]$. We shall show that for all x and t in $[a, b]$ we have $|F(x) - F(t)| \leq K|x - t|$. Let x and t

belong to $[a, b]$, and assume, without loss of generality, that $t \leq x$. Then we have

$$|F(x) - F(t)| = \left| \int_a^x f - \int_a^t f \right| = \left| \int_t^x f \right| \leq \int_t^x |f| \leq \int_t^x K = K|x - t|. \qquad \blacksquare$$

8.17.3 Theorem. Suppose f is Riemann-integrable on $[a, b]$, where $a < b$; and for every $x \in [a, b]$, define

$$F(x) = \int_a^x f.$$

Suppose that $c \in [a, b]$ and that f is continuous at c. Then F is differentiable at c, and $F'(c) = f(c)$.

■ *Proof.* Note first that if $x \in [a, b]$ and $x \neq c$, then

$$\frac{F(x) - F(c)}{x - c} - f(c) = \frac{1}{x - c} \int_c^x (f - f(c)).$$

We shall show that the latter expression tends to zero as $x \to c$. Let $\varepsilon > 0$; and choose $\delta > 0$ such that whenever $x \in [a, b]$ and $|x - c| < \delta$, we have $|f(x) - f(c)| < \varepsilon$. Now suppose $x \in [a, b] \backslash \{c\}$ and $|x - c| < \delta$. In order to show that $|1/(x - c) \int_c^x (f - f(c))| \leq \varepsilon$, we shall consider separately the cases $x < c$ and $x > c$. Suppose first that $x < c$. Then

$$\left| \frac{1}{x - c} \int_c^x (f - f(c)) \right| \leq \frac{1}{|x - c|} \int_x^c |f - f(c)| \leq \frac{1}{|x - c|} \int_x^c \varepsilon = \varepsilon.$$

Similarly, when $x > c$, we have

$$\left| \frac{1}{x - c} \int_c^x (f - f(c)) \right| \leq \frac{1}{|x - c|} \int_c^x |f - f(c)| \leq \frac{1}{|x - c|} \int_c^x \varepsilon = \varepsilon. \qquad \blacksquare$$

8.17.4 The Change-of-Variable Theorem. Suppose that ϕ is differentiable on $[a, b]$, and suppose further that ϕ' is Riemann-integrable on $[a, b]$. Suppose f is continuous on the range of ϕ. Then

$$\int_a^b f(\phi(t))\phi'(t) \, dt = \int_{\phi(a)}^{\phi(b)} f.$$

■ *Proof.* The first thing we need to observe is that since $f \circ \phi$ is the composition of two continuous functions, it is continuous on $[a, b]$ and, therefore, Riemann-integrable on $[a, b]$. Therefore, since ϕ' is Riemann-integrable on $[a, b]$, it follows from Theorem 8.16.5 that the product of these two functions is integrable. This means that the left side of the required identity is meaningful. Now the range of ϕ

is, of course, an interval. Call it $[\alpha, \beta]$. For every $x \in [\alpha, \beta]$, define

$$F(x) = \int_\alpha^x f.$$

Then since f is continuous on $[\alpha, \beta]$, it follows from Theorem 8.17.3 that for every $x \in [\alpha, \beta]$ we have $F'(x) = f(x)$. Now define $h(t) = F(\phi(t))$ for all $t \in [a, b]$. From the chain rule we see that for every $t \in [a, b]$ we have

$$h'(t) = F'(\phi(t))\phi'(t) = f(\phi(t))\phi'(t);$$

and it follows from Theorem 8.17.1 that

$$\int_a^b f(\phi(t))\phi'(t) \, dt = \int_a^b h' = h(b) - h(a) = F(\phi(b)) - F(\phi(a))$$

$$= \int_\alpha^{\phi(b)} f - \int_\alpha^{\phi(a)} f = \int_{\phi(a)}^{\phi(b)} f. \qquad \blacksquare$$

8.17.5   Theorem: Integration by Parts.   Suppose that f and g are differentiable on $[a, b]$, and suppose further that f' and g' are Riemann-integrable on $[a, b]$. Then

$$\int_a^b fg' = f(b)g(b) - f(a)g(a) - \int_a^b f'g.$$

▪ **Proof.**   We observe, first, that the functions $f'g$ and fg', being products of Riemann-integrable functions, are Riemann-integrable. Now define $h = fg$. Then since $h' = f'g + fg'$, it follows that h' is Riemann-integrable on $[a, b]$; and it follows from Theorem 8.17.1 that $\int_a^b h' = h(b) - h(a)$. In other words,

$$\int_a^b (f'g + fg') = f(b)g(b) - f(a)g(a),$$

and this is precisely what we had to prove. $\qquad \blacksquare$

8.18   EXERCISES

▪ **1.** Prove that if f is continuous on $[-1, 1]$, then

$$\int_0^{2\pi} f(\sin x) \cos x \, dx = 0;$$

and that if f is continuous on $[0, 1]$, then

$$\int_0^{\pi/2} f(\sin x) \, dx = \int_{\pi/2}^\pi f(\sin x) \, dx.$$

2. Given $\alpha > 0$, prove that $\int_0^{\pi/2} \sin^\alpha x \, dx = 2^\alpha \int_0^{\pi/2} \sin^\alpha x \cos^\alpha x \, dx$.

3. Suppose that ϕ is differentiable on $[a, b]$, and suppose further that ϕ' is Riemann-integrable on $[a, b]$ and $\phi(a) = \phi(b)$. Prove that if f is any function which is continuous on the range of ϕ, then we have $\int_a^b f(\phi(t))\phi'(t) \, dt = 0$.

4. Given that f is Riemann-integrable on $[a, b]$ and that c is any number, prove that $\int_a^b f(t)\, dt = \int_{a+c}^{b+c} f(t - c)\, dt$.

5. Given any numbers a, b, and c such that $ac < bc$, and given that f is continuous on the interval $[ac, bc]$, prove that $\int_{ac}^{bc} f(t)\, dt = c \int_a^b f(ct)\, dt$.

■ 6. Use exercise 5 to show that if a and b are positive numbers, then

$$\int_1^a \frac{1}{t}\, dt + \int_1^b \frac{1}{t}\, dt = \int_1^{ab} \frac{1}{t}\, dt.$$

Another proof of this identity will be given in Section 9.2.3.

7. Suppose that f is continuous on $[a, b]$, that g is a nonnegative, Riemann-integrable function on $[a, b]$, and that $\int_a^b g > 0$. Prove that the ratio $(\int_a^b fg)(\int_a^b g)^{-1}$ lies between the minimum and maximum values of the function f. Deduce that there exists a number $c \in [a, b]$ such that

$$\int_a^b fg = f(c) \int_a^b g.$$

This result is known as the **mean value theorem for integrals**.

8. Given f continuous on $[a, b]$, prove that there exists a number $c \in [a, b]$ such that $\int_a^b f = f(c)(b - a)$.

9. Given that f is a nonnegative, continuous function on $[a, b]$ and that $\int_a^b f = 0$, prove that f is the constant zero function.

10. Given that f is continuous on $[a, b]$ and that $\int_a^x f = 0$ for every $x \in [a, b]$, prove that f is the constant zero function.

*11. Given that f is a strictly increasing continuous function on $[a, b]$, prove that

$$\int_a^b f + \int_{f(a)}^{f(b)} f^{-1} = bf(b) - af(a).$$

*12. Given that f is a strictly increasing, continuous, unbounded function on $[0, \infty)$ and that $f(0) = 0$, prove that for every pair of positive numbers a and b

$$ab \le \int_0^a f + \int_0^b f^{-1}.$$

*13. Suppose that a and b are positive numbers and that p and q are positive numbers satisfying $1/p + 1/q = 1$. Prove that $ab \le a^p/p + b^q/q$. *Hint:* Apply exercise 8 with an appropriate function f. This inequality is known as W. H. Young's inequality.

*14. Given f Riemann-integrable on $[a, b]$ and $\alpha > 0$, prove that the function $|f|^\alpha$ is Riemann-integrable on $[a, b]$. *Hint:* Use Theorem 8.16.6.

*15. Prove that if f is nonnegative and Riemann-integrable on $[a, b]$, and if $\alpha > 0$, then we have $\int_a^b f = 0$ if and only if $\int_a^b f^\alpha = 0$. *Hint:* Use Exercise 8.15(6).

*16. Given f Riemann-integrable on $[a, b]$ and a number $p > 1$, the *p-norm* $\|f\|_p$ of f on $[a, b]$ is defined to be $(\int_a^b |f|^p)^{1/p}$. Prove that if f and g are Riemann-integrable on $[a, b]$ and p and q are positive numbers satisfying $1/p + 1/q = 1$, then

$$\int_a^b |fg| \le \|f\|_p \|g\|_q.$$

This inequality is known as Hölder's inequality. *Hint:* If the right side of this inequality is zero, use exercise 15 to show that the left side is also zero. Otherwise, apply exercise 13 to the numbers $|f(x)|/\|f\|_p$ and $|g(x)|/\|g\|_q$ for each $x \in [a, b]$, and integrate.

8.19* THE ROLE OF THE MESH OF A PARTITION; DARBOUX'S THEOREM

The material of this section will be used in Section 9.8.8 in the proof of Fichtenholz's beautiful theorem on the inversion of repeated integrals. Except for this, the section can be omitted without loss of continuity.

The concept of mesh of a partition which we defined in Section 8.2.1 plays a very important role in the theory of Riemann integration; for roughly speaking, if f is Riemann-integrable on an interval $[a, b]$, then the function $\omega(\mathcal{P}, f)$ will be small for *all* partitions $\mathcal{P}$ which have sufficiently small mesh. This fact, which was hinted at during the proof of Theorem 8.16.2, is a dramatic improvement over Theorem 8.14.5, which says only that if f is Riemann-integrable on $[a, b]$, then there must *exist* a partition $\mathcal{P}$ for which $\omega(\mathcal{P}, f)$ is small.

The key to the theorems that follow is the fact that if $\mathcal{Q}$ is the partition $(q_0, q_1, \ldots, q_N)$ of an interval $[a, b]$, and $\mathcal{P}$ is any other partition $(x_0, \ldots, x_n)$ of $[a, b]$, then not more than $2N$ of the intervals $[x_{i-1}, x_i]$ of $\mathcal{P}$ can contain a point of $\mathcal{Q}$. Therefore, writing the union of these intervals as U, we have $m(U) \le 2N\|\mathcal{P}\|$. Furthermore, given any point $x \in [a, b]\backslash U$, x must lie in some interval $[x_{i-1}, x_i]$ that does not contain a point of $\mathcal{Q}$ and which must therefore be included in some interval (q_{j-1}, q_j):

It follows that whenever $x \in [a, b]\backslash U$, we have

$$\ell(\mathcal{Q}, f)(x) \le \ell(\mathcal{P}, f)(x) \le f(x) \le u(\mathcal{P}, f)(x) \le u(\mathcal{Q}, f)(x);$$

and consequently, $\omega(\mathcal{P}, f)(x) \le \omega(\mathcal{Q}, f)(x)$.

8.19.1* Theorem. Suppose f is Riemann-integrable on $[a, b]$. Then given $\varepsilon > 0$, there exists a number $\delta > 0$ such that for every partition $\mathcal{P}$ of $[a, b]$ satisfying $\|\mathcal{P}\| < \delta$, we have $\int_a^b \omega(\mathcal{P}, f) < \varepsilon$.

▪ *Proof.* Let $\varepsilon > 0$; choose a positive number K such that $|f| \le K$; and using Theorem 8.14.5, choose a partition $\mathcal{Q} = (q_0, q_1, \ldots, q_N)$ of $[a, b]$ such that

$\int_a^b \omega(\mathcal{Q}, f) < \varepsilon/2$. Now suppose that $\mathcal{P}$ is any partition $(x_0, \ldots, x_n)$ of $[a, b]$, define U as above, and define $V = [a, b]\backslash U$. Then

$$\int_a^b \omega(\mathcal{P}, f) = \int_U \omega(\mathcal{P}, f) + \int_V \omega(\mathcal{P}, f) \leq \int_U 2K + \int_V \omega(\mathcal{Q}, f)$$

$$\leq \int_U 2K + \int_a^b \omega(\mathcal{Q}, f)$$

$$\leq 2Km(U) + \frac{\varepsilon}{2} \leq 4KN\|\mathcal{P}\| + \frac{\varepsilon}{2},$$

and the latter expression will be less than ε provided $\|\mathcal{P}\| < \varepsilon/8kN$.   ∎

8.19.2*   Theorem.   Suppose f is Riemann-integrable on $[a, b]$. Then given $\varepsilon > 0$, there exists a number $\delta > 0$ such that for every partition $\mathcal{P}$ of $[a, b]$ satisfying $\|\mathcal{P}\| < \delta$, if we define E to be the set $\{x \in [a, b] \mid \omega(\mathcal{P}, f)(x) \geq \varepsilon\}$, then we have $m(E) < \varepsilon$.

■ *Proof.*   Let $\varepsilon > 0$; choose a positive number K such that $|f| \leq K$; and using Theorem 8.14.5, choose a partition $\mathcal{Q} = (q_0, q_1, \ldots, q_N)$ of $[a, b]$ such that if $A = \{x \in [a, b] \mid \omega(\mathcal{Q}, f)(x) \geq \varepsilon/2\}$, then $m(A) < \varepsilon/2$. Now suppose that $\mathcal{P}$ is any partition $(x_0, \ldots, x_n)$ of $[a, b]$; define U as above, and define $E = \{x \in [a, b] \mid \omega(\mathcal{P}, f)(x) \geq \varepsilon\}$. Then we clearly have $E \subseteq A \cup U$, and so $m(E) \leq m(A) + m(U) < \varepsilon/2 + 2N\|\mathcal{P}\|$. The latter expression will be less than ε as long as $\|\mathcal{P}\| < \varepsilon/4N$.   ■

8.19.3*   Darboux's Theorem.   Suppose f is Riemann-integrable on $[a, b]$. Then given $\varepsilon > 0$, there exists a number $\delta > 0$ such that for every partition $\mathcal{P}$ of the form $(x_0, \ldots, x_n)$ satisfying $\|\mathcal{P}\| < \delta$ and for every choice of numbers t_i in the intervals $[x_{i-1}, x_i]$ for $i = 1, 2, \ldots, n$, we have

$$\left| \sum_{i=1}^n f(t_i)(x_i - x_{i-1}) - \int_a^b f \right| < \varepsilon.$$

■ *Proof.*   Let $\varepsilon > 0$; choose a positive number K such that $|f| \leq K$; and using Theorem 8.14.5, choose a partition $\mathcal{Q} = (q_0, q_1, \ldots, q_N)$ of $[a, b]$ such that $\int_a^b \omega(\mathcal{Q}, f) < \varepsilon/4$. Now suppose that $\mathcal{P}$ is any partition $(x_0, \ldots, x_n)$ of $[a, b]$, and that for each $i = 1, 2, \ldots, n$, t_i is a point in the interval $[x_{i-1}, x_i]$. In order to estimate the value of the sum $\sum_{i=1}^n f(t_i)(x_i - x_{i-1})$, we shall express it as the integral of a step function. Define g to be the step function that takes the value $f(x_i)$ at each point x_i of $\mathcal{P}$ and which takes the constant value $f(t_i)$ in each interval (x_{i-1}, x_i). Then $\sum_{i=1}^n f(t_i)(x_i - x_{i-1}) = \int_a^b g$.

We shall show that $|\int_a^b f - \int_a^b g|$ will be less than ε provided that $\|\mathcal{P}\|$ is small enough; and for this purpose we shall compare each of the integrals $\int_a^b f$ and $\int_a^b g$ with $\int_a^b u(\mathcal{Q}, f)$.

Now on the one hand, $\ell(\mathcal{Q}, f) \le f \le u(\mathcal{Q}, f)$; and therefore,

$$\left| \int_a^b u(\mathcal{Q}, f) - \int_a^b f \right| \le \int_a^b [u(\mathcal{Q}, f) - \ell(\mathcal{Q}, f)] < \frac{\varepsilon}{4}.$$

On the other hand, defining U as in the previous results of this section and $V = [a, b]\setminus U$, we see that $\ell(\mathcal{Q}, f)(x) \le g(x) \le u(\mathcal{Q}, f)(x)$ for all points $x \in V$. Therefore,

$$\left| \int_a^b u(\mathcal{Q}, f) - \int_a^b g \right| \le \int_a^b |u(\mathcal{Q}, f) - g| = \int_U |u(\mathcal{Q}, f) - g| + \int_V |u(\mathcal{Q}, f) - g|$$

$$\le \int_U 2K + \int_V [u(\mathcal{Q}, f) - \ell(\mathcal{Q}, f)] < 4KN\|\mathcal{P}\| + \frac{\varepsilon}{4}.$$

From this it follows that $|\int_a^b f - \int_a^b g| < 4KN\|\mathcal{P}\| + \varepsilon/2$, and the latter expression will be less than ε provided that $\|\mathcal{P}\| < \varepsilon/8KN$. ∎

8.19.4*   A Variation of Darboux's Theorem.   Suppose f is a bounded function on an interval $[a, b]$. Suppose that $\mathcal{I}$ is some number, and for every natural number n suppose $\mathcal{P}_n$ is the regular n-partition of $[a, b]$. The ith point of $\mathcal{P}_n$ is, of course, $a + i(b - a)/n$, but for simplicity we shall write this as x_{ni}. The following two conditions are equivalent:

(1) f is Riemann-integrable on $[a, b]$ and $\int_a^b f = \mathcal{I}$.

(2) For every choice of numbers t_{ni} in the intervals $[x_{n\,i-1}, x_{ni}]$, where $n \in \mathbf{Z}^+$ and $i = 1, 2, \ldots, n$, we have

$$\sum_{i=1}^n f(t_{ni})(x_{ni} - x_{n\,i-1}) \to \mathcal{I} \qquad \text{as} \qquad n \to \infty.$$

■ **Proof.**   The fact that (1) $\Rightarrow$ (2) follows directly from Theorem 8.19.3. Now to prove that (2) $\Rightarrow$ (1), assume that (2) holds. In order to prove that f is Riemann-integrable on $[a, b]$, we shall show that for every $\varepsilon > 0$ there is a partition $\mathcal{P}$ of $[a, b]$ such that $\int_a^b \omega(\mathcal{P}, f) < \varepsilon$. Let $\varepsilon > 0$. For each natural number n and each $i = 1, 2, \ldots, n$, choose numbers s_{ni} and t_{ni} in $(x_{n\,i-1}, x_{ni})$ such that if α_{ni} and β_{ni} are the constant values of $\ell(\mathcal{P}_n, f)$ and $u(\mathcal{P}_n, f)$ in the interval $(x_{n\,i-1}, x_{ni})$, then

$$f(s_{ni}) - \frac{\varepsilon}{4(b - a)} < \alpha_{ni} \qquad \text{and} \qquad f(t_{ni}) + \frac{\varepsilon}{4(b - a)} > \beta_{ni}.$$

Using the fact that both

$$\sum_{i=1}^n f(s_{ni})(x_{ni} - x_{n\,i-1}) \qquad \text{and} \qquad \sum_{i=1}^n f(t_{ni})(x_{ni} - x_{n\,i-1})$$

approach $\mathscr{I}$ as $n \to \infty$, choose n such that

$$\left| \sum_{i=1}^{n} f(s_{ni})(x_{ni} - x_{n\,i-1}) - \mathscr{I} \right| < \frac{\varepsilon}{4} \quad \text{and} \quad \left| \sum_{i=1}^{n} f(t_{ni})(x_{ni} - x_{n\,i-1}) - \mathscr{I} \right| < \frac{\varepsilon}{4}.$$

From this we see that

$$\int_{a}^{b} \mathscr{W}(\mathscr{P}_n, f) = \sum_{i=1}^{n} (\beta_{ni} - \alpha_{ni})(x_{ni} - x_{n\,i-1})$$

$$\leq \sum_{i=1}^{n} \left[f(t_{ni}) - f(s_{ni}) + \frac{\varepsilon}{2(b-a)} \right] (x_{ni} - x_{n\,i-1})$$

$$= \sum_{i=1}^{n} f(t_{ni})(x_{ni} - x_{n\,i-1}) - \mathscr{I} + \mathscr{I} - \sum_{i=1}^{n} f(s_{ni})(x_{ni} - x_{n\,i-1}) + \frac{\varepsilon}{2}$$

$$< \varepsilon. \qquad \blacksquare$$

8.20 THE BOUNDED CONVERGENCE THEOREM

8.20.1 Pointwise Convergence of a Sequence of Functions. If (f_n) is a sequence of functions defined on a set S, then we say that (f_n) **converges pointwise** on S if for every point $x \in S$ the sequence $(f_n(x))$ is convergent. In such a case, if we define $f(x) = \lim_{n \to \infty} f_n(x)$ for each point $x \in S$, then we call f the **limit function** of the sequence (f_n); and we say that the sequence (f_n) *converges pointwise to f on S*, and we write $f_n \to f$ *pointwise on S*.

The following examples illustrate this definition.

(1) For each natural number n, define $f_n(x) = x^n$ for $0 \leq x \leq 1$ (see Figure 8.15). Then since $f_n(x) \to 0$ whenever $0 \leq x < 1$ and $f_n(1) \to 1$, we see that the sequence (f_n) converges pointwise on $[0, 1]$ and that the limit function of this sequence is the function $f : [0, 1] \to \mathbf{R}$, where $f(x) = 0$ for $x < 1$ and $f(1) = 1$. It might be of interest to observe that although each function f_n is continuous, the limit function f is not. We shall have more to say about this phenomenon in Chapter 11.

(2) For each natural number n, define $f_n(x) = \sin nx / (1 + nx)$ for $x \geq 0$ (see Figure 8.16). It is easy to see that (f_n) converges pointwise on $[0, \infty)$ to the constant function 0.

(3) For each natural number n, define $f_n(x) = (1 + x/n)^n$ for all $x \in \mathbf{R}$. Using L'Hospital's rule one may show easily that (f_n) converges pointwise on $\mathbf{R}$ to the exponential function.

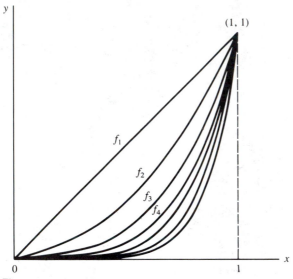

Figure 8.15

8.20.2 The Inadequacy of Pointwise Convergence. If (f_n) is a sequence of Riemann-integrable functions on an interval $[a, b]$ and (f_n) converges pointwise on $[a, b]$ to a function f, then it might seem reasonable to expect that $\int_a^b f_n \to \int_a^b f$ as $n \to \infty$; but unfortunately, even in some of the simplest situations, this is asking for too much. In the first place, f might not be integrable; but even when f is integrable, the result we have suggested can fail. The following three examples point to this failure.

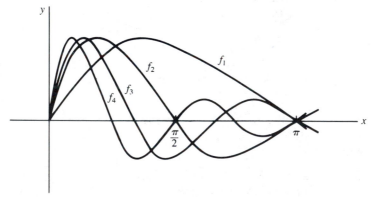

Figure 8.16

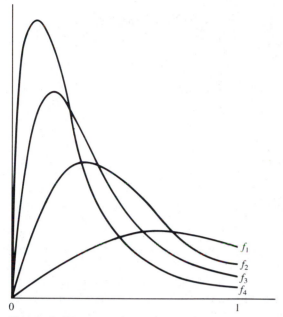

Figure 8.17

(1) For every natural number n, we define the function f_n on the interval $[0, 1]$ by

$$f_n(x) = \frac{2n^2x}{(1 + n^2x^2)^2} \qquad \text{for all} \qquad x \in [0, 1].$$

Using the results of Section 8.17, we see that

$$\int_0^1 f_n = \int_0^1 \frac{2n^2x}{(1 + n^2x^2)^2}\, dx = 1 - \frac{1}{1 + n^2} \to 1 \qquad \text{as} \qquad n \to \infty.$$

But on the other hand, for every $x \in [0, 1]$ we have $f_n(x) \to 0$, and, of course, $\int_0^1 0 \neq 1$. Figure 8.17 illustrates the graph of f_n for a few values of n.

(2) In the previous example all the functions were continuous. We now show that the pathology that occurred in that example can even occur with step functions. For each natural number n we define the function f_n on the interval $[0, 1]$ by $f_n(x) = n$ whenever $0 < x < 1/n$, and $f_n(x) = 0$ whenever $x = 0$ or $1/n \leq x \leq 1$ (see Figure 8.18). Just as in example (1), we have $f_n(x) \to 0$ as $n \to \infty$ for every $x \in [0, 1]$, and this time $\int_0^1 f_n = 1$ for every $n \in \mathbf{Z}^+$.

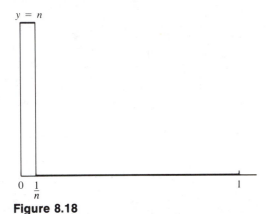

Figure 8.18

(3) We define f to be the nonintegrable function of Example 8.11.6; in other words, for each point $x \in [0, 1]$, $f(x) = 1$ if x is rational and $f(x) = 0$ if x is irrational. For each natural number n we define the function $f_n : [0, 1] \rightarrow$ $\boldsymbol{R}$ as follows: $f_n(x) = 1$ if x is rational with reduced form i/j with $j \leq n$, and $f_n(x) = 0$ for all other points $x \in [0, 1]$. It is easy to see that (f_n) converges pointwise to f. Now for each n the function f_n is zero at all but finitely many points of $[0, 1]$; and therefore, f_n, being a step function, must be Riemann-integrable on $[0, 1]$. This shows that a sequence of Riemann-integrable functions can converge pointwise to a nonintegrable function. One might observe that in this example we have $0 \leq f_n \leq 1$ and $\int_0^1 f_n = 0$ for each n.

These three examples show that in order to guarantee the condition $\int_a^b f_n \rightarrow$ $\int_a^b f$, we need to know more than just the pointwise convergence of (f_n) to f. We need to know that f is integrable, and in addition, we need some sort of boundedness condition that will prevent the existence of the high peaks that we saw in the graphs of the functions f_n in examples (1) and (2). As we saw in these examples, it is not good enough to know merely that each individual function f_n is bounded. To eliminate the possibility that the graph of f_n will peak at higher and higher values as n increases, we need to know that there is a single number K which "dominates" all of the functions f_n in the sense that $|f_n| \leq K$ for all n. Roughly speaking, the bounded convergence theorem tells us that if such a number K exists, then all will be well.

8.20.3 Definition of Bounded Convergence. A sequence (f_n) which converges pointwise on a set S is said to **converge boundedly** on S if there exists a number K such that for every natural number n and every point $x \in S$ we have $|f_n(x)| \leq K$. If (f_n) converges boundedly on S and f is its limit function, then we write $f_n \rightarrow f$ *boundedly on S.*

8.20.4 The Bounded Convergence Theorem. Suppose (f_n) is a sequence of Riemann-integrable functions on an interval $[a, b]$ which converges boundedly on $[a, b]$ to a Riemann-integrable function f. Then $\int_a^b f_n \to \int_a^b f$.

In addition to the form of the bounded convergence theorem which is stated here, you will also see a number of its variations and extensions. A variation for derivatives appears as Theorem 8.20.6, and a sharper form of the theorem appears as Theorem 8.22.1. In Chapter 9 you will see an extension of the theorem known as the **dominated convergence theorem**, which applies to improper integrals, and yet another extension of the theorem can be found in Appendix B.

The heart of the bounded convergence theorem is contained in the following important theorem, which we shall prove first.

8.20.5 Theorem: The Contracting-Sequences Theorem.[5] Suppose that (A_n) is a sequence of bounded subsets of R, that $A_{n+1} \subseteq A_n$ for every n, and that $\bigcap_{n=1}^{\infty} A_n = \emptyset$. For each n, define

$$\alpha_n = \sup\{m(E) \mid E \text{ is an elementary subset of } A_n\}.$$

Then $\alpha_n \to 0$ as $n \to \infty$.

■ **Proof.** The sequence (α_n) is clearly decreasing. To obtain a contradiction, assume that this sequence does not converge to 0; and choose a number $\delta > 0$ such that $\alpha_n > \delta$ for every n. For each n, choose an elementary subset E_n of A_n such that $m(E_n) > \alpha_n - \delta/2^n$. Now for each n, using Theorem 8.7.5(1), we choose a closed elementary subset F_n of E_n such that $m(F_n) > \alpha_n - \delta/2^n$; and we define

$$H_n = \bigcap_{i=1}^{n} F_i.$$

Observe that for every n the set H_n is a closed, bounded (elementary) set and that $H_{n+1} \subseteq H_n$. If we could show that every one of the sets H_n must be nonempty, then it would follow from the Cantor intersection theorem, Theorem 4.13.3, that $\bigcap_{n=1}^{\infty} H_n \neq \emptyset$ in spite of the fact that the larger sets A_n have an empty intersection. This would be our desired contradiction. So what we have to do in order to complete the proof is to show that every one of the sets H_n must be nonempty. For this purpose, we make two observations.

First, given any natural number n and any elementary subset E of $A_n \backslash F_n$, we have

$$m(E) + m(F_n) = m(E \cup F_n) \leq \alpha_n;$$

and therefore, since $m(F_n) > \alpha_n - \delta/2^n$, we have $m(E) < \delta/2^n$.

---

[5] The last important occasion on which we saw contracting sequences of sets was in Section 4.13 when we studied the Cantor intersection theorem, and as a matter of fact, it would be a good idea to review that section now. We are about to make use of it.

Second, given any natural number n and any elementary subset E of $A_n\backslash H_n$, since

$$E = (E\backslash F_1) \cup (E\backslash F_2) \cup \cdots \cup (E\backslash F_n),$$

and since $E\backslash F_i$ is an elementary subset of $A_i\backslash F_i$ for each $i = 1, 2, \ldots, n$, we have

$$m(E) \le \sum_{i=1}^{n} m(E\backslash F_i) < \sum_{i=1}^{n} \frac{\delta}{2^i} = \delta\left(1 - \frac{1}{2^n}\right) < \delta.$$

Look, now, at the two sets A_n and $A_n\backslash H_n$. The set A_n *must* have an elementary subset of measure greater than δ, but the set $A_n\backslash H_n$ can *not*. So for each n the sets A_n and $A_n\backslash H_n$ cannot be the same, and it follows that the set H_n is not empty. ∎

■ **Proof of the Bounded Convergence Theorem.** Choose a number K such that $|f_n(x)| \le K$ for all natural numbers n and all points $x \in [a, b]$. Then given $x \in [a, b]$, we also have $|f(x)| \le K$ because $f_n(x) \to f(x)$. For each n, define $g_n = |f_n - f|$. Then for every $x \in [a, b]$ we have $g_n(x) \to 0$; and given n and x, we have

$$0 \le g_n(x) \le |f_n(x)| + |f(x)| \le 2K.$$

Furthermore, for each n we have

$$\left|\int_a^b f_n - \int_a^b f\right| = \left|\int_a^b (f_n - f)\right| \le \int_a^b |f_n - f| = \int_a^b g_n;$$

and therefore, in order to show that $\int_a^b f_n \to \int_a^b f$, all we have to show is that $\int_a^b g_n \to 0$.

Let $\varepsilon > 0$. For each n, define A_n to be the set of all those points $x \in [a, b]$ for which $g_i(x) \ge \varepsilon/2(b - a)$ for at least one natural number $i \ge n$. It is easy to see that for each n we have $A_{n+1} \subseteq A_n$; in other words, the sequence (A_n) is contracting. Furthermore, given any number $x \in [a, b]$, since $g_i(x) \to 0$ as $i \to \infty$, there must exist a natural number n such that $g_i(x) < \varepsilon/2(b - a)$ for all $i \ge n$; in other words, $x \notin A_n$. From this it follows that $\bigcap_{n=1}^{\infty} A_n = \emptyset$. Therefore, we may use the contracting-sequences theorem to choose a natural number N such that whenever $n \ge N$ and E is an elementary subset of A_n, we have $m(E) < \varepsilon/4K$.

We shall now complete the proof by showing that whenever $n \ge N$, we have $\int_a^b g_n \le \varepsilon$. Let $n \ge N$. Since the integral of a Riemann-integrable function is the same as its lower integral, if we want to show that $\int_a^b g_n \le \varepsilon$, then all we have to show is that whenever s is a step function and $0 \le s \le g_n$, we have $\int_a^b s \le \varepsilon$. Suppose, then, that s is a step function and that $0 \le s \le g_n$. Define

$$E = \left\{x \in [a, b] \,\middle|\, s(x) \ge \frac{\varepsilon}{2(b - a)}\right\} \quad \text{and} \quad F = [a, b]\backslash E.$$

Then E and F are elementary subsets of $[a, b]$; and since $E \subseteq A_n$, we have $m(E) < \varepsilon/4K$. Therefore,

$$\int_a^b s = \int_E s + \int_F s \le \int_E 2K + \int_F \frac{\varepsilon}{2(b-a)} \le \int_E 2K + \int_a^b \frac{\varepsilon}{2(b-a)}$$

$$= 2Km(E) + \frac{\varepsilon}{2(b-a)}(b-a) < \varepsilon. \qquad \blacksquare$$

8.20.6 A Bounded Convergence Theorem for Derivatives.

Given a sequence (f_n) of differentiable functions on an interval, and given that $f_n(x) \to f(x)$ for all x in the interval, there is no reason to expect $f_n'(x) \to f'(x)$. For example, if $f_n(x) = (\sin nx)/\sqrt{n}$ for all real numbers x and natural numbers n, then obviously, $f_n(x) \to 0$ for every x. But $f_n'(x) = \sqrt{n} \cos nx$ for all x, and so $f_n'(0) \to \infty$ as $n \to \infty$; and as a matter of fact, for most values of x the sequence $(f_n'(x))$ is unbounded and has no limit at all. In the light of this example, the following theorem is interesting.

Theorem. Suppose that (f_n) is a sequence of differentiable functions on an interval $[a, b]$, that f_n' is continuous on $[a, b]$ for each n, that the sequence (f_n') converges boundedly on $[a, b]$ to a continuous function g, and that for some number c in $[a, b]$ the sequence $(f_n(c))$ converges. Then the sequence (f_n) converges pointwise on $[a, b]$. Furthermore, if $f_n \to f$, then f is differentiable on $[a, b]$ and $f' = g$.

■ **Proof.** Assume that $f_n(c) \to \alpha$ as $n \to \infty$. For every natural number n and every point $x \in [a, b]$ we have $\int_c^x f_n' = f_n(x) - f_n(c)$; and therefore, since $f_n(x) = f_n(c) + \int_c^x f_n'$, it follows from the bounded convergence theorem that $f_n(x) \to \alpha + \int_c^x g$ for every $x \in [a, b]$. So if we define $f(x) = \lim_{n\to\infty} f_n(x)$ for each x, it follows from Theorem 8.17.3 that $f' = g$.

For a different form of this theorem that does not require the functions f_n' to be continuous, see Theorem 7.17 of Rudin [13].

8.21 EXERCISES

■ **1.** For every natural number n and every point $x \in [0, 1]$ we define

$$f_n(x) = \frac{2n^2x}{(1 + n^2x^2)^2} \quad \text{and} \quad g_n(x) = \frac{2nx}{(1 + n^2x^2)^2}.$$

Prove that both (f_n) and (g_n) converge pointwise on $[0, 1]$ to the constant function 0. Prove that (g_n) converges boundedly, but (f_n) does not. What do you notice about the limits $\lim_{n\to\infty} \int_0^1 f_n$ and $\lim_{n\to\infty} \int_0^1 g_n$? If you would like to see a solution of this exercise, look at Examples 11.3.5(1) and 11.3.5(2).

2. Repeat exercise 1 for the sequences (f_n) and (g_n), where for every natural number n and every point $x \in [0, 1]$ we have

$$f_n(x) = (2n^2x)e^{-n^2x^2} \quad \text{and} \quad g_n(x) = (2nx)e^{-n^2x^2}.$$

3. For every natural number n we define $f_n : [0, 1] \to R$ by $f_n(x) = 0$ if $x < 1/n$ and $f_n(x) = 1/nx$ if $x \geq 1/n$. Prove that (f_n) converges boundedly on $[0, 1]$ to the constant function 0; and by applying the bounded convergence theorem to this sequence, deduce that $(\log n)/n \to 0$ as $n \to \infty$.

■ 4. Prove that if $f_n \to f$ and $g_n \to g$ pointwise on S, then $f_n + g_n \to f + g$ and $f_n g_n \to fg$ pointwise on S. Is the same true for bounded convergence?

■ 5. Suppose that (f_n) is a sequence of differentiable functions on $[a, b]$, that for each n the function f_n' is Riemann-integrable on $[a, b]$, and that $|f_n'| \leq 1$. Prove that if $f_n' \to 0$ pointwise on $[a, b]$, then there exists a sequence (α_n) such that for every point $x \in [a, b]$ we have $f_n(x) - \alpha_n \to 0$ as $n \to \infty$.

6. Suppose (p_n) is a strictly increasing sequence of natural numbers. Prove that there exists a point $x \in (0, 2\pi)$ such that the sequence $(\sin p_n x)$ diverges. *Hint:* Define

$$f_n(x) = (\sin p_n x - \sin p_{n+1}x)^2$$

for all n and x. What is $\int_0^{2\pi} f_n$? Show that it is not the case that $f_n(x) \to 0$ as $n \to \infty$ for every point x.

*7. Prove that there exists a sequence of step functions that converges boundedly on R to χ_Q.

8.22 THE INADEQUACY OF RIEMANN INTEGRATION

In a nutshell, the principal defect of the theory of Riemann integration is that it is possible for a sequence of Riemann-integrable functions to converge boundedly to a function which is not integrable. [See Example 8.20.2(3).] Consequently, we find ourselves faced with an interesting question: Suppose that a sequence (f_n) of Riemann-integrable functions on an interval $[a, b]$ converges boundedly on $[a, b]$ to a function f, and suppose that f is not Riemann-integrable. What will happen to $\int_a^b f_n$ as $n \to \infty$? Can we be sure that the sequence $(\int_a^b f_n)$ converges, or is it perhaps possible that the nonintegrability of the limit function f might cause this sequence of integrals to oscillate badly enough to pick up more than one partial limit? As the following variation of the bounded convergence theorem shows, this sequence of integrals must converge.

8.22.1 Theorem: A Variation of the Bounded Convergence Theorem.

Suppose that (f_n) is a sequence of Riemann-integrable functions on an interval

$[a, b]$ which converges boundedly on $[a, b]$. Then the sequence $(\int_a^b f_n)$ must converge.

■ **Proof.** In order to show that the sequence $(\int_a^b f_n)$ converges, what we shall actually show is that this sequence is a Cauchy sequence. Its convergence will then follow from Theorem 4.11.3. Let $\varepsilon > 0$. Choose a positive number K such that $|f_n(x)| \leq K$ for all $x \in [a, b]$ and $n \in \mathbf{Z}^+$; and for every n, define A_n to be the set of all those points $x \in [a, b]$ for which $|f_i(x) - f_j(x)| \geq \varepsilon/2(b - a)$ for at least one pair of natural numbers i and j such that $i \geq n$ and $j \geq n$. The sequence (A_n) is obviously contracting, and from the fact that $(f_i(x))$ is a Cauchy sequence for each point x in $[a, b]$, it is easy to see that $\bigcap_{n=1}^\infty A_n = \emptyset$. Using the contracting-sequences theorem, we now choose a natural number N such that whenever $n \geq N$ and E is an elementary subset of A_n, we have $m(E) < \varepsilon/4K$.

We shall now complete the proof by showing that whenever $i \geq N$ and $j \geq N$, we have

$$\left| \int_a^b f_i - \int_a^b f_j \right| \leq \varepsilon.$$

Let $i \geq N$ and $j \geq N$. Since

$$\left| \int_a^b f_i - \int_a^b f_j \right| = \left| \int_a^b (f_i - f_j) \right| \leq \int_a^b |f_i - f_j|,$$

we need only show that the latter integral does not exceed ε. Since the integral of a Riemann-integrable function is the same as its lower integral, if we want to show that $\int_a^b |f_i - f_j| \leq \varepsilon$, then all we have to show is that whenever s is a step function and $0 \leq s \leq |f_i - f_j|$, we have $\int_a^b s \leq \varepsilon$. Suppose, then, that s is a step function and that $0 \leq s \leq |f_i - f_j|$. Define

$$E = \left\{ x \in [a, b] \mid s(x) \geq \frac{\varepsilon}{2(b - a)} \right\} \quad \text{and} \quad F = [a, b] \backslash E.$$

Then E and F are elementary subsets of $[a, b]$; and since $E \subseteq A_n$, we have $m(E) < \varepsilon/4K$. Therefore,

$$\int_a^b s = \int_E s + \int_F s \leq \int_E 2K + \int_F \frac{\varepsilon}{2(b - a)}$$

$$\leq 2Km(E) + \frac{\varepsilon}{2(b - a)} (b - a) < \varepsilon. \qquad ■$$

8.22.2 Theorem. Suppose that (f_n) and (g_n) are two sequences of Riemann-integrable functions which converge boundedly on $[a, b]$ to the same function f. Then the sequences $(\int_a^b f_n)$ and $(\int_a^b g_n)$ must have the same limit.

■ *Proof.* Since the sequence $(f_n - g_n)$ converges boundedly to the constant function zero, it follows from the bounded convergence theorem that $\int_a^b (f_n - g_n) \to 0$; and therefore $\lim_{n \to \infty} \int_a^b f_n = \lim_{n \to \infty} \int_a^b g_n$. ■

8.22.3 Concluding Remarks. Suppose that f is a bounded function on an interval $[a, b]$ and that f is the limit of a boundedly convergent sequence (f_n) of integrable functions on $[a, b]$. As we have said, there is no guarantee that f has to be integrable on $[a, b]$; but nevertheless, the sequence $(\int_a^b f_n)$ converges. Furthermore, we know from Theorem 8.22.2 that $\lim_{n \to \infty} \int_a^b f_n$ depends only on the function f and is independent of the sequence (f_n). There is therefore a compelling reason to suggest that whether or not f is Riemann-integrable on $[a, b]$, the integral $\int_a^b f$ should exist in any satisfactory theory and should be defined to be $\lim_{n \to \infty} \int_a^b f_n$.

Let us apply this principle to the function f of Section 8.11.6. Here we have a function which, from the point of view of Riemann integration, is too wild to be integrated. Now in spite of this, we saw in Example 8.20.2(3) that f is the limit of a boundedly convergent sequence (f_n) of step functions on $[0, 1]$ such that $\int_0^1 f_n = 0$ for each n. Therefore, given *any* sequence (g_n) of Riemann-integrable functions such that $g_n \to f$ boundedly on $[0, 1]$, we have $\int_0^1 g_n \to 0$; and this leads us to conclude that in a satisfactory theory of integration we should have $\int_0^1 f = 0$. In the light of this observation, the very fact that f is 0 at every irrational point in the interval and 1 at every rational suggests that one ought to be able to ignore the set Q of rational numbers as being too small to have any significance. In other words, one ought to be able to treat this function f just as if it were the constant function zero. This would be a good time to recall the famous quotation from Cantor that appears in Chapter 2:

> The rationals are spotted in the line like stars in a black sky while the dense blackness is the firmament of the irrationals.

The trouble with Riemann integration is that this theory doesn't seem to be able to distinguish between the dense blackness and a few unimportant spots.[6]

8.23 EXERCISES

■ **1.** Suppose that (f_n) is a sequence of differentiable functions on an interval $[a, b]$, that for each n the function f_n' is Riemann-integrable on $[a, b]$, and that the sequence (f_n') converges boundedly on $[a, b]$. Prove that either the sequence $(f_n(x))$ converges for every point $x \in [a, b]$ or $(f_n(x))$ diverges for every $x \in [a, b]$. Give an example to show that the second of these options can actually occur.

[6] For some other ways in which the set Q can be seen as a very small subset of R, see Appendixes A and B.

■ **2.** Prove that Theorem 8.22.1 is, in fact, a sharper form of the bounded convergence theorem than the form stated as Theorem 8.20.4.

*■ **3.** (This exercise is starred because it makes use of the idea of countability which is introduced in Appendix A.) Given a countable subset E of an interval $[a, b]$, explain why in any satisfactory theory of integration we should have $\int_a^b \chi_E = 0$.

This concludes the mainstream of Chapter 8. At this stage you can choose to read Appendix B on sets of measure zero, or you can omit this appendix and proceed right away with Chapter 9.

Appendix *B* $\genfrac{}{}{0pt}{}{\textit{(Optional Appendix}}{\textit{to Chapter 8)}}$

Sets of
Measure Zero

B.1 INTRODUCTION

During our study of the Riemann integral in Chapter 8 there were several occasions when the thrust of what we were saying seemed to revolve around the "smallness" of certain sets of numbers. Perhaps the most striking example was provided by Theorem 8.16.3, where we learned that a bounded function f is Riemann-integrable if the set of points where f is discontinuous is "small." In the context of that theorem a set is "small" if it can be placed inside elementary sets of arbitrarily small measure, but this notion of smallness is not the only one we saw in Chapter 8. For example, the set $Q \cap [0, 1]$ is not small in this way. The set $Q \cap [0, 1]$ cannot be placed inside an elementary set of measure less than one, but in spite of this, the ruler function which is discontinuous everywhere in $Q \cap [0, 1]$ still manages to be Riemann-integrable. [See Section 5.8, Exercise 6.5(15), and Exercises 8.15(4) and 8.15(5).] The integrability of the ruler function depends upon another kind of smallness of the set $Q \cap [0, 1]$. If you have read Appendix A, then you know that this set is countable; but even if you have not yet studied countability, you still know that $Q \cap [0, 1]$ does not contain any of the "dense blackness'' of the irrationals. The smallness of the set $\boldsymbol{Q} \cap [0, 1]$ was important to us again in Section 8.22.3, where this time we were concerned with its characteristic function. The characteristic function of $\boldsymbol{Q} \cap [0, 1]$ is in some ways highly pathological, being discontinuous everywhere in $[0, 1]$ and failing to be Riemann-integrable. But in spite of this, we saw in Section 8.22.3 how the theory of integration seems to allow us to think of this function as being the constant function zero except in the ''small'' set $\boldsymbol{Q} \cap [0, 1]$.

With these examples in mind, one can begin to understand that the concept of smallness of a set is of fundamental importance in the theory of integration. There are a number of variations of this concept, the most general and the most useful of these being the concept of *measure zero* which forms the subject matter of this appendix. In this appendix you will find a precise definition of a set of measure zero, and you will see how, with a good understanding of this idea, one can prove a number of interesting theorems which were beyond our reach in Chapter 8.

B.2 THE BASIC DEFINITIONS

What should we mean by a set of measure zero? If we look at Theorem 8.16.3, we might be inclined to define the concept in the following way.

B.2.1 Nineteenth-Century-Style Definition of a Set of Measure Zero.
A bounded set A is said to have *measure zero* if for every number $\varepsilon > 0$ there exists an elementary set E such that $A \subseteq E$ and $m(E) < \varepsilon$.

This definition seems to be tailor-made for Theorem 8.16.3, but is it? For one thing, that theorem was only the *junior* version of the Lebesgue criterion for Riemann integrability. It was inadequate in the sense that all it provided was a *sufficient* condition for a given function to be Riemann-integrable, while the true Lebesgue criterion is both necessary and sufficient. And second, according to the previous definition, the ''small'' set $\boldsymbol{Q} \cap [0, 1]$ is not small at all; for given any elementary set E which includes $\boldsymbol{Q} \cap [0, 1]$, it is clear that $m(E) \geq 1$. Having thus seen that the definition of measure zero given in this subsection is not adequate, we are ready for the one that is.

B.2.2 Definition of a Set of Measure Zero.
Given $A \subseteq \boldsymbol{R}$, we say that A is a *set of measure zero*—or alternatively, that A *has measure zero*—if for every number $\varepsilon > 0$ there exists an expanding sequence (E_n) of elementary sets such that $A \subseteq \cup_{n=1}^{\infty} E_n$ and $m(E_n) < \varepsilon$ for every $n \in \boldsymbol{Z}^+$.

Note that in order for a set to have measure zero, there is no need for the set to be bounded. Note also that any set satisfying Definition B.2.1 will certainly have measure zero; for if A satisfies Definition B.2.1 and $\varepsilon > 0$, we can choose an elementary set $E \supseteq A$ such that $m(E) < \varepsilon$ and then define $E_n = E$ for every n. On the other hand, the converse of this is not true. As we shall see soon, the set $Q \cap [0, 1]$ has measure zero even though it obviously fails to satisfy Definition B.2.1. This brings us to our first theorem, which tells us that for closed, bounded sets the two ideas are the same.

B.2.3 Theorem: Equivalence of the Two Definitions for Closed, Bounded Sets.

Suppose A is a closed, bounded set of real numbers. Then A has measure zero if and only if A satisfies the conditions of Definition B.2.1.

■ **Proof.** The *if* part of this theorem is trivial. What we have to do is show that if A has measure zero, then A satisfies the conditions of Definition B.2.1. Suppose that A has measure zero and that $\varepsilon > 0$. Choose an expanding sequence (E_n) of elementary sets such that $A \subseteq \cup_{n=1}^{\infty} E_n$ and $m(E_n) < \varepsilon/4$ for every $n \in \mathbf{Z}^+$. Now using Theorem 8.7.5, we first choose an open elementary set U_1 such that $E_1 \subseteq U_1$ and $m(U_1) < \varepsilon/4$. Then for each $n \geq 2$ we choose an open elementary set $U_n \supseteq E_n \backslash E_{n-1}$ such that

$$m(U_n) < m(E_n \backslash E_{n-1}) + \frac{\varepsilon}{2^{n+1}}.$$

For each n, define $V_n = \cup_{i=1}^{n} U_i$. Then (V_n) is an expanding sequence of open elementary sets; and for each n

$$m(V_n) \leq \sum_{i=1}^{n} m(U_i) < \frac{\varepsilon}{4} + \sum_{i=2}^{n} m(E_i \backslash E_{i-1}) + \sum_{i=2}^{n} \frac{\varepsilon}{2^{i+1}}$$

$$< \frac{\varepsilon}{2} + m(E_n \backslash E_1) \leq \frac{\varepsilon}{2} + m(E_n) < \varepsilon.$$

Since $E_n \subseteq V_n$ for every n, it follows that $A \subseteq \cup_{n=1}^{\infty} V_n$; and we deduce from Theorem 4.13.4 that for some $n \in \mathbf{Z}^+$ we have $A \subseteq V_n$. This is all we need to show that A satisfies the conditions of Definition B.2.1. ■

An important corollary of Theorem B.2.3 is the fact that there exist sets which don't have measure zero. For example, $[0, 1]$ is a closed, bounded set which does not satisfy the conditions of Definition B.2.1. Therefore, $[0, 1]$ does not have measure zero, and similarly, no interval of positive length can have measure zero. This fact is not as trivial as it may look. Did you notice that we used the completeness of $\mathbf{R}$ in the proof of the theorem?

Our next theorem is one of the punch lines of this appendix. Obviously, the smaller a set is, the more likely it is to have measure zero (e.g., a subset of a set of measure zero must have measure zero); but the following theorem shows how to make some larger sets of measure zero.

B.3 THEOREM: UNION OF A SEQUENCE OF SETS OF MEASURE ZERO

Suppose (A_n) is a sequence of sets of measure zero. Then $\cup_{n=1}^{\infty} A_n$ must have measure zero.

■ **Proof.** Let $\varepsilon > 0$. For every $n \in Z^+$, choose a sequence (E_{ni}) of elementary sets such that $E_{ni} \subseteq E_{n\,i+1}$ and $m(E_{ni}) < \varepsilon/2^n$ for every $i \in Z^+$, and $A_n \subseteq \cup_{i=1}^{\infty} E_{ni}$. See Figure B.1.

For every $i \in Z^+$, define $E_i = \cup_{n=1}^{i} E_{ni}$. Thus

$$E_1 = E_{11}$$

$$E_2 = E_{12} \cup E_{22}$$

$$E_3 = E_{13} \cup E_{23} \cup E_{33}$$

$$E_4 = E_{14} \cup E_{24} \cup E_{34} \cup E_{44}$$

$$E_5 = E_{15} \cup E_{25} \cup E_{35} \cup E_{45} \cup E_{55}$$

$$\vdots$$

We see easily that (E_i) is an expanding sequence of elementary sets and that for each i we have

$$m(E_i) \le \sum_{n=1}^{i} m(E_{ni}) < \sum_{n=1}^{i} \frac{\varepsilon}{2^n} < \varepsilon.$$

All we need to do now to complete the proof is show that $\cup_{n=1}^{\infty} A_n \subseteq \cup_{i=1}^{\infty} E_i$. Let $x \in \cup_{n=1}^{\infty} A_n$. Choose a natural number N such that $x \in A_N$. Using the fact that $A_N \subseteq \cup_{i=1}^{\infty} E_{Ni}$, choose j such that $x \in E_{Nj}$. Then whenever i exceeds the larger of the two numbers j and N, we have

$$x \in \bigcup_{n=1}^{i} E_{ni} = E_i. \qquad ■$$

B.4 AN IMPORTANT EXAMPLE

The set $Q \cap [0, 1]$ has measure zero. As a matter of fact, so does the set Q of all rational numbers—truly a set of stars in a black sky.

■ **Proof.** For every $n \in Z^+$, define A_n to be the subset of $Q \cap [0, 1]$ consisting of those rational numbers with reduced form i/j, where $j \le n$. Since each set A_n is finite, we see that A_n has measure zero for every n. But $Q \cap [0, 1] = \cup_{n=1}^{\infty} A_n$, and so it follows at once from Theorem B.3 that $Q \cap [0, 1]$ has measure zero. Now define $Q_n = Q \cap [-n, n]$ for every integer n. Arguing as above, we see that Q_n has measure zero for every n; and since $Q = \cup_{n=1}^{\infty} Q_n$, it follows that Q has measure zero. ■

$$
\begin{array}{ccccccccccccccc}
\boxed{E_{11}} & \subseteq & E_{12} & \subseteq & E_{13} & \subseteq & E_{14} & \subseteq & E_{15} & \subseteq & E_{16} & \subseteq \cdots \subseteq & E_{1i} & \subseteq \cdots & \text{and} \quad m(E_{1i}) < \frac{\varepsilon}{2^1} \\
E_{21} & \subseteq & E_{22} & \subseteq & E_{23} & \subseteq & E_{24} & \subseteq & E_{25} & \subseteq & E_{26} & \subseteq \cdots \subseteq & E_{2i} & \subseteq \cdots & \text{and} \quad m(E_{2i}) < \frac{\varepsilon}{2^2} \\
E_{31} & \subseteq & E_{32} & \subseteq & E_{33} & \subseteq & E_{34} & \subseteq & E_{35} & \subseteq & E_{36} & \subseteq \cdots \subseteq & E_{3i} & \subseteq \cdots & \text{and} \quad m(E_{3i}) < \frac{\varepsilon}{2^3} \\
E_{41} & \subseteq & E_{42} & \subseteq & E_{43} & \subseteq & E_{44} & \subseteq & E_{45} & \subseteq & E_{46} & \subseteq \cdots \subseteq & E_{4i} & \subseteq \cdots & \text{and} \quad m(E_{4i}) < \frac{\varepsilon}{2^4} \\
E_{51} & \subseteq & E_{52} & \subseteq & E_{53} & \subseteq & E_{54} & \subseteq & E_{55} & \subseteq & E_{56} & \subseteq \cdots \subseteq & E_{5i} & \subseteq \cdots & \text{and} \quad m(E_{5i}) < \frac{\varepsilon}{2^5} \\
E_{61} & \subseteq & E_{62} & \subseteq & E_{63} & \subseteq & E_{64} & \subseteq & E_{65} & \subseteq & E_{66} & \subseteq \cdots \subseteq & E_{6i} & \subseteq \cdots & \text{and} \quad m(E_{6i}) < \frac{\varepsilon}{2^6} \\
 & & & & & & & & & \vdots & & & & & \vdots \\
E_{n1} & \subseteq & E_{n2} & \subseteq & E_{n3} & \subseteq & E_{n4} & \subseteq & E_{n5} & \subseteq & E_{n6} & \subseteq \cdots \subseteq & E_{ni} & \subseteq \cdots & \text{and} \quad m(E_{ni}) < \frac{\varepsilon}{2^n} \\
 & & & & & & & & & \vdots & & & & & \vdots
\end{array}
$$

Figure B.1

B.5 THE CONCEPTS *ALMOST EVERY* AND *ALMOST EVERYWHERE*

B.5.1 Definition. Suppose A is a set of real numbers. A subset B of A is said to contain *almost every* point of A if the set $A \backslash B$ has measure zero. For example, we would say that a given function f defined on a set A is *continuous at almost every point of A*—or, alternatively, *is continuous almost everywhere in A*—if the set of points of A at which f is not continuous has measure zero. As another example, we might say that *almost every* real number is irrational.

B.5.2 Theorem. Suppose that f and g are Riemann-integrable on an interval $[a, b]$ and that $f(x) = g(x)$ for almost every point $x \in [a, b]$. Then

$$\int_a^b f = \int_a^b g.$$

■ **Proof.** We shall show that $\int_a^b |f - g| = 0$; and to do this, we need only show that whenever s is a step function and $0 \leq s \leq |f - g|$, we have $\int_a^b s = 0$. Assume, then, that s is such a step function, and choose a partition $\mathcal{P} = (x_0, \ldots, x_n)$ of $[a, b]$ such that s steps within $\mathcal{P}$. Then given any $i = 1, \ldots, n$, since the interval (x_{i-1}, x_i) does not have measure zero, the constant value of s in (x_{i-1}, x_i) must be zero; and it follows at once that $\int_a^b s = 0$, as required. ■

We are now ready to state a sharper version of the bounded convergence theorem.

B.5.3 *Almost Everywhere* Version of the Bounded Convergence Theorem. Suppose (f_n) is a sequence of functions which are Riemann-integrable on a given interval $[a, b]$. Suppose that K is some positive number and that for every $n \in \mathbf{Z}^+$ the function f_n satisfies the inequality $|f_n(x)| \leq K$ for almost every point $x \in [a, b]$.

(1) If the sequence $(f_n(x))$ is convergent for almost every point $x \in [a, b]$, then the sequence of integrals $\int_a^b f_n$ must also be convergent.

(2) If f is a Riemann-integrable function on $[a, b]$ and $f_n(x) \to f(x)$ for almost every point $x \in [a, b]$, then we have

$$\int_a^b f_n \to \int_a^b f.$$

■ **Proof.** For each n, define $A_n = \{x \in [a, b] \mid |f_n(x)| > K\}$. Then each set A_n has measure zero. Define $A = \bigcup_{n=1}^{\infty} A_n$. We conclude from Theorem B.3 that A has measure zero; and given any point $x \in [a, b] \backslash A$, the inequality $|f_n(x)| \leq K$ holds for every n. For each n, define $g_n = [(-K) \vee f_n] \wedge K$ [see Exercise 8.4(4) for the meaning of $\vee$ and $\wedge$]. Note that each function g_n is Riemann-integrable on $[a, b]$, that $|g_n| \le K$, and that since g_n and f_n are equal almost everywhere, we have $\int_a^b f_n = \int_a^b g_n$.

■ **Proof of Part (1).**   Define B to be the set of points $x \in [a, b]$ for which the sequence $(f_n(x))$ fails to be convergent. Then B has measure zero. Define $C = A \cup B$, and note that C has measure zero. Now we want to show that the sequence of integrals $\int_a^b f_n$ is a Cauchy sequence. Let $\varepsilon > 0$. Choose an expanding sequence (E_n) of elementary sets such that $m(E_n) < \varepsilon/4K$ for every n and such that $C \subseteq \cup_{n=1}^\infty E_n$. Now define

$$h_n = g_n(1 - \chi_{E_n}).$$

Given any point $x \in \cup_{n=1}^\infty E_n$, the sequence of numbers $h_n(x)$ is eventually zero; and so $h_n(x) \to 0$. On the other hand, if $x \notin \cup_{n=1}^\infty E_n$, then $x \notin C$; and so $h_n(x) = f_n(x)$ for every n; and again, the sequence $(h_n(x))$ converges. Therefore, the sequence $(h_n(x))$ converges for every x; and since $|h_n| \le K$ for every n, it follows from the bounded convergence theorem that the sequence of integrals $\int_a^b h_n$ must converge. Choose a natural number N such that whenever $i \ge N$ and $j \ge N$, we have $|\int_a^b h_i - \int_a^b h_j| < \varepsilon/4$. Then whenever $i \ge N$ and $j \ge N$, we have

$$\left| \int_a^b f_i - \int_a^b f_j \right| = \left| \int_a^b g_i - \int_a^b g_j \right| = \left| \int_a^b (h_i + g_i \chi_{E_i}) - \int_a^b (h_j + g_j \chi_{E_j}) \right|$$

$$\le \left| \int_a^b h_i - \int_a^b h_j \right| + \int_a^b |g_i| \chi_{E_i} + \int_a^b |g_j| \chi_{E_j}$$

$$< \frac{\varepsilon}{4} + \int_a^b K \chi_{E_i} + \int_a^b K \chi_{E_j} \le \frac{\varepsilon}{4} + Km(E_i) + Km(E_j) < \varepsilon.$$

■ **Proof of Part (2).**   This is very similar to part (1). This time we define B to be the set of points $x \in [a, b]$ for which the sequence $(f_n(x))$ fails to be convergent to the number $f(x)$. Once again, B has measure zero, and we define $C = A \cup B$. Define $g = [(-K) \vee f] \wedge K$, and note that g is Riemann-integrable on $[a, b]$ and $|g| \le K$; and since the functions f and g are equal at every point of $[a, b] \backslash A$, we have $\int_a^b f = \int_a^b g$. Now let $\varepsilon > 0$, and choose the sequence (E_n) as in the proof of part (1). For every n, define

$$h_n = (g - g_n)(1 - \chi_{E_n}).$$

Since $h_n(x) = f(x) - f_n(x)$ whenever $x \in [a, b] \backslash \cup_{n=1}^\infty E_n$, we see that $h_n(x) \to 0$ for every x; and therefore, since $|h_n| \le K$ for every n, it follows from the bounded convergence theorem that $\int_a^b h_n \to 0$. Choose a natural number N such that whenever $n \ge N$, we have $|\int_a^b h_n| < \varepsilon/2$. Then whenever $n \ge N$, we have

$$\left| \int_a^b (f - f_n) \right| = \left| \int_a^b (g - g_n) \right| = \left| \int_a^b [h_n + (g - g_n) \chi_{E_n}] \right|$$

$$\le \left| \int_a^b h_n \right| + \int_a^b 2K \chi_{E_n} < \varepsilon. \qquad \blacksquare$$

B.5.4 An Application to Decreasing Sequences of Nonnegative Functions. Suppose (f_n) is a decreasing sequence of nonnegative, Riemann-integrable functions on an interval $[a, b]$; and for each point $x \in [a, b]$, define $f(x) = \lim_{n \to \infty} f_n(x)$. Then the following two conditions are equivalent:

(1) $\int_a^b f_n \to 0$ as $n \to \infty$.

(2) $f(x) = 0$ for almost every point $x \in [a, b]$.

■ **Proof.** Since (f_n) converges almost everywhere on $[a, b]$ to the constant function zero and f_1 is bounded, the fact that $(2) \Rightarrow (1)$ follows at once from Theorem B.5.3. Now assume (1); and to obtain a contradiction, assume that the set $\{x \in [a, b] \mid f(x) > 0\}$ does not have measure zero. Using the fact that

$$\{x \in [a, b] \mid f(x) > 0\} = \bigcup_{k=1}^{\infty} \left\{ x \in [a, b] \mid f(x) > \frac{1}{k} \right\}$$

and Theorem B.3, choose a natural number k such that the set $A = \{x \in [a, b] \mid f(x) > 1/k\}$ does not have measure zero. For each n, choose a step function $S_n \geq f_n$ such that $\int_a^b S_n < \int_a^b f_n + 1/n$. We note that $\int_a^b S_n \to 0$ as $n \to \infty$. Now for each n, define $E_n = \{x \in [a, b] \mid S_n(x) \geq 1/k\}$. Then each set E_n is elementary; and for each n, $A \subseteq E_n$. But from the inequality

$$\int_a^b S_n \geq \int_a^b S_n \chi_{E_n} \geq \frac{1}{k} m(E_n),$$

we deduce that $m(E_n) \to 0$ as $n \to \infty$; and so it follows that not only does the set A have measure zero (contradicting our choice of k) but this set actually satisfies the more restrictive condition of Section B.2.1.   ■

B.6   THE LEBESGUE CRITERION FOR RIEMANN INTEGRABILITY

We now begin our approach to the important criterion of Lebesgue which sharpens Theorem 8.16.3 by providing a necessary and sufficient condition for a function to be Riemann-integrable. Our first theorem in this section is of interest in its own right.

B.6.1 Theorem. Suppose f is a bounded function on an interval $[a, b]$. For each $n \in \mathbf{Z}^+$, suppose $\mathcal{P}_n$ is the regular 2^n-partition of $[a, b]$. For every $x \in [a, b]$, define $\omega(x) = \lim_{n \to \infty} \omega(\mathcal{P}_n, f)(x)$. Then the following two conditions are equivalent:

(1) f is Riemann-integrable on $[a, b]$.

(2) $\omega(x) = 0$ for almost every point $x \in [a, b]$.

■ **Proof.** We mention first that since $\mathcal{P}_{n+1}$ is a refinement of $\mathcal{P}_n$ for every n, the sequence of functions $\omega(\mathcal{P}_n, f)$ is decreasing; and therefore, the definition of $\omega(x)$

for each x makes sense. Now by Theorem B.5.4, condition (2) is equivalent to the condition that $\int_a^b \mathcal{W}(\mathcal{P}_n, f) \to 0$; and it follows from Theorem 8.14.5 that (2) $\Rightarrow$ (1). On the other hand, if f is Riemann-integrable on $[a, b]$, since $\|\mathcal{P}_n\| \to 0$, it follows from Theorem 8.19.1 that $\int_a^b \mathcal{W}(\mathcal{P}_n, f) \to 0$; and therefore, (1) $\Rightarrow$ (2). ∎

B.6.2   Lemma.   With the notation of Theorem B.6.1, if we define A_n to be the set of points of the partition $\mathcal{P}_n$, and $A = \cup_{n=1}^{\infty} A_n$, then the set A has measure zero. Furthermore, given any point $x \in [a, b] \backslash A$, the following two conditions are equivalent:

(1) f is continuous at x.

(2) $\mathcal{W}(x) = 0$.

∎ **Proof.**   Since each set A_n, being finite, must have measure zero, it follows from Theorem B.3 that A has measure zero. Now suppose that $x \in [a, b] \backslash A$; and in order to prove that (1) $\Rightarrow$ (2), assume that f is continuous at x. We shall prove that $\mathcal{W}(x) = 0$ by showing that for every $\varepsilon > 0$ we have $\mathcal{W}(x) < \varepsilon$. Let $\varepsilon > 0$. Choose a number $\delta > 0$ such that for every $t \in [a, b] \cap (x - \delta, x + \delta)$ we have $|f(x) - f(t)| < \varepsilon/4$. Choose a natural number n such that $\|\mathcal{P}_n\| < \delta$. Since x is not a point of $\mathcal{P}_n$, x must lie between two successive points of $\mathcal{P}_n$, say $x_{i-1} < x < x_i$. Now since for all points r and t in the interval (x_{i-1}, x_i) we have

$$|f(r) - f(t)| \leq |f(r) - f(x)| + |f(x) - f(t)| < \frac{\varepsilon}{4} + \frac{\varepsilon}{4} = \frac{\varepsilon}{2},$$

it follows that the constant value of $\mathcal{W}(\mathcal{P}_n, f)$ in (x_{i-1}, x_i) cannot exceed $\varepsilon/2$; and so $\mathcal{W}(x) \leq \varepsilon/2 < \varepsilon$.

Finally, to prove that (2) $\Rightarrow$ (1), assume that $\mathcal{W}(x) = 0$. Let $\varepsilon > 0$. Choose a natural number n such that $\mathcal{W}(\mathcal{P}_n, f)(x) < \varepsilon$. As before, since x is not a point of $\mathcal{P}_n$, x must lie between two successive points x_{i-1} and x_i of $\mathcal{P}_n$. It follows directly from the definition of $\mathcal{W}(\mathcal{P}_n, f)$ that for every $t \in (x_{i-1}, x_i)$ we have $|f(x) - f(t)| < \varepsilon$. ∎

B.6.3   Theorem: The Lebesgue Criterion for Riemann Integrability.   Suppose f is a bounded function on an interval $[a, b]$. Then the following two conditions are equivalent:

(1) f is Riemann-integrable on $[a, b]$.

(2) f is continuous at almost every point of $[a, b]$.

∎ **Proof.**   The theorem follows at once from Theorem B.6.1 and Lemma B.6.2. ∎

Now that we have the Lebesgue criterion for Riemann integrability, we can keep the promise we made in Chapter 8 and prove a sharper form of the composition theorem (Theorem 8.16.6).

B.7   THEOREM: A SHARPER FORM OF THE COMPOSITION THEOREM

Suppose f is Riemann-integrable on an interval $[a, b]$ and that ϕ is continuous on the range of f. Then the composition $\phi \circ f$ of ϕ with f is Riemann-integrable on $[a, b]$.

■ **Proof.**   Since $\phi \circ f$ is continuous at every point of $[a, b]$ at which f is continuous, the result follows at once from the Lebesgue criterion for Riemann integrability.   ■

B.8   COUNTABILITY AND MEASURE ZERO

This section makes use of the material of Appendix A.

Up to this point, we have seen two ways of saying that a set A of numbers is very small: the concept of measure zero and the concept of countable set that appeared in Section A.7. The two concepts have very similar properties; for example, a subset of a set of measure zero has measure zero, and a subset of a countable set is countable. Furthermore, if (A_n) is a sequence of sets of measure zero, then we saw in Theorem B.3 that $\cup_{n=1}^{\infty} A_n$ has measure zero; and if (A_n) is a sequence of countable sets, then we saw in Theorem A.7.8 that $\cup_{n=1}^{\infty} A_n$ is countable. In this, our last section, we compare those two concepts. In our first theorem we observe that countable sets always have measure zero.

B.8.1   Theorem.   Every countable set has measure zero.

■ **Proof.**   We remark first that if a set A is finite, then A is an elementary set of measure zero. Suppose now that a given set A is countable and infinite, and write A in the form $\{x_n \mid n \in \mathbf{Z}^+\}$. For every n the set $A_n = \{x_1, x_2, \ldots, x_n\}$ is finite and therefore has measure zero; and since $A = \cup_{n=1}^{\infty} A_n$, the fact that A has measure zero follows at once from Theorem B.3.   ■

Now we ask the obvious question. Is the converse of this theorem true? The answer is *no*. As a matter of fact, not only can one find uncountable sets of measure zero, but we shall even find a set C such that $C \sim \mathbf{R}$; and C satisfies the stronger type of "measure zero" criterion which in Section B.2.1 was termed the "nineteenth-century" concept of measure zero. The set C that we have in mind is called the *Cantor set*. Strictly speaking, a precise definition of the Cantor set depends upon the notion of an infinite series, which is introduced in Chapter 10. Until you reach that chapter, we suggest that you read the definition unofficially, making use of the knowledge of infinite series that you gained in elementary calculus. The definition also appears in Exercise 10.3(7), and it is recommended that you do the first five parts of this exercise now.

B.8.2 The Cantor Set. The *Cantor set* C is defined to be the set of all those numbers that can be expressed as a ternary decimal $\sum_{n=1}^{\infty} a_n/3^n$, where (a_n) is any sequence in $\{0, 2\}$. Since there is an obvious one–one function from C onto $\{0, 2\}^{Z^+}$, we see that $C \sim \mathbf{R}$. Now the smallest number in the Cantor set is obviously 0, which we obtain by taking $a_n = 0$ for each n in the above expression; and at the other extreme we can take $a_n = 2$ for every n and obtain

$$\sum_{n=1}^{\infty} \frac{a_n}{3^n} = \sum_{n=1}^{\infty} \frac{2}{3^n} = 1.$$

This shows that $C \subseteq [0, 1]$.

We now observe that no point of C lies in the interval $(\frac{1}{3}, \frac{2}{3})$. To see this, notice that if $a_1 = 2$, then the least possible value of $\sum_{n=1}^{\infty} a_n/3^n$ is $\frac{2}{3}$; if $a_1 = 0$, then the greatest possible value of $\sum_{n=1}^{\infty} a_n/3^n$ is $\sum_{n=2}^{\infty} 2/3^n = \frac{1}{3}$:

By a similar argument, one can see that in the event that $a_1 = 0$, and if $a_2 = 0$, then the greatest possible value of $\sum_{n=1}^{\infty} a_n/3^n$ is $\frac{1}{9}$; and if $a_2 = 2$, then the least possible value of $\sum_{n=1}^{\infty} a_n/3^n$ is $\frac{2}{9}$. So no point of C lies in the interval $(\frac{1}{9}, \frac{2}{9})$; and similarly, no point of C lies in the interval $(\frac{7}{9}, \frac{8}{9})$:

By continuing this process, we may see that for every natural number n the Cantor set C is a subset of an elementary set of Lebesgue measure $(\frac{2}{3})^n$. It follows that C satisfies the conditions of Definition B.2.1.

B.8.3 Cantor's Continuum Hypothesis. We have seen that not only can a set of measure zero be uncountable, but even a set which satisfies the more demanding conditions of Definition B.2.1 can be equivalent to $\mathbf{R}$. This might suggest the question: Can an uncountable set of real numbers be so small that it is not equivalent to $\mathbf{R}$, or is every uncountable set of real numbers equivalent to $\mathbf{R}$?

This question was considered by Cantor, who was unable to resolve it and who finally suggested as an hypothesis that every uncountable set of real numbers should be equivalent to $\mathbf{R}$. For a number of years this hypothesis was a major focus of attention, as mathematicians tried either to prove it or to disprove it; but we know today that the question is impossible to answer. With the techniques of modern mathematical logic it can be shown that it is impossible to prove Cantor's hypothesis true and that it is also impossible to prove that the hypothesis is false. If you would like to read further on this interesting topic, look in your library for a good book on the theory of sets.

Chapter 9

Convergence of Integrals

9.1 IMPROPER INTEGRALS

9.1.1 Introduction. Some of the integrals that were most useful to you in your elementary calculus courses do not fall within the scope of the previous chapter because they run into difficulty at one of the two endpoints of the interval of integration. To remind ourselves how this can happen, let us take a look at some examples.

(1) The integral $\int_0^1 (1/\sqrt{x})\, dx$ should, of course, be equal to $2\sqrt{x}\,|_0^1 = 2$; but the trouble is that its integrand (the function we are integrating) is not defined at 0, and even worse, this integrand is not bounded. As you know, the theory of Riemann integration that we studied in the previous chapter was confined strictly to bounded functions. So we have to ask our usual question: What does it *mean* to say that $\int_0^1 (1/\sqrt{x})\, dx = 2$? In elementary calculus you answered this question by observing that whenever $0 < w \le 1$, the integral $\int_w^1 (1/\sqrt{x})\, dx$ exists and is equal to $2 - 2\sqrt{w}$, and that the latter expression tends to 2 as $w \to 0$ (from the right). This is the sense in which we might say that $\int_0^1 (1/\sqrt{x})\, dx = 2$.

(2) The integral $\int_0^1 (1/\sqrt{1 - x^2})\, dx$ equals $\pi/2$; but once again, the integrand is unbounded. The point is that whenever $0 \le w < 1$, we have

$$\int_0^w \frac{1}{\sqrt{1 - x^2}}\, dx = \arcsin x \Big|_0^w = \arcsin w,$$

which approaches $\pi/2$ as $w \to 1$ (from the left).

(3) The integral $\int_1^\infty (1/x^2)\, dx$ does not fit into the previous chapter even though its integrand is bounded, because in this example the interval of integration is illegal. As you may remember from elementary calculus, this integral may be defined by saying that

$$\int_1^\infty \frac{1}{x^2}\, dx = \lim_{w \to \infty} \int_1^w \frac{1}{x^2}\, dx = \lim_{w \to \infty} \left(1 - \frac{1}{w}\right) = 1.$$

These examples suggest the definition of an improper integral given in the next subsection.

9.1.2 Definition of an Improper Integral.
We define an improper integral of the type $\int_a^{\to b} f(x)\, dx$ as follows: If $-\infty < a < b \le \infty$ and $f : [a, b) \to \mathbf{R}$, and if the function f is Riemann-integrable on every interval $[a, w]$ for which $a < w < b$, and the limit $\lim_{w \to b} \int_a^w f(x)\, dx$ exists, then

$$\int_a^{\to b} f(x)\, dx = \lim_{w \to b} \int_a^w f(x)\, dx.$$

Similarly, we define an improper integral of the type $\int_{a \leftarrow}^b f(x)\, dx$: If $-\infty \le a < b < \infty$ and $f : (a, b] \to \mathbf{R}$, and if the function f is Riemann-integrable on the interval $[w, b]$ whenever $a < w < b$, and the limit $\lim_{w \to a} \int_w^b f(x)\, dx$ exists, then

$$\int_{a \leftarrow}^b f(x)\, dx = \lim_{w \to a} \int_w^b f(x)\, dx.$$

If an improper integral of the form $\int_a^{\to b} f(x)\, dx$ or $\int_{a \leftarrow}^b f(x)\, dx$ exists and is a real number (in other words, if it exists and is not $\pm\infty$), then we say that the improper integral **converges** and that f is **improper Riemann-integrable** on $[a, b)$ or $(a, b]$. An improper integral which does not converge is said to **diverge**. This means that there are two distinct ways for an improper integral $\int_{a \leftarrow}^b f(x)\, dx$ or $\int_a^{\to b} f(x)\, dx$ to diverge; either the appropriate limit $\lim_{w \to a} \int_w^b f(x)\, dx$ or $\lim_{w \to b} \int_a^w f(x)\, dx$ is $\pm\infty$, or this limit does not exist. Notice how similar this use of the words *convergent* and *divergent* is to our use of these words for sequences. To remind yourself how we used these words for sequences, take another look at Section 4.3.10.

9.1.3 Some Examples of Improper Integrals.
In our first two examples we look at some convergent improper integrals.

(1) We begin with the simplest kind of improper integral, an integral that exists in the ordinary Riemann sense. Suppose f is Riemann-integrable on $[a, b]$. Then f is Riemann-integrable on $[a, x]$ for every $x \in [a, b]$, and it is easy to see that $\int_a^x f \to \int_a^b f$ as $x \to b$ (from the left). We conclude that if a function f is Riemann-integrable on an interval $[a, b]$, then f is improper Riemann-integrable on $[a, b)$; and we have $\int_a^{\to b} f = \int_a^b f$. In the same way, if f is Riemann-integrable on $[a, b]$, then f is improper Riemann-integrable on $(a, b]$ and $\int_{a\leftarrow}^b f = \int_a^b f$.

(2) For some examples of convergent improper integrals that do not exist in the ordinary Riemann sense, we return to the integrals we looked at in Section 9.1.1. From the discussion given there we see that these improper integrals converge, and that

$$\int_{0\leftarrow}^1 \frac{1}{\sqrt{x}}\, dx = 2, \quad \int_0^{\to 1} \frac{1}{\sqrt{1-x^2}}\, dx = \frac{\pi}{2}, \quad \text{and} \quad \int_1^{\to\infty} \frac{1}{x^2}\, dx = 1.$$

The following four examples exhibit some of the ways in which an improper integral might diverge.

(3) Whenever $w \geq 1$, we have $\int_1^w (1/\sqrt{x})\, dx = 2\sqrt{w} - 2$; and since the latter expression tends to infinity as $w \to \infty$, we see that $\int_1^{\to\infty} (1/\sqrt{x})\, dx = \infty$ and that the integral diverges. In the same way, one can show that the integral $\int_{0\leftarrow}^1 (1/x^2)\, dx$ also diverges.

(4) Whenever $w \geq 1$, we have $\int_0^w \cos x\, dx = \sin w$; and since the latter expression does not have a limit as $w \to \infty$, we see that the integral $\int_0^{\to\infty} \cos x\, dx$ diverges.

(5) Referring to Figure 9.1, we define $f: [0, \infty) \to \mathbf{R}$ as follows: $f(x) = 1$ if $n - 1 \leq x < n$ for some odd natural number n, and $f(x) = -1$ if $n - 1 \leq x < n$ for some even natural number n. If we like, we can combine these two statements and say that if n is any natural number, and $n - 1 \leq x < n$, then $f(x) = (-1)^{n-1}$. Since $\int_0^n f(x)\, dx$ is 1 when n is odd and is 0 when n is even, we see that $\lim_{w\to\infty} \int_0^w f(x)\, dx$ does not exist; and it follows that the integral $\int_0^{\to\infty} f(x)\, dx$ diverges.

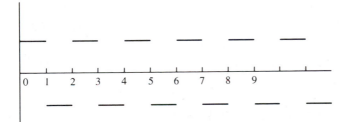

Figure 9.1

(6) We define $f: (0, 1] \to \mathbf{R}$ as follows: $f(1) = 1$ and $f(x) = n(n + 1)(-1)^{n+1}$ whenever n is a natural number and $1/(n + 1) \leq x < 1/n$. It is not hard to see that the integral $\int_{1/n}^{1} f$ is 1 when n is even and is 0 when n is odd. It follows that $\lim_{w \to 0} \int_{w}^{1} f$ does not exist; and therefore, the integral $\int_{0 \leftarrow}^{1} f$ diverges.

9.1.4 More General Improper Integrals; an Extension of Our Notation.
As you may recall from elementary calculus, we sometimes work with integrals which have two or more points of "improper" behavior, and we deal with such integrals by taking their points of improper behavior one at a time. For example, if $a < b$, an improper integral of the form $\int_{a \leftrightarrow}^{\to b} f$ is understood to be the sum of two improper integrals $\int_{a \leftarrow}^{c} f$ and $\int_{c}^{\to b} f$, where c is any chosen point of the interval (a, b); and $\int_{a \leftrightarrow}^{\to b} f$ is understood to be convergent when both of the integrals $\int_{a \leftarrow}^{c} f$ and $\int_{c}^{\to b} f$ converge. When it is clear that a given integral may be improper at some points, we often write it just as if it were a regular Riemann integral. For example, the integral

$$\int_{0}^{\infty} \frac{1}{\sqrt{|x(x - 3)(x - 5)|}} \, dx$$

contains six bits of improper behavior, these being at 0, at ∞, and on each side of the points 3 and 5. Strictly speaking, this integral means the sum

$$\int_{0 \leftarrow}^{1} + \int_{1}^{\to 3} + \int_{3 \leftarrow}^{4} + \int_{4}^{\to 5} + \int_{5 \leftarrow}^{6} + \int_{6}^{\to \infty} \frac{1}{\sqrt{|x(x - 3)(x - 5)|}} \, dx,$$

and in order to show that it converges, we must establish the convergence of all six of the latter integrals. As a matter of fact, all six of these integrals do converge. This will be clear when we come to Section 9.4.

9.1.5 Theorem: Linearity of Convergent Improper Integrals.
Suppose f and g are improper Riemann-integrable on $[a, b)$.

(1) $f + g$ is improper Riemann-integrable on $[a, b)$, and we have

$$\int_{a}^{\to b} (f + g) = \int_{a}^{\to b} f + \int_{a}^{\to b} g.$$

(2) For every number α the function αf is improper Riemann-integrable on $[a, b)$, and we have

$$\int_{a}^{\to b} \alpha f = \alpha \int_{a}^{\to b} f.$$

■ **Proof.** To prove part (1), we notice that for every $w \in [a, b)$ we have

$$\int_{a}^{w} (f + g) = \int_{a}^{w} f + \int_{a}^{w} g,$$

and the result follows at once by letting $w \to b$. The proof of part (2) is similar.

∎

9.1.6 Theorem: Additivity of Convergent Improper Integrals. Suppose f is improper Riemann-integrable on $[a, b)$. Then given any number $c \in [a, b)$, we have $\int_a^{\to b} f = \int_a^c f + \int_c^{\to b} f$.

∎ **Proof.** The result follows at once from the fact that whenever $c < w < b$, we have $\int_a^w f = \int_a^c f + \int_c^w f$.

∎

Our last theorem in this section is an analogue for integrals of the monotone sequence theorem (Theorem 4.10.1).

9.1.7 Theorem. Suppose that $-\infty < a < b \leq \infty$, that f is a nonnegative function defined on $[a, b)$, and that f is Riemann-integrable on the interval $[a, w]$ whenever $a < w < b$. Then either $\int_a^{\to b} f(x)\, dx$ converges or $\int_a^{\to b} f(x)\, dx = \infty$.

∎ **Proof.** For each number $w \in [a, b)$, define $F(w) = \int_a^w f(x)\, dx$. If w_1 and w_2 are points of $[a, b)$ and $w_1 < w_2$, then

$$F(w_2) - F(w_1) = \int_{w_1}^{w_2} f(x)\, dx \geq 0,$$

and so the function F is increasing on $[a, b)$. The existence of $\lim_{w \to b} F(w)$ now follows at once from Theorem 6.10.6.

∎

In a similar fashion, one can obtain an analogue of Theorem 9.1.7 for integrals of the form $\int_{a \leftarrow}^b f$. We leave the task of doing this as an exercise.

9.1.8 Some Simple Exercises

1. Evaluate each of the following improper integrals, when possible, and specify those that diverge.

(a) $\displaystyle \int_0^{\to \infty} \frac{1}{(1 + x^2)^{3/2}}\, dx$

(b) $\displaystyle \int_2^{\to \infty} \frac{1}{x\sqrt{x^2 - 1}}\, dx$

(c) $\displaystyle \int_{1 \leftarrow}^2 \frac{1}{x\sqrt{x^2 - 1}}\, dx$

(d) $\displaystyle \int_{1 \leftarrow}^{\to \infty} \frac{1}{x\sqrt{x^2 - 1}}\, dx$

(e) $\displaystyle \int_0^{\to \pi/2} \tan x\, dx$

(f) $\displaystyle \int_0^{\to \pi/2} \sqrt{\tan x \sin x}\, dx$

(g) $\displaystyle \int_{0 \leftarrow}^{\pi/2} \frac{x \cos x - \sin x}{x^2}\, dx$

2. Interpret the integral $\int_0^2 1/(x-1)^{1/3}\,dx$ as the sum of two improper integrals and evaluate it.

3. Justify the assertion that was made in Section 9.1.3 that if f is Riemann-integrable on $[a, b]$, then $\int_a^x f \to \int_a^b f$ as $x \to b$ (from the left).

4. Prove that if f is bounded on $[a, b]$ and is improper Riemann-integrable on $[a, b)$, then f is Riemann-integrable on $[a, b]$ and $\int_a^{\to b} f = \int_a^b f$.

5. The purpose of this exercise is to justify the definition of an integral of the type $\int_{a\leftarrow}^{\to b}$ given in Section 9.1.4. Suppose that $-\infty \leq a < c < b \leq \infty$ and that the improper integrals $\int_{a\leftarrow}^c f$ and $\int_c^{\to b} f$ are both convergent. Prove that for every $d \in (a, b)$ we have

$$\int_{a\leftarrow}^d f + \int_d^{\to b} f = \int_{a\leftarrow}^c f + \int_c^{\to b} f.$$

9.2 AN IMPORTANT DIVERGENT INTEGRAL; DEFINITION OF THE FUNCTION log[1]

9.2.1 Introduction. As you know, the logarithmic and exponential functions and the trigonometric functions play a vitally important role in calculus. While we have used them occasionally in our examples and exercises, we haven't yet defined them at the level at which we are working, and it is clear that sooner or later we must do so. We shall leave the trigonometric functions to be defined after we have spoken about power series in Chapter 11. In this section we shall define the logarithmic and exponential functions.

How should these functions be defined? Once again, we find ourselves asking what something *means*. What we want, of course, is a pair of functions, called log and exp, which are inverse functions of one another and which satisfy a set of rules like the following:

(1) $\log : (0, \infty) \to R$ and $\exp : R \to (0, \infty)$. Furthermore, $\log 1 = 0$ and $\exp 0 = 1$.

(2) Given any positive numbers a and b, we have $\log(ab) = \log a + \log b$; and given any numbers a and b, we have $\exp(a + b) = (\exp a)(\exp b)$.

(3) For every positive number x we have $\log' x = 1/x$.

(4) For every number x we have $\exp' x = \exp x$.

(5) For every number x we have $\exp x = \sum_{n=0}^{\infty} x^n/n!$.

Notice how (5) tells us exactly what $\exp x$ has to be for every real number x, and in some presentations of calculus this rule is used as the *definition* of the

---

[1] Scientists and engineers often refer to the natural logarithm as ln, reserving the symbol log for $\log_{10}$. In keeping with the usage in most texts in pure mathematics, we shall reserve the symbol log for the natural logarithm.

exponential function. We obviously can't do this here because we haven't yet studied series. Instead, we shall make our definition of log in the traditional way, using the requirement that log $1 = 0$ and that log$'$ $x = 1/x$ for every positive number x.

Specifically, we define

$$\log x = \int_1^x \frac{1}{t}\, dt \qquad \text{for every positive number } x.$$

9.2.2 Theorem: Uniqueness of the log Function. Given any function $f:(0, \infty) \to \mathbf{R}$, the following two conditions are equivalent:

(1) $f = \log$.

(2) $f(1) = 0$ and $f'(x) = 1/x$ for every positive number x.

■ *Proof.*   The fact that (1) $\Rightarrow$ (2) follows at once from Theorem 8.17.3. To prove that (2) $\Rightarrow$ (1), assume that (2) holds. Define $g = f - \log$. Then since $g'(x) = 1/x - 1/x = 0$ for every $x \in (0, \infty)$, we see that the function g is constant; and since $g(1) = 0$, we see that this constant must be zero. Therefore, $f(x) = \log x$ for every $x \in (0, \infty)$. ■

9.2.3 Theorem: The Algebraic Properties of log

(1) Given any positive numbers a and b, we have

$$\log(ab) = \log a + \log b.$$

(2) Given any positive number a and any integer n, we have

$$\log(a^n) = n \log a.$$

■ *Proof.*   To prove part (1), suppose a and b are positive numbers. For every positive number x, define $f(x) = \log(ax) - \log x$. Differentiating, we obtain $f'(x) = a/ax - 1/x = 0$ for every $x > 0$; and it follows that f must be constant on $(0, \infty)$. Therefore, since $f(1) = \log a$, we have $f(b) = \log a$; in other words, $\log(ab) - \log b = \log a$.

Now to prove (2), suppose n is some integer; and for $x > 0$, define $f(x) = \log(x^n) - n \log x$. To see that this function f is the constant zero, we note that $f(1) = 0$ and that for every $x \in (0, \infty)$ we have $f'(x) = nx^{n-1}/x^n - n/x = 0$. ■

9.2.4 Theorem

(1) log is a strictly increasing function on $(0, \infty)$.

(2) $\log x \to \infty$ as $x \to \infty$, and $\log x \to -\infty$ as $x \to 0$ (from the right).

(3) The range of log is $\mathbf{R}$.

■ *Proof.* Part (1) follows at once from the fact that the derivative of log is positive, and it follows that both $\lim_{x \to \infty} \log x$ and $\lim_{x \to 0} \log x$ must exist. We now complete the proof of (2) by observing that since $\log 2 > 0$, we have $\log(2^n) = n \log 2 \to \infty$ as $n \to \infty$ and $\log(2^n) \to -\infty$ as $n \to -\infty$. Part (3) now follows at once from Theorem 6.8.1. ■

It is worth noting that the fact that $\log x \to \infty$ as $x \to \infty$ is just the fact that the integral $\int_1^{\to \infty} 1/x \, dx$ diverges.

9.2.5 Definition of the Exponential Function.
The function exp is defined to be the inverse function of the function log.

Note that $\exp : \mathbf{R} \to (0, \infty)$, that exp is strictly increasing on $\mathbf{R}$, that $\exp x \to \infty$ as $x \to \infty$, and that $\exp x \to 0$ as $x \to -\infty$.

9.2.6 Theorem: The Derivative of exp.
The function exp is differentiable on $\mathbf{R}$, and $\exp' = \exp$.

■ *Proof.* Let x be any real number, and define $t = \exp x$. Then $x = \log t$, and we deduce from Theorem 7.7.1 that

$$\exp' x = \frac{1}{\log' t} = t = \exp x.$$ ■

9.2.7 Theorem: The Algebraic Properties of exp.
Given any numbers a and b and any integer n, we have the following:

(1) $\exp(a + b) = (\exp a)(\exp b)$.

(2) $(\exp a)^n = \exp(na)$.

■ *Proof.* This theorem follows at once from Theorem 9.2.3: Put $u = \exp a$ and $v = \exp b$. Then

$$\exp(a + b) = \exp(\log u + \log v) = \exp[\log(uv)] = uv.$$

Furthermore, $\exp(na) = \exp(n \log u) = \exp[\log(u^n)] = u^n = (\exp a)^n$. ■

9.2.8 Extension to Other Bases.
We now extend the idea of exponents and logarithms to general bases in the traditional way, motivated by the fact that if $a > 0$ and n is any integer, then $a^n = \exp[\log(a^n)] = \exp(n \log a)$.

(1) Given $a > 0$ and $x \in \mathbf{R}$, we define $a^x = \exp(x \log a)$.

(2) Given any positive number $a \neq 1$ and any $x > 0$, we define

$$\log_a x = \frac{\log x}{\log a}.$$

Note that if $y = \log_a x$, then since $y = \log x/\log a$, we have $\log x = y \log a$, and therefore, $x = \exp(y \log a) = a^y$.

Finally, we define $e = \exp 1$. Note that $\log e = 1$, that $\exp x = e^x$ for every $x \in R$, and that $\log_e = \log$.

In the exercises that follow you will have an opportunity to verify some of the standard properties of the exponential and logarithmic functions.

9.3   EXERCISES

1. Prove that if $a > 0$ and x and y are any real numbers, then $a^x \cdot a^y = a^{x+y}$.

2. Prove that if $a > 0$ and x and y are any real numbers, then $(a^x)^y = a^{xy}$. Deduce that if m and n are integers and $n \neq 0$, then $a^{m/n} = \sqrt[n]{a^m}$.

3. Prove that if a and b are positive numbers, then for any x we have $a^x \cdot b^x = (ab)^x$.

4. Suppose that α is any real number and that $f(x) = x^\alpha$ for all $x > 0$. Prove that for every x we have $f'(x) = \alpha x^{\alpha-1}$.

5. Suppose that $a > 0$ and that $f(x) = a^x$ for all $x \in R$. Prove that for every x we have $f'(x) = a^x \cdot \log a$.

6. Prove that if $f: R \to R, f(0) = 1$, and $f' = f$, then $f = \exp$. *Hint:* Look at the function $g = f/\exp$.

7. Take another look at the proof of Theorem 9.2.3(2). Now find a different proof which is purely algebraic and uses part (1) of the theorem.

8. Suppose that $f: R \to R$ and that for all numbers x and t we have $f(x + t) = f(x)f(t)$.
   (a) Prove that either $f(x) = 0$ for every number x or $f(x) \neq 0$ for every number x.
   (b) Prove that if f is not the constant zero, then $f(0) = 1$ and $f(x) > 0$ for every number x.
   (c) Prove that if f is not the constant zero and $a = f(1)$, then for every rational number x we have $f(x) = a^x$. Deduce that in the event that f is continuous, the equation $f(x) = a^x$ holds for all numbers x.

9. Make a rough sketch of the curve

$$y = (\log x)^3 - 4(\log x)^2$$

showing its local maximum and minimum points and points of inflection.

10. Prove that if $\alpha \in R$, then $(1 + \alpha x)^{1/x} \to e^\alpha$ as $x \to 0$. *Hint:* First look at the limit of $\log(1 + \alpha x)/x$.

11. Prove that if $\alpha \in R$, then $(1 + \alpha/x)^x \to e^\alpha$ as $x \to \infty$.

12. Suppose p is some number; and for all $x \in (0, 1]$, define $f(x) = 1/x^p$. Prove that the integral $\int_{0\leftarrow}^{1} f$ converges when $p < 1$ and diverges when $p \geq 1$.

13. Suppose p is some number; and for all $x \in [1, \infty)$, define $f(x) = 1/x^p$. Prove that the integral $\int_{1}^{\rightarrow\infty} f$ converges when $p > 1$ and diverges when $p \leq 1$.

14. Find the values of p for which the integral $\int_{2}^{\rightarrow\infty} 1/x(\log x)^p \, dx$ converges and the values of p for which the integral $\int_{1\leftarrow}^{2} 1/x(\log x)^p \, dx$ converges.

15. Show that the integral $\int_{0}^{\rightarrow\infty} e^{-x} \sin x \, dx$ converges, and evaluate it. *Hint:* Use integration by parts.

16. Suppose $\alpha \in \mathbf{R}\setminus\{0\}$, and define $S = \{x \mid 1 + \alpha x > 0\}$.
    (a) Prove that if $g(x) = \alpha x - (1 + \alpha x) \log(1 + \alpha x)$ for all $x \in S$, then $g(x) < 0$ for all $x \in S\setminus\{0\}$.
    (b) Prove that if $f(x) = \log(1 + \alpha x)/x$ for all $x \in S\setminus\{0\}$ and $f(0) = \alpha$, then f is continuous and strictly decreasing on S. Deduce that for every positive member x of S we have $f(x) < \alpha$.
    (c) Prove that for every positive member x of S we have $(1 + \alpha x)^{1/x} < e^{\alpha}$.
    (d) Prove that if $x > |\alpha|$, then $(1 + \alpha/x)^x < e^{\alpha}$.

17. By applying the Taylor mean value theorem (Theorem 7.13.6), to exp on $[0, 1]$, show that for every natural number n there is a number $c \in (0, 1)$ such that

$$e = 1 + 1 + \frac{1}{2!} + \frac{1}{3!} + \cdots + \frac{1}{n!} + \frac{e^c}{(n + 1)!}.$$

Deduce that for every natural number n we have

$$0 < e - \sum_{j=0}^{n} \frac{1}{j!} < \frac{e}{(n + 1)!}.$$

Putting $n = 2$ in this inequality, show that $e < 3$, deduce that

$$0 < e - \sum_{j=0}^{n} \frac{1}{j!} < \frac{3}{(n + 1)!},$$

and deduce finally that $\sum_{j=0}^{n} 1/j! \to e$ as $n \to \infty$.

■ 18. Using the last inequality of exercise 17, prove that the number e is irrational.

9.4 THE COMPARISON TESTS FOR INTEGRALS OF NONNEGATIVE FUNCTIONS

As we saw in Theorem 9.1.7, if a function f is nonnegative on an interval $[a, b)$ and is Riemann-integrable on $[a, w]$ whenever $a < w < b$, then the only way the integral $\int_{a}^{\rightarrow b} f$ can fail to converge is that $\int_{a}^{\rightarrow b} f = \infty$. From this it follows that as long as we are talking about nonnegative functions, the integral of a smaller function is more likely to converge than the integral of a larger function. This is stated precisely in the results that follow, which we call the **comparison tests**. We shall state them for improper integrals of the form $\int_a^{\to b}$. Their statements and proofs for other types of improper integrals are similar.

9.4.1   Theorem: The Fundamental Comparison Test.   Suppose that f and g are nonnegative functions defined on an interval $[a, b)$, where $-\infty < a < b \le \infty$. Suppose that the improper integral $\int_a^{\to b} g$ converges and that f is Riemann-integrable on $[a, w]$ for every $w \in [a, b)$; and suppose that there exists a point $c \in [a, b)$ such that $f(x) \le g(x)$ whenever $x \in [c, b)$. Then the improper integral $\int_a^{\to b} f$ must converge.

■ *Proof.*   Whenever $w \in [c, b)$, we have

$$\int_a^w f = \int_a^c f + \int_c^w f \le \int_a^c f + \int_c^w g \le \int_a^c f + \int_c^{\to b} g.$$

Therefore, it is impossible to have $\lim_{w \to b} \int_a^w f = \infty$, and the result follows at once from Theorem 9.1.7.   ■

9.4.2   Theorem: The Limit Comparison Test.   Suppose that f and g are nonnegative functions defined on an interval $[a, b)$, where $-\infty < a < b \le \infty$. Suppose that the improper integral $\int_a^{\to b} g$ converges and that f is Riemann-integrable on $[a, w]$ for every $w \in [a, b)$; and suppose that for some number $\alpha \in [0, \infty)$ we have $f(x)/g(x) \to \alpha$ as $x \to b$. Then the improper integral $\int_a^{\to b} f$ must converge.

■ *Proof.*   Choose a point $c \in [a, b)$ such that whenever $c < w < b$, we have $f(x)/g(x) < \alpha + 1$. Now whenever $w \in [c, b)$, we have

$$\int_a^w f = \int_a^c f + \int_c^w f \le \int_a^c f + (\alpha + 1) \int_c^w g \le \int_a^c f + (\alpha + 1) \int_c^{\to b} g,$$

and we complete the proof just as we did in Theorem 9.4.1.   ■

9.5   ABSOLUTE AND CONDITIONAL CONVERGENCE

9.5.1   Definition of Absolute Convergence.   Suppose that $-\infty < a < b \le \infty$, that $f: [a, b) \to R$, and that the function f is Riemann-integrable on every interval $[a, w]$ for which $a < w < b$. Of course, $|f|$ is also Riemann-integrable on $[a, w]$ whenever $a < w < b$. If the integral $\int_a^{\to b} |f|$ converges, then we say that the integral $\int_a^{\to b} f$ **converges absolutely**.

9.5.2   Theorem.   Every absolutely convergent improper integral is convergent. Furthermore, if $\int_a^{\to b} f$ is absolutely convergent, then $|\int_a^{\to b} f| \le \int_a^{\to b} |f|$.

■ **Proof.** Suppose $\int_a^{\to b} f$ is absolutely convergent. In order to prove that $\int_a^{\to b} f$ converges, we shall express f as the difference of two nonnegative functions f_1 and f_2 whose sum is $|f|$. The functions f_1 and f_2 that satisfy these requirements are

$$f_1 = \frac{|f| + f}{2} \quad \text{and} \quad f_2 = \frac{|f| - f}{2}.$$

Since $0 \le f_1 \le |f|$, and $0 \le f_2 \le |f|$, it follows from the comparison test, Theorem 9.4.1, that the integrals $\int_a^{\to b} f_1$ and $\int_a^{\to b} f_2$ are convergent; and therefore, since $f = f_1 - f_2$, the convergence of $\int_a^{\to b} f$ follows from Theorem 9.1.5. The final assertion of the theorem now follows from the fact that whenever $w \in [a, b)$, we have $|\int_a^w f| \le \int_a^w |f|$. ■

9.5.3 Definition of Conditional Convergence.

It can happen that a given improper integral converges but does not converge absolutely, and in such a case we say that the integral is **conditionally convergent**.

When a given integral is not absolutely convergent, it is often quite hard to determine its convergence or divergence because we cannot use the comparison test in this situation. As a result, the concept of conditional convergence is somewhat more difficult than the concept of absolute convergence. But even when the comparison test is not available to us, there is still an important technique that allows us to conclude that certain integrals are convergent. This technique is provided by the theorem that follows, which is known as Dirichlet's test.

9.5.4 Theorem: Dirichlet's Test.

Suppose $-\infty < a < b \le \infty$. Suppose that f is a positive decreasing function on $[a, b)$ with a continuous derivative and that $f(x) \to 0$ as $x \to b$. Suppose that g is continuous on $[a, b)$ and that there is a number K such that $|\int_a^x g| \le K$ for all $x \in [a, b)$. Then the integral $\int_a^{\to b} fg$ converges.

■ **Proof.** For each $x \in [a, b)$, define $G(x) = \int_a^x g$. Then $G(a) = 0$, and for every $x \in [a, b)$ we have $G'(x) = g(x)$. Therefore, for each $x \in [a, b)$ we can integrate by parts to obtain

$$\int_a^x fg = \int_a^x fG' = f(x)G(x) - f(a)G(a) - \int_a^x Gf' = f(x)G(x) - \int_a^x Gf'.$$

Choose a number K such that $|G(x)| \le K$ for all $x \in [a, b)$. Then for each x we have $|f(x)G(x)| \le Kf(x)$; and since $f(x) \to 0$ as $x \to b$, it follows from the sandwich theorem that $f(x)G(x) \to 0$ as $x \to b$. So to complete the proof, all we need to do is show that the integral $\int_a^{\to b} Gf'$ converges. To this end, we observe that for each $x \in [a, b)$ we have $|G(x)f'(x)| \le K(-f'(x))$ and that since $\int_a^x K(-f') = Kf(a) - Kf(x)$ for all $x \in [a, b)$, we have $\int_a^x K(-f') \to Kf(a)$ as $x \to b$. Therefore, the convergence of the integral $\int_a^{\to b} |Gf'|$ follows from the comparison test; and since the integral $\int_a^{\to b} Gf'$ converges absolutely, it converges. ■

The final theorem in this section tells us something about the size of the integral whose convergence has been established by Dirichlet's test.

9.5.5 Theorem. Suppose $-\infty < a < b \leq \infty$. Suppose that f is a positive decreasing function on $[a, b)$ with a continuous derivative and that $f(x) \to 0$ as $x \to b$. Suppose that g is continuous on $[a, b)$ and that there is a number K such that $|\int_a^x g| \leq K$ for all $x \in [a, b)$. Then for every $x \in [a, b)$ we have $|\int_a^x fg| \leq Kf(a)$, and consequently, $|\int_a^{\to b} fg| \leq Kf(a)$.

■ ***Proof.*** For each $x \in [a, b)$, define $G(x) = \int_a^x g$, and note that $|G(x)| \leq K$. Now given $x \in [a, b)$, we have

$$\int_a^x fg = f(x)G(x) - \int_a^x Gf';$$

and therefore, since $f' \leq 0$,

$$\left|\int_a^x fg\right| \leq |f(x)G(x)| + \int_a^x |G|(-f') \leq Kf(x) + \int_a^x K(-f')$$

$$= Kf(x) + Kf(a) - Kf(x) = Kf(a).$$ ■

9.5.6 Example. Suppose a is any positive number. Whenever $x \geq a$, we have

$$\left|\int_a^x \sin t \, dt\right| = |\cos x - \cos a| \leq 2,$$

and it therefore follows from Dirichlet's test that the integral $\int_a^{\to\infty} (\sin x)/x \, dx$ converges. One may also deduce from Theorem 9.5.5 that $|\int_a^{\to\infty} (\sin x)/x \, dx| \leq 2/a$.

It is also worth remarking that since $(\sin x)/x \to 1$ as $x \to 0$, the integral $\int_0^1 (\sin x)/x \, dx$ is not improper; and therefore, the integral $\int_0^{\to\infty} (\sin x)/x \, dx$ converges. We shall see soon that the latter integral has a value of $\pi/2$. Another interesting fact about this integral is that it is conditionally convergent; in other words, the integral $\int_0^{\to\infty} |(\sin x)/x| \, dx$ diverges. We shall prove this using series in the next chapter.

9.6 EXERCISES

■ **1.** Determine the convergence or divergence of the following integrals.

(a) $\int_1^{\to\infty} \frac{\sqrt{x}}{x^2 - x + 1} \, dx$

(b) $\int_{0\leftarrow}^1 \frac{1}{x + x^2} \, dx$

(c) $\int_1^{\to\infty} \frac{\sin x}{x^2} \, dx$

(d) $\int_{0\leftarrow}^{\to\infty} \frac{\sin x}{x\sqrt{x}} \, dx$

(e) $\int_0^{\to \pi/2} \sqrt{\tan x} \, dx$ **(f)** $\int_{1\leftarrow}^2 \frac{1}{\log x} \, dx$

(g) $\int_{0\leftarrow}^{\pi/2} \log(\sin x) \, dx$ **(h)** $\int_2^{\to \infty} \frac{1}{(\log x)^{\log x}} \, dx$

(i) $\int_3^{\to \infty} \frac{1}{(\log \log x)^{\log x}} \, dx$ **(j)** $\int_{30}^{\to \infty} \frac{1}{(\log \log \log x)^{\log x}} \, dx$

(k) $\int_3^{\to \infty} \frac{1}{(\log x)^{\log \log x}} \, dx$ **(l)** $\int_1^{\to \infty} \frac{1}{\exp(\sqrt{\log x})} \, dx$

2. (a) Prove that $\int_1^{\to \infty} x^{\alpha-1} e^{-x} \, dx$ converges for all numbers α. *Hint:* Look at the limit $\lim_{x\to\infty} x^{\alpha-1} e^{-x}/x^{-2}$, and use Theorem 9.4.2.

(b) Prove that $\int_{0\leftarrow}^1 x^{\alpha-1} e^{-x} \, dx$ converges if and only if $\alpha > 0$, and deduce that the integral $\int_{0\leftarrow}^\infty x^{\alpha-1} e^{-x} \, dx$ converges if and only if $\alpha > 0$. This integral defines the value at α of the **gamma function** and is usually denoted as $\Gamma(\alpha)$.

3. Prove that the integral $\int_{0\leftarrow}^{\to 1} (1-t)^{\alpha-1} t^{\beta-1} \, dt$ converges if and only if both α and β are positive. This integral defines the value at (α, β) of the **beta function** and is usually denoted as $B(\alpha, \beta)$. We shall explore some of the important properties of the gamma and beta functions in Exercises 9.9(11) and 9.9(12).

■**4.** Determine the convergence or divergence of the following integrals.

(a) $\int_0^{\to \infty} \frac{\sin x}{\sqrt{x}} \, dx$ **(b)** $\int_0^{\to \infty} \frac{\sin^2 x}{x} \, dx$

(c) $\int_2^{\to \infty} \frac{\sin x}{\log x} \, dx$ **(d)** $\int_1^{\to \infty} \frac{e^x \sin(e^x)}{x} \, dx$

5. Integrate by parts to obtain the identity

$$\int_0^w \frac{\sin^2 x}{x^2} \, dx = \frac{-\sin^2 w}{w} + \int_0^w \frac{2 \sin x \cos x}{x} \, dx.$$

Now show that each of the following integrals equals $\int_0^{\to\infty} (\sin x)/x \, dx$.

(a) $\int_0^{\to \infty} \frac{2 \sin x \cos x}{x} \, dx$ **(b)** $\int_0^{\to \infty} \frac{\sin^2 x}{x^2} \, dx$

(c) $\int_0^{\to \infty} \frac{2 \sin^2 x \cos^2 x}{x^2} \, dx$ **(d)** $\int_0^{\to \infty} \frac{2 \sin^4 x}{x^2} \, dx$

9.7 THE DOMINATED CONVERGENCE THEOREM

The dominated convergence theorem is a useful extension of the bounded convergence theorem to include improper integrals. Stated roughly for improper integrals of the type $\int_a^{\to b}$, the dominated convergence theorem tells us that if $f_n(x) \to f(x)$ for every point x in $[a, b)$, and if the functions f_n are "dominated" by some improper Riemann-integrable function g, then $\int_a^{\to b} f_n \to \int_a^{\to b} f$.

A precise statement of the theorem follows.

9.7.1 Dominated Convergence Theorem. Suppose that $-\infty < a < b \le \infty$, that (f_n) is a sequence of improper Riemann-integrable functions on $[a, b)$, and that (f_n) converges pointwise on $[a, b)$ to an improper Riemann-integrable function f. If there exists an improper Riemann-integrable function g on $[a, b)$ such that $|f_n| \le g$ for all n, then we have

$$\int_a^{\to b} f_n \to \int_a^{\to b} f.$$

■ *Proof.* From the comparison test (Theorem 9.4.1) we see that the integrals $\int_a^{\to b} f_n$ and $\int_a^{\to b} f$ converge absolutely. The key to the proof now lies in the fact that whenever $w \in [a, b)$, we have

$$\left| \int_a^{\to b} f_n - \int_a^{\to b} f \right| \le \int_a^{\to b} |f_n - f| = \int_a^w |f_n - f| + \int_w^{\to b} |f_n - f|$$

$$\le \int_a^w |f_n - f| + \int_w^{\to b} 2g.$$

Let $\varepsilon > 0$. Using the fact that $\int_a^{\to b} g$ converges, choose a number w in $[a, b)$ such that $\int_a^{\to b} g - \int_a^w g < \varepsilon/4$; in other words, $\int_w^{\to b} g < \varepsilon/4$. Since g is bounded on the interval $[a, w]$ and $|f_n| \le g$ for all n, we may apply the bounded convergence theorem to (f_n) on $[a, w]$ to choose a number N such that whenever $n \ge N$, we have $\int_a^w |f_n - f| < \varepsilon/2$. From the previous inequalities it follows that whenever $n \ge N$, we have

$$\left| \int_a^{\to b} f_n - \int_a^{\to b} f \right| \le \int_a^w |f_n - f| + \int_w^{\to b} 2g < \frac{\varepsilon}{2} + \frac{\varepsilon}{2} = \varepsilon. \qquad ■$$

9.7.2 Some Examples. The three examples given in this subsection illustrate the dominated convergence theorem.

(1) In this example we show that for improper integrals, if we want to transpose a limit and an integral, it is not good enough to assume that all the functions are bounded by some constant. For each natural number n, let f_n be the function defined on $[0, \infty)$ by $f_n(x) = 1/n$ if $0 \le x \le n$ and $f_n(x) = 0$ if $x > n$. (See Figure 9.2.) It is clear that $0 \le f_n \le 1$ for each n and that $f_n(x) \to 0$ for every point $x \in [0, \infty)$. But since $\int_0^{\to \infty} f_n = 1$ for every n, we do not have $\int_0^{\to \infty} f_n \to 0$.

(2) In this example we see that the dominated convergence theorem can sometimes be used even when the functions are unbounded. For each natural number n and for each $x \in (0, 1]$, we define $f_n(x) = e^{-nx}/\sqrt{x}$.

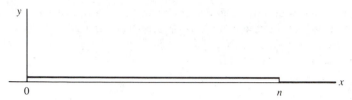

Figure 9.2

Since $f_n(x) \to 0$ for every $x \in (0, 1]$, we may use the dominated convergence theorem to conclude that $\int_{0\leftarrow}^{1} f_n \to 0$. This application of the theorem is valid even though the functions f_n are unbounded because $f_n(x) \leq 1/\sqrt{x}$ for all n and x; and, of course, the integral $\int_{0\leftarrow}^{1} 1/\sqrt{x} \, dx$ converges.

(3)* Using the dominated convergence theorem, we can deduce that whenever $\alpha > -1$, we have

$$\lim_{n\to\infty} \int_{1/n}^{n} \left(1 - \frac{x}{n}\right)^n x^\alpha \, dx = \int_{0}^{\to\infty} e^{-x} x^\alpha \, dx$$

To see this, we have to make use of Exercises 9.3(11) and 9.3(16)(d). For each n we define $f_n(x) = (1 - x/n)^n x^\alpha$ when $1/n \leq x \leq n$ and $f_n(x) = 0$ otherwise. Then for each $x \in (0, \infty)$ we have $f_n(x) \to e^{-x} x^\alpha$ as $n \to \infty$; and whenever $x \in (0, \infty)$ and $n \in \mathbf{Z}^+$, we have $|f_n(x)| \leq e^{-x} x^\alpha$. The previous assertion can now be proved by applying the dominated convergence theorem to each of the integrals $\int_0^1 f_n$ and $\int_1^{\to\infty} f_n$.

9.8 INTEGRALS THAT DEPEND ON A PARAMETER

9.8.1 Introduction. In this section we shall be looking at integrals of the type $\int_a^b f(x, y) \, dx$, where the function f is a *function of two variables*—in other words, when $f(x, y)$ is defined at each of a certain set of ordered pairs (x, y) of real numbers. As an example of an integral of this type, let us consider $\int_0^\pi e^{-xy} \sin x \, dx$. Given any number y, the function "of x" that sends each number x to the number $e^{-xy} \sin x$ is continuous on $\mathbf{R}$; and integrating by parts, one can show that

$$\int_0^\pi e^{-xy} \sin x \, dx = \frac{1 + e^{-\pi y}}{y^2 + 1}.$$

This allows us to define a function $\phi : \mathbf{R} \to \mathbf{R}$ by $\phi(y) = \int_0^\pi e^{-xy} \sin x \, dx$ for every $y \in \mathbf{R}$, and there are a number of interesting questions one might ask about functions ϕ which are defined in this way. We might ask, for example, whether

$$\phi'(y) = \int_0^\pi \frac{\partial}{\partial y} (e^{-xy} \sin x) \, dx = \int_0^\pi -xe^{-xy} \sin x \, dx,$$

and as a matter of fact, if you care to evaluate these expressions, you will see that this equation is true. There are times when it is extremely valuable to be able to differentiate a function of this type. Let us look, for example, at the function $\phi : R \to R$ defined by

$$\phi(y) = \int_0^{\to\infty} e^{-\alpha x} \frac{\sin xy}{x} \, dx \qquad \text{for every } y \in R,$$

where α is some given positive number. We deduce at once from Dirichlet's test (or even from the comparison test) that this integral converges for every y, and so this definition of the function ϕ makes sense. Now although the integral $\int_0^{\to\infty} e^{-\alpha x} [(\sin xy)/x] \, dx$ is not as easy to evaluate as the one we looked at earlier, differentiation with respect to y yields an integral that can be evaluated easily by parts. If we knew that $\phi'(y)$ could be found by writing

$$\phi'(y) = \int_0^{\to\infty} \frac{\partial}{\partial y} \left(e^{-\alpha x} \frac{\sin xy}{x} \right) dx = \int_0^{\to\infty} e^{-\alpha x} \cos xy \, dx,$$

then by evaluating the latter integral, we would obtain $\phi'(y) = \alpha/(\alpha^2 + y^2)$. From this it would follow that for some constant c we have $\phi(y) = c + \arctan(y/\alpha)$; and using the fact that $\phi(0) = 0$, we could deduce that $c = 0$. So if we could justify this method of finding ϕ', we would know that whenever $\alpha > 0$ and $y \in R$, we have

$$\int_0^{\to\infty} e^{-\alpha x} \frac{\sin xy}{x} \, dx = \arctan \frac{y}{\alpha}.$$

Putting $y = 1$ in this identity yields

$$\int_0^{\to\infty} e^{-\alpha x} \frac{\sin x}{x} \, dx = \arctan \frac{1}{\alpha} = \frac{\pi}{2} - \arctan \alpha;$$

and now letting $\alpha \to 0$ suggests the identity $\int_0^{\to\infty} (\sin x)/x \, dx = \pi/2$.

In order to place all this on a sound footing, we need a theory of integrals of the form $\int_a^b f(x, y) \, dx$. We begin this theory now.

9.8.2 Functions of Two Variables.

Suppose A and B are sets of real numbers. The **Cartesian product** $A \times B$ of A and B is defined to be the set of all ordered pairs (x, y) that can be formed by taking $x \in A$ and $y \in B$.

Now suppose f is a function with domain $A \times B$; in other words, f associates to each point (x, y) in $A \times B$ a number which is written as $f(x, y)$. Given any point $x \in A$, the function that sends each point $y \in B$ to the number $f(x, y)$ is a function whose domain is B; and we write this function as $f(x, \cdot)$. Similarly, for each point $y \in B$ the function that sends each point $x \in A$ to the number $f(x, y)$ is a function whose domain is A; and we write this function as $f(\cdot, y)$.

Given any point $(x, y) \in A \times B$, the derivative (if it exists) of the function $f(\cdot, y)$ at x is written as $D_1 f(x, y)$; and similarly, the derivative of $f(x, \cdot)$ at y is written as $D_2 f(x, y)$. These derivatives are, of course, the partial derivatives that are often written as $\partial f / \partial x$ and $\partial f / \partial y$ in elementary calculus.

Now suppose that the set A is an interval; for example, let us suppose that $A = [a, b)$. If for a given point $y \in B$ the function $f(\cdot, y)$ is improper Riemann-integrable on $[a, b)$, then we write the integral $\int_a^{\to b} f(\cdot, y)$ as $\int_a^{\to b} f(x, y) \, dx$.[2] If $f(\cdot, y)$ is improper Riemann-integrable on $[a, b)$ for every $y \in B$, then the function $\phi : B \to \mathbf{R}$ defined by $\phi(y) = \int_a^{\to b} f(x, y) \, dx$ for every $y \in B$ is written as $\int_a^{\to b} f(x, \cdot) \, dx$.

Similarly, if B is an interval—for example, if $B = [c, d)$—and if $f(x, \cdot)$ is improper Riemann-integrable on $[c, d)$ for every $x \in A$, then we give the obvious meaning to the function $\int_c^{\to d} f(\cdot, y) \, dy$.

Finally, if $A = [a, b)$ and $B = [c, d)$, and if the function $\int_a^{\to b} f(x, \cdot) \, dx$ is improper Riemann-integrable on $[c, d)$, then we write the integral of this function as $\int_c^{\to d} \int_a^{\to b} f(x, y) \, dx \, dy$. This is called a **repeated** or **iterated improper Riemann integral**.

The repeated integral $\int_a^{\to b} \int_c^{\to d} f(x, y) \, dy \, dx$ has a similar meaning.

9.8.3 Theorem: Continuity with Respect to a Parameter. Suppose $f : [a, b) \times S \to \mathbf{R}$, where $-\infty < a < b \le \infty$, and $S \subseteq \mathbf{R}$. Suppose that for every point $x \in [a, b)$ the function $f(x, \cdot)$ is continuous on S, and that for every point $y \in S$ the function $f(\cdot, y)$ is improper Riemann-integrable on $[a, b)$. Suppose, finally, that there exists an improper Riemann-integrable function g on $[a, b)$ such that $|f(x, y)| \le g(x)$ whenever $x \in [a, b)$ and $y \in S$. Then the function $\int_a^{\to b} f(x, \cdot) \, dx$ is continuous on S.

▪ **Proof.** Let $y \in S$. To show that $\int_a^{\to b} f(x, \cdot) \, dx$ is continuous at y, we shall use Theorem 6.3.2. Let (y_n) be a sequence in S converging to y. For each $x \in S$ the continuity of the function $f(x, \cdot)$ at y guarantees that $f(x, y_n) \to f(x, y)$, and we may therefore deduce from the dominated convergence theorem that $\int_a^{\to b} f(x, y_n) \, dx \to \int_a^{\to b} f(x, y) \, dx$. ▪

9.8.4 Example. For each number $y \ge 0$, define $\phi(y) = \int_{0 \leftarrow}^1 (x - y)/(x + y)^3 \, dx$. Then $\phi(0) = \infty$, and for all $y > 0$ we have $\phi(y) = -1/(1 + y)^2$. So in spite of the fact that the integrand is continuous in y for every point $x \in (0, 1]$, the function ϕ is not continuous at 0. Why doesn't this contradict Theorem 9.8.3?

9.8.5 Theorem: Differentiation with Respect to a Parameter. Suppose $f : [a, b) \times S \to \mathbf{R}$, where $-\infty < a < b \le \infty$ and S is an interval. Suppose that for every point $x \in [a, b)$ the function $f(x, \cdot)$ is differentiable on S, and that for every point $y \in S$ the functions $f(\cdot, y)$ and $D_2 f(\cdot, y)$ are improper Riemann-integrable on $[a, b)$. Suppose, finally, that there exists an improper Riemann-integrable function g on $[a, b)$ such that $|D_2 f(x, y)| \le g(x)$ whenever $x \in [a, b)$ and $y \in S$. Then the

---

[2] There is no reason why the integral $\int_a^{\to b} f(x, y) \, dx$ has to be improper. We have shown it as an improper integral to allow for the possibility that it *might* be.

function $\phi = \int_a^{\to b} f(x, \cdot)\, dx$ is differentiable on S, and for each point $y \in S$ we have $\phi'(y) = \int_a^{\to b} D_2 f(x, y)\, dx$.

■ **Proof.**   Let $y \in S$. To show that $\phi'(y) = \int_a^{\to b} D_2 f(x, y)\, dx$, we shall use Theorem 7.5.1. Let (y_n) be a sequence in $S \backslash \{y\}$ converging to y. For every point $x \in [a, b)$ we have

$$\frac{f(x, y_n) - f(x, y)}{y_n - y} \to D_2 f(x, y)$$

as $n \to \infty$, and the idea of the proof is to use the dominated convergence theorem to obtain

$$\frac{\phi(y_n) - \phi(y)}{y_n - y} = \int_a^{\to b} \frac{f(x, y_n) - f(x, y)}{y_n - y}\, dx \to \int_a^{\to b} D_2 f(x, y)\, dx.$$

All we have to demonstrate to make this argument valid is that whenever $x \in [a, b)$ and $n \in \mathbf{Z}^+$, we have

$$\left| \frac{f(x, y_n) - f(x, y)}{y_n - y} \right| \le g(x).$$

But given such x and n, we can use the first mean value theorem (Theorem 7.12.6) to choose a number t between y_n and y such that

$$\frac{f(x, y_n) - f(x, y)}{y_n - y} = D_2 f(x, t),$$

and this yields the inequality at once.   ■

9.8.6   A Return to the Integrals of Section 9.8.1.   With the help of Theorems 9.8.3 and 9.8.5, we now return to the integral

$$\phi(y) = \int_0^{\to \infty} e^{-\alpha x} \frac{\sin xy}{x}\, dx$$

which was discussed in Section 9.8.1. Suppose $\alpha > 0$, and define $f(x, y) = e^{-\alpha x} (\sin xy)/x$ for all $x > 0$ and all y. Then for all $x > 0$ and all y we have $D_2 f(x, y) = e^{-\alpha x} \cos xy$; and because of the inequality $|D_2 f(x, y)| \le e^{-\alpha x}$, we can conclude from Theorem 9.8.5 that the identity $\phi'(y) = \int_0^{\to \infty} e^{-\alpha x} \cos xy\, dx$ does indeed hold. This justifies the identities

$$\int_0^{\to \infty} e^{-\alpha x} \frac{\sin xy}{x}\, dx = \arctan \frac{y}{\alpha}$$

and

$$\int_0^{\to \infty} e^{-\alpha x} \frac{\sin x}{x}\, dx = \frac{\pi}{2} - \arctan \alpha$$

which were predicted in Section 9.8.1.

In Section 9.8.1 we also suggested that one might take the limit as $\alpha \to 0$ in the latter identity and obtain $\int_0^{\to \infty} (\sin x)/x \, dx = \pi/2$, but we do not yet have the tools to do this. One might have hoped that we could obtain this limit with the help of Theorem 9.8.3, using the fact that

$$\left| e^{-\alpha x} \frac{\sin xy}{x} \right| \leq \left| \frac{\sin x}{x} \right|,$$

but this technique does not work because (as we shall see in Section 10.6.8) the integral $\int_0^{\to \infty} |(\sin x)/x| \, dx$ diverges. The tools we need are provided by the theorem of Niels Abel that follows.

9.8.7 Abel's Theorem for Integrals.   Suppose a is a real number and that f is improper Riemann-integrable on $[a, \infty)$. For every number $\alpha > 0$, define $\phi(\alpha) = \int_a^{\to \infty} e^{-\alpha x} f(x) \, dx$. Then $\phi(\alpha) \to \int_a^{\to \infty} f(x) \, dx$ as $\alpha \to 0$.

Before proving Abel's theorem, we should note that if we had assumed that $|f|$ is improper Riemann-integrable on $[a, \infty)$, then the theorem would have followed at once from Theorem 9.8.3. The whole point of Abel's theorem is that the integral $\int_a^{\to \infty} f(x) \, dx$ might be conditionally convergent.

■ *Proof of Abel's Theorem.*   It follows from Dirichlet's test that $\phi(\alpha)$ exists for every $\alpha > 0$. Let $\varepsilon > 0$. Choose a number u such that whenever $v \geq u$, we have $|\int_a^{\to \infty} f - \int_a^v f| < \varepsilon/8$; in other words, $|\int_v^{\to \infty} f| < \varepsilon/8$. Then whenever $v \geq u$, we have

$$\left| \int_u^v f \right| = \left| \int_u^{\to \infty} f - \int_v^{\to \infty} f \right| \leq \left| \int_u^{\to \infty} f \right| + \left| \int_v^{\to \infty} f \right| < \frac{\varepsilon}{4}.$$

Now given $\alpha > 0$, since $e^{-\alpha u} < 1$, it follows from Theorem 9.5.5 that $|\int_u^{\to \infty} e^{-\alpha x} f(x) \, dx| < \varepsilon/4$. Therefore, for $\alpha > 0$

$$\left| \int_a^{\to \infty} e^{-\alpha x} f(x) \, dx - \int_a^{\to \infty} f \right|$$

$$\leq \left| \int_a^u e^{-\alpha x} f(x) \, dx - \int_a^u f \right| + \left| \int_u^{\to \infty} e^{-\alpha x} f(x) \, dx \right| + \left| \int_u^{\to \infty} f \right|$$

$$< \left| e^{-\alpha a} - 1 \right| \int_a^u |f| + \frac{\varepsilon}{2},$$

and we can make the latter expression less than ε by making α small enough. ■

Our final topic in this section concerns the equality of repeated Riemann integrals. We begin with a beautiful theorem of Fichtenholz that says roughly that as long as we are dealing with ordinary Riemann integrals (not improper integrals), nothing can go wrong.

9.8.8 Fichtenholz's Theorem on Inversion of Iterated Riemann Integrals: First Form of the Theorem.   If f is a bounded real function on the rectangle $[a, b] \times [c, d]$, then the identity

$$\int_a^b \int_c^d f(x, y) \; dy \; dx = \int_c^d \int_a^b f(x, y) \; dx \; dy$$

will hold as long as both sides exist as repeated Riemann integrals.

Second (and Slightly Stronger) Form of the Theorem. Suppose f is a bounded function on the rectangle $[a, b] \times [c, d]$. Suppose that for every point $x \in [a, b]$ the function $f(x, \cdot)$ is Riemann-integrable on $[c, d]$, and that for every point $y \in [c, d]$ the function $f(\cdot, y)$ is Riemann-integrable on $[a, b]$.

(1) The function $\phi : [a, b] \to \mathbf{R}$ defined by $\phi(x) = \int_c^d f(x, y) \; dy$ for all $x \in [a, b]$ is Riemann-integrable on $[a, b]$.

(2) The function $\psi : [c, d] \to \mathbf{R}$ defined by $\psi(y) = \int_a^b f(x, y) \; dx$ for all $y \in [c, d]$ is Riemann-integrable on $[c, d]$.

(3) $\int_a^b \phi(x) \; dx = \int_c^d \psi(y) \; dy$; in other words,

$$\int_a^b \int_c^d f(x, y) \; dy \; dx = \int_c^d \int_a^b f(x, y) \; dx \; dy.$$

We mention that there is a stronger version of the second form of Fichtenholz's theorem that permits some of the integrals to be the more general Lebesgue integrals which we are not studying here. Details may be found in Lewin [10]. Some abstract generalizations of Fichtenholz's theorem may be found in Luxemburg [11].

■ *Proof of the First Form.* Let ϕ and ψ be defined as in the second form of the theorem; and for each natural number n, denote as $\mathcal{P}_n$ the regular n-partition of $[c, d]$. For $i = 1, \ldots, n$ the ith point of $\mathcal{P}_n$ is, of course, $c + i(d - c)/n$, but for simplicity we shall denote this as y_{ni}. For each natural number n and each $x \in [a, b]$, define

$$\phi_n(x) = \sum_{i=1}^n f(x, y_{ni})(y_{ni} - y_{n\,i-1}).$$

Since the function $f(x, \cdot)$ is Riemann-integrable for every $x \in [a, b]$, and since $\|\mathcal{P}_n\| \to 0$, it follows from Darboux's theorem (Theorem 8.19.3) that $\phi_n(x) \to \phi(x)$ as $n \to \infty$ for each $x \in [a, b]$. Since f is bounded, it follows easily that $\phi_n \to \phi$ boundedly on $[a, b]$; and since we have also assumed that ϕ is Riemann-integrable on $[a, b]$, we may deduce from the bounded convergence theorem that

$$\int_a^b \phi_n(x) \; dx \to \int_a^b \phi(x) \; dx = \int_a^b \int_c^d f(x, y) \; dy \; dx \qquad \text{as} \qquad n \to \infty.$$

But for each n we have

$$\int_a^b \phi_n(x) \; dx = \int_a^b \sum_{i=1}^n f(x, y_{ni})(y_{ni} - y_{n\,i-1}) \; dx$$

$$= \sum_{i=1}^{n} \left[\int_a^b f(x, y_{ni}) \, dx \right] (y_{ni} - y_{n\,i-1})$$

$$= \sum_{i=1}^{n} \psi(y_{ni})(y_{ni} - y_{n\,i-1});$$

and since ψ is Riemann-integrable on $[c, d]$, we can use Darboux's theorem once again to show that the latter expression approaches $\int_c^d \psi(y) \, dy$ as $n \to \infty$. This shows that $\int_a^b \phi(x) \, dx = \int_c^d \psi(y) \, dy$, which is what we had to prove. ∎

■ Proof of the Second Form. The difference between this second form of the theorem and the first form is that the Riemann integrability of the functions ϕ and ψ is now part of the conclusion. What we have to show therefore is that ϕ and ψ are automatically Riemann-integrable on $[a, b]$ and $[c, d]$, respectively. As before let $\mathcal{P}_n$ be the regular n-partition of $[c, d]$ for each natural number n, and denote the ith point of $\mathcal{P}_n$ as y_{ni}. To show that ψ is Riemann-integrable on $[c, d]$, we shall use Theorem 8.19.4. We shall show that there is a number $\mathcal{I}$ such that for every possible choice of numbers t_{ni} in the intervals $[y_{n\,i-1}, y_{ni}]$, we have

$$\sum_{i=1}^{n} \psi(t_{ni})(y_{ni} - y_{n\,i-1}) \to \mathcal{I} \qquad \text{as} \qquad n \to \infty.$$

Let us look for the moment at any one possible choice of the numbers t_{ni}. For each natural number n and $x \in [a, b]$, define

$$\phi_n(x) = \sum_{i=1}^{n} f(x, t_{ni})(y_{ni} - y_{n\,i-1}).$$

Since the function $f(x, \cdot)$ is Riemann-integrable for every $x \in [a, b]$, and since $\|\mathcal{P}_n\| \to 0$, it follows from Darboux's theorem that $\phi_n(x) \to \phi(x)$ as $n \to \infty$ for each $x \in [a, b]$. It therefore follows from Theorem 8.22.1 that the sequence of integrals $\int_a^b \phi_n(x) \, dx$ converges. The limit of this sequence of integrals is obviously independent of the choice of numbers t_{ni}; for if t_{ni}^* is another choice, and the functions ϕ_n^* are defined analogously by

$$\phi_n^*(x) = \sum_{i=1}^{n} f(x, t_{ni}^*)(y_{ni} - y_{n\,i-1}) \qquad \text{for } x \in [a, b],$$

then we also have $\phi_n^*(x) \to \phi(x)$ for all $x \in [a, b]$, and the bounded convergence theorem implies that $\int_a^b [\phi_n(x) - \phi_n^*(x)] \, dx \to 0$. Now for each n we have

$$\int_a^b \phi_n(x) \, dx = \sum_{i=1}^{n} \psi(t_{ni})(y_{ni} - y_{n\,i-1});$$

and therefore, the latter expression tends to a limit as required. This shows that ψ is Riemann-integrable on $[c, d]$. The proof that ϕ is Riemann-integrable on $[a, b]$ is similar. ∎

9.8.9   The Failure of Fichtenholz's Theorem for Improper Integrals.   To see that the analogue of Fichtenholz's theorem for improper integrals is false, all we have to notice is that

$$\int_{0\leftarrow}^{1} \int_{0}^{1} \frac{x - y}{(x + y)^3} \, dx \, dy = -\frac{1}{2} \quad \text{and} \quad \int_{0\leftarrow}^{1} \int_{0}^{1} \frac{x - y}{(x + y)^3} \, dy \, dx = \frac{1}{2}.$$

Among the other well-known examples of repeated improper integrals which depend upon the order in which the integration is performed are the integrals

$$\int_{1}^{\to\infty} \int_{1}^{\to\infty} \frac{x - y}{(x + y)^3} \, dx \, dy \quad \text{and} \quad \int_{1}^{\to\infty} \int_{1}^{\to\infty} \frac{x^2 - y^2}{(x^2 + y^2)^2} \, dx \, dy.$$

Notice, however, that in all three of these integrals the integrand changes sign. In our final theorem we shall see that an analogue of Fichtenholz's theorem for improper integrals *can* be found when the integrand is nonnegative.

9.8.10   Fichtenholz's Theorem for Improper Integrals.   If f is a nonnegative real function on the rectangle $[a, b) \times [c, d)$, then the identity

$$\int_{a}^{\to b} \int_{c}^{\to d} f(x, y) \, dy \, dx = \int_{c}^{\to d} \int_{a}^{\to b} f(x, y) \, dx \, dy$$

will hold as long as both sides exist as repeated improper Riemann integrals.

Note that this statement includes the assertion that if either side of the identity is ∞, then so is the other.

■ *Proof.*   We shall prove the desired equality by showing that neither side can be smaller than the other. Let us assume, for example, that

$$\int_{a}^{\to b} \int_{c}^{\to d} f(x, y) \, dy \, dx < \int_{c}^{\to d} \int_{a}^{\to b} f(x, y) \, dx \, dy.$$

Using the fact that

$$\int_{c}^{\to d} \int_{a}^{\to b} f(x, y) \, dx \, dy = \lim_{v \to d} \int_{c}^{v} \int_{a}^{\to b} f(x, y) \, dx \, dy,$$

choose a point $v \in [c, d)$ such that

$$\int_{a}^{\to b} \int_{c}^{\to d} f(x, y) \, dy \, dx < \int_{c}^{v} \int_{a}^{\to b} f(x, y) \, dx \, dy.$$

Choose a sequence (u_n) in $[a, b)$ such that $u_n \to b$. Define

$$\phi(y) = \int_{a}^{\to b} f(x, y) \, dx \quad \text{and} \quad \phi_n(y) = \int_{a}^{u_n} f(x, y) \, dx$$

for all $y \in [c, v]$ and $n \in \mathbf{Z}^+$. Then since $\phi_n \leq \phi$ for every n, and $\phi_n(y) \to \phi(y)$ for every point $y \in [c, v]$, and since the function ϕ is Riemann-integrable on $[c, v]$, it follows from the bounded convergence theorem that

$$\int_c^v \phi_n(y) \, dy \to \int_c^v \phi(y) \, dy = \int_c^v \int_a^{\to b} f(x, y) \, dx \, dy.$$

Therefore, for n sufficiently large,

$$\int_a^{\to b} \int_c^{\to d} f(x, y) \, dy \, dx < \int_c^v \phi_n(y) \, dy.$$

We may therefore choose a point $u \in [a, b)$ such that

$$\int_a^{\to b} \int_c^{\to d} f(x, y) \, dy \, dx < \int_c^v \int_a^u f(x, y) \, dx \, dy.$$

But using Fichtenholz's theorem for ordinary Riemann integrals, we obtain

$$\int_c^v \int_a^u f(x, y) \, dx \, dy = \int_a^u \int_c^v f(x, y) \, dy \, dx$$

$$\leq \int_a^u \int_c^{\to d} f(x, y) \, dy \, dx \leq \int_a^{\to b} \int_c^{\to d} f(x, y) \, dy \, dx$$

and this contradicts our choice of u and v. ∎

9.9 EXERCISES

1. Given

$$f(x) = \int_0^x \exp(-t^2) \, dt,$$

$$g(x) = \int_0^1 \frac{\exp[-x^2(t^2 + 1)]}{t^2 + 1} \, dt,$$

and

$$h(x) = [f(x)]^2 + g(x)$$

for all $x \in R$, prove that the function h must be constant. What is the value of the constant? Use this to show that $\int_0^{\to\infty} \exp(-x^2) \, dx = \sqrt{\pi}/2$.

2. Given $\phi(y) = \int_0^{\to\infty} \exp(-x^2) \cos 2xy \, dx$ and $f(y) = \exp(y^2) \phi(y)$ for all $y \in R$, prove that the function f must be constant. What is the value of this constant? *Hint:* Use exercise 1.

■ **3.** Find an explicit formula for the integral $\int_0^{\to\infty} \exp(-x^2) \cos 2xy \, dx$.

■ **4.** Evaluate the integral $\int_0^{\to\infty} \int_0^{\to\infty} \exp(-x^2) \cos 2xy \, dx$. What happens in this integral if we invert the order of integration?

■ **5.** Apply Fichtenholz's theorem to the integral $\int_0^{\to\infty} \int_0^{\to\infty} f(x, y) \, dx \, dy$, where we define $f(x, y) = \exp(-y^3)$ when $x < y^2$ and $f(x, y) = 0$ otherwise. Now evaluate the integral $\int_0^{\to\infty} \int_{\sqrt{x}}^{\to\infty} \exp(-y^3) \, dy \, dx$.

6. Express $1/(1 + t^2x^2)(1 + t^2y^2)$ in partial fractions; and show that if

$$f(x, y) = \int_0^{\to\infty} \frac{1}{(1 + t^2x^2)(1 + t^2y^2)} \, dt, \quad \text{then} \quad f(x, y) = \frac{\pi}{2(x + y)}.$$

Now apply Fichtenholz's theorem (more than once) to the integral $\int_{0\leftarrow}^1 \int_0^1 f(x, y) \, dx \, dy$ and deduce that

$$\int_0^{\to\infty} \frac{(\arctan x)^2}{x^2} \, dx = \pi \log 2.$$

■ 7. Use the results obtained in exercise 6 and evaluate the integrals

$$\int_0^{\pi/2} \frac{x^2}{\sin^2 x} \, dx, \quad \int_0^{\pi/2} x \cot x \, dx, \quad \text{and} \quad \int_{0\leftarrow}^{\pi/2} \log \sin x \, dx.$$

8. Prove that if f is a function on the rectangle $[a, b) \times [c, d)$, then the identity

$$\int_a^{\to b} \int_c^{\to d} f(x, y) \, dy \, dx = \int_c^{\to d} \int_a^{\to b} f(x, y) \, dx \, dy$$

will hold as long as both sides exist as repeated improper Riemann integrals and the integrals on the left side of the identity converge absolutely. *Hint:* Apply Theorem 9.8.10 to the functions $|f|$ and $|f| + f$.

9. Given that f is improper Riemann-integrable on $[0, \infty)$, that $a \geq 0$, and that $g(u) = f(u - a)$ for all $u \geq a$, prove that g is improper Riemann-integrable on $[a, \infty)$ and that $\int_0^{\to\infty} f(x) \, dx = \int_a^{\to\infty} g(u) \, du$. *Hint:* Use Exercise 8.18(4).

10. In this exercise f and g are nonnegative, improper Riemann-integrable functions on $[0, \infty)$; given any point (x, y) in $[0, \infty) \times [0, \infty)$, we define $h(x, y) = f(x - y)g(y)$ when $y \leq x$ and $h(x, y) = 0$ when $y > x$.
(a) Prove that

$$\int_0^{\to\infty} \int_0^{\to\infty} h(x, y) \, dx \, dy = \int_0^{\to\infty} \int_y^{\to\infty} f(x - y)g(y) \, dx \, dy$$

$$= \left(\int_0^{\to\infty} f \right) \left(\int_0^{\to\infty} g \right)$$

Hint: Use exercise 9.
(b) Apply Fichtenholz's theorem for improper integrals to the integral $\int_0^{\to\infty} \int_0^{\to\infty} h(x, y) \, dx \, dy$, and deduce that this integral equals $\int_0^{\to\infty} \int_0^x f(x - y)g(y) \, dy \, dx$.

■ 11. The purpose of this exercise is to explore some of the properties of the gamma and beta functions that were defined in Exercises 9.6(2) and 9.6(3). Suppose that α and β are positive numbers.
(a) Apply exercise 10 to the functions f and g defined by $f(x) = x^{\alpha-1}e^{-x}$ and $g(x) = x^{\beta-1}e^{-x}$ for all $x > 0$, and deduce that

$$\Gamma(\alpha)\Gamma(\beta) = \int_0^{\to\infty} \int_0^x e^{-x}(x - y)^{\alpha-1}y^{\beta-1} \, dy \, dx.$$

(b) By making the substitution $y = ux$ in the integral $\int_0^x (x - y)^{\alpha-1} y^{\beta-1} \, dy$, deduce that $\Gamma(\alpha)\Gamma(\beta) = \Gamma(\alpha + \beta)B(\alpha, \beta)$.

(c) Using integration by parts, show that $\Gamma(\alpha + 1) = \alpha\Gamma(\alpha)$. Deduce that if α is any natural number, then $\Gamma(\alpha) = (\alpha - 1)!$.

(d) By making the substitution $t = \sin^2 \theta$ in the definition of the beta function, show that $B(\alpha, \beta) = 2 \int_0^{\pi/2} \sin^{2\alpha-1} \theta \cos^{2\beta-1} \theta \, d\theta$.

(e) Use part (d) to evaluate $B(\tfrac{1}{2}, \tfrac{1}{2})$, and then using part (b), show that $\Gamma(\tfrac{1}{2}) = \sqrt{\pi}$.

(f) By making the substitution $x = u^2$ in the definition of the gamma function, show that $\Gamma(\alpha) = 2 \int_{0\leftarrow}^{\to\infty} u^{2\alpha-1} \exp(-u^2) \, du$. Now using part (e), find another way to evaluate the integral in exercise 1.

(g) Given $p > -1$, show that

$$\int_0^\pi \sin^p \theta \, d\theta = 2 \int_0^{\pi/2} \sin^p \theta \, d\theta = \frac{\sqrt{\pi}\Gamma[(p + 1)/2]}{\Gamma(p/2 + 1)}.$$

Hint: Use Exercise 8.18(1).

■ **(h)** Prove that $B(\alpha, \alpha) = 2^{1-2\alpha}B(\alpha, \tfrac{1}{2})$, and deduce that

$$\sqrt{\pi}\, \Gamma(2\alpha) = 2^{2\alpha-1}\Gamma(\alpha)\Gamma(\alpha + \tfrac{1}{2}).$$

Hint: Use Exercise 8.18(2).

■ **(i)** Prove that $\int_0^{\to\pi/2} \sqrt{\tan x} \, dx = \pi/\sqrt{2}$. *Hint:* Use parts (d) and (h).

*12. The purpose of this exercise is to encourage you to read a proof of an interesting result known as **Stirling's formula**, which states that

$$\frac{\Gamma(x + 1)}{x^x e^{-x}\sqrt{x}} \to \sqrt{2\pi} \quad \text{as} \quad x \to \infty.$$

Note that if n is a large natural number, then Stirling's formula suggests that an approximate value for $n!$ is $\sqrt{2\pi n}\, n^n e^{-n}$. Although there are many proofs of this theorem in the literature, you might like to look at Section 8.22 on page 194 of Rudin [13], where the proof is sketched in the form of a sequence of exercises. In the proof as it appears in [13], you are asked to do five simple exercises which appear on page 195 marked (a), (b), (c), (d), and (e). You should *omit* (b) and do the other four. You should also ignore the reference made in [13] to "Exercise 12 of Chapter 7." Instead of using that exercise, you should complete your proof of Stirling's formula by using the dominated convergence theorem (Theorem 9.7.1).

Chapter *10*

Infinite Series

10.1 INTRODUCTION

In this chapter we shall study the important type of sequence which is formed when the terms of a given sequence are added together. If (a_n) is a given sequence, then the **infinite series** $\Sigma\ a_n$, written alternatively as $\Sigma_n\ a_n$, is defined to be the sequence whose value at each natural number n is $\Sigma_{j=1}^{n}\ a_j$. To understand this definition, let us look at some examples.

(1) Let $a_n = 1$ for every $n \in \mathbf{Z}^+$. Then the value of $\Sigma\ a_n$ at each natural number n is $\Sigma_{j=1}^{n} 1 = n$.

(2) Let $a_n = (-1)^n$ for every $n \in \mathbf{Z}^+$. Then the value of $\Sigma\ a_n$ at each n is -1 if n is odd and is 0 if n is even.

(3) Let $a_n = \log(1 + 1/n)$ for every $n \in \mathbf{Z}^+$. Then the value of $\Sigma\ a_n$ at each n is

$$\sum_{j=1}^{n} \log\left(1 + \frac{1}{j}\right) = \sum_{j=1}^{n} [\log(j + 1) - \log j]$$

$$= (\log 2 - \log 1) + (\log 3 - \log 2) + \cdots$$

$$+ [\log n - \log(n - 1)] + [\log(n + 1) - \log n]$$

$$= \log(n + 1).$$

(4) Let $a_n = 1/n(n + 1)$ for every $n \in \mathbf{Z}^+$. Then the value of Σa_n at each n is

$$\sum_{j=1}^{n} \left[\frac{1}{j(j + 1)} \right] = \sum_{j=1}^{n} \left(\frac{1}{j} - \frac{1}{j + 1} \right) = 1 - \frac{1}{n + 1}.$$

(5) Suppose x is any number unequal to 1; and for each $n \in \mathbf{Z}^+$, define $a_n = x^{n-1}$. Then it follows from the algebraic identity

$$(1 - x)(1 + x + x^2 + \cdots + x^{n-1}) = 1 - x^n$$

that the value of Σa_n at each n is $(1 - x^n)/(1 - x)$.

In the preceding five examples we were able to give a simple formula for the value of Σa_n at each n, but most of the time, this isn't possible. For example:

(6) Let $a_n = 1/n^2$ for every $n \in \mathbf{Z}^+$. Then the value of Σa_n at each natural number n is

$$\sum_{j=1}^{n} \frac{1}{j^2} = \frac{1}{1^2} + \frac{1}{2^2} + \frac{1}{3^2} + \frac{1}{4^2} + \cdots + \frac{1}{n^2},$$

but there is no known elementary formula for this expression.

Given a sequence (a_n), the series Σa_n is called the *series with nth term* a_n, and the value of Σa_n at a given natural number n is called the **nth partial sum** of the series Σa_n. In other words, the nth partial sum of the series Σa_n is defined to be the number $\Sigma_{j=1}^{n} a_j$. In the event that the sequence (a_n) starts at an integer N, instead of starting at the number 1, then the series Σa_n is sometimes denoted as $\Sigma_{n \geq N} a_n$; and in this case the nth partial sum of the series is defined to be $\Sigma_{j=N}^{n} a_j$ for every integer $n \geq N$.

Now rephrasing the preceding six examples in the language of partial sums, we have the following:

(1) The nth partial sum of the series $\Sigma 1$ is n.

(2) The nth partial sum of the series $\Sigma(-1)^n$ is -1 if n is odd and is 0 if n is even.

(3) The nth partial sum of the series $\Sigma \log(1 + 1/n)$ is $\log(n + 1)$.

(4) The nth partial sum of the series

$$\sum \frac{1}{n(n + 1)} \quad \text{is} \quad 1 - \frac{1}{n + 1}.$$

(5) If $x \neq 1$, then the nth partial sum of the series Σx^{n-1} is $(1 - x^n)/(1 - x)$. As you may know, this series is called a **geometric series.**

(6) The nth partial sum of the series $\Sigma 1/n^2$ is

$$\sum_{j=1}^{n} \frac{1}{j^2} = \frac{1}{1^2} + \frac{1}{2^2} + \frac{1}{3^2} + \frac{1}{4^2} + \cdots + \frac{1}{n^2}.$$

10.2 CONVERGENT AND DIVERGENT SERIES

As we have said, if (a_n) is a given sequence, then $\Sigma\, a_n$ means the sequence whose value at each natural number n is $\Sigma_{j=1}^{n}\, a_j$. Therefore, $\Sigma\, a_n$ is **convergent** when there exists a real number α such that $\Sigma_{j=1}^{n}\, a_j \to \alpha$ as $n \to \infty$. Just as we did with sequences and improper integrals, we use the word *divergence* to indicate the absence of convergence. A series is said to be **divergent** when it is not convergent. This means that there are two ways in which a given series $\Sigma\, a_n$ might diverge: Either $\Sigma_{j=1}^{n}\, a_j \to \pm\infty$ as $n \to \infty$, or $\Sigma_{j=1}^{n}\, a_j$ does not tend to a limit at all, even in $[-\infty, \infty]$. If $\Sigma_{j=1}^{n}\, a_j$ does tend to a limit, then this limit is called the *sum* of the series $\Sigma\, a_n$ and is written as $\Sigma_{j=1}^{\infty}\, a_j$; and since the symbol j is clearly unimportant here, we could just as well write the sum as $\Sigma_{n=1}^{\infty}\, a_n$.

Let us return once again to the examples of Section 10.1.

(1) $\Sigma_{j=1}^{n} 1 = n \to \infty$ as $n \to \infty$. Therefore, $\Sigma_{n=1}^{\infty} 1 = \infty$, and the series $\Sigma\, 1$ diverges.

(2) The nth partial sum of the series $\Sigma(-1)^n$ does not tend to a limit as $n \to \infty$. Therefore, the sum of the series $\Sigma(-1)^n$ is undefined, and the series diverges.

(3) The nth partial sum of the series $\Sigma \log(1 + 1/n)$ is $\log(n + 1)$, which tends to infinity as $n \to \infty$. Therefore, $\Sigma_{n=1}^{\infty} \log(1 + 1/n) = \infty$, and the series $\Sigma \log(1 + 1/n)$ diverges.

(4) $\Sigma_{j=1}^{n} 1/j(j + 1) = 1 - 1/(n + 1) \to 1$ as $n \to \infty$. Therefore, the series $\Sigma\, 1/n(n + 1)$ converges, and $\Sigma_{n=1}^{\infty} 1/n(n + 1) = 1$.

(5) Suppose $|x| < 1$. From Example 4.10.2(2), we know that $x^n \to 0$ as $n \to \infty$. Therefore,

$$\sum_{j=1}^{n} x^{j-1} = \frac{1 - x^n}{1 - x} \to \frac{1}{1 - x} \qquad \text{as} \qquad n \to \infty.$$

We have therefore proved that when $|x| < 1$, the geometric series $\Sigma\, x^{n-1}$ converges and that its sum is $1/(1 - x)$.

(6) Since we are not in possession of a simple formula for $\Sigma_{j=1}^{n} 1/j^2$, it is not at all obvious what happens to this expression as $n \to \infty$. Certainly, if $s_n = \Sigma_{j=1}^{n} 1/j^2$ for each n, then the sequence (s_n) is increasing; and the question of whether $\Sigma\, 1/n^2$ is convergent or divergent therefore reduces to the question of whether or not the sequence (s_n) is bounded. Later in this chapter, when we know a little more about series, we shall prove easily that the series $\Sigma\, 1/n^2$ *does* converge, and in Chapter 11 we shall show that the sum of this series is $\pi^2/6$.

Our next example gives another series whose nth partial sum cannot be expressed by a simple formula; but this time, we shall be able to see that the series converges, and we shall find its sum.

(7) For every $x \in [0, 1)$ and every natural number n, we define $f_n(x)$ to be the nth partial sum of the series $\Sigma(-x)^{n-1}$; in other words, we define

$$f_n(x) = 1 - x + x^2 - \cdots + (-x)^{n-1} = \frac{1 - (-x)^n}{1 + x};$$

and for each n we define $f_n(1) = \frac{1}{2}$. For each n the function f_n is continuous at every point of $[0, 1]$, except at 1; and therefore, f_n is Riemann-integrable on $[0, 1]$. Since $0 \leq f_n \leq 1$ for each n, and since $f_n(x) \to 1/(1 + x)$ as $n \to \infty$ for every $x \in [0, 1]$, it follows from the bounded convergence theorem that

$$\int_0^1 f_n \to \int_0^1 \frac{1}{1 + x} \, dx = \log 2.$$

But for every n we have

$$\int_0^1 f_n = \int_0^1 [1 - x + x^2 - \cdots + (-x)^{n-1}] \, dx$$

$$= 1 - \frac{1}{2} + \frac{1}{3} - \cdots + \frac{(-1)^{n-1}}{n} = \sum_{j=1}^n \frac{(-1)^{j-1}}{j},$$

and it follows that $\sum_{j=1}^n (-1)^{j-1}/j \to \log 2$ as $n \to \infty$. We have therefore shown that the series $\Sigma(-1)^{n-1}/n$ converges, and that

$$\sum_{n=1}^{\infty} \frac{(-1)^{n-1}}{n} = \log 2.$$

(8) In our final example of this section we return to Example 4.10.2(4), where we looked at decimals. What we saw in that example was that if (a_n) is any sequence in the set $\{0, 1, 2, 3, 4, 5, 6, 7, 8, 9\}$, and if for each n, x_n is the nth partial sum of the series $\Sigma a_n/10^n$, in other words, if

$$x_n = \frac{a_1}{10^1} + \frac{a_2}{10^2} + \frac{a_3}{10^3} + \cdots + \frac{a_n}{10^n},$$

then (x_n) is convergent, and its limit lies in $[0, 1]$. This means that every decimal of the form $\Sigma a_n/10^n$ converges to a number in $[0, 1]$.

10.3 EXERCISES

1. Find the nth partial sum of the series $\Sigma(n - 3)/n(n + 1)(n + 3)$. Deduce that this series is convergent and find its sum. *Hint:* Use partial fractions.

2. Given $x \neq 1$, find the nth partial sum of the series Σnx^{n-1}. *Hint:* The required expression is the derivative of the nth partial sum of the series Σx^n.

3. Given $|x| < 1$, prove that the series Σnx^{n-1} is convergent, and find its sum.

4. For every $x \in [0, 1)$ and every natural number n, we define $f_n(x)$ to be the nth partial sum of the series $\sum_{n \geq 0}(-x^2)^{n-1}$; and for each n we define $f_n(1) = \frac{1}{2}$.
(a) Prove that for $n \in \mathbf{Z}^+$ and $x \in [0, 1)$ we have

$$f_n(x) = \frac{1 - (-x^2)^n}{1 + x^2};$$

and deduce that for each n we have $0 \leq f_n \leq 1$. Prove that for every point $x \in [0, 1]$ we have $f_n(x) \to 1/(1 + x^2)$ as $n \to \infty$.
(b) Say why each function f_n is Riemann-integrable on $[0, 1]$, and show that

$$\int_0^1 f_n = \sum_{j=1}^n \frac{(-1)^{j-1}}{2j - 1}.$$

(c) Applying the bounded convergence theorem to (f_n), deduce that the series $\sum(-1)^{n-1}/(2n - 1)$ is convergent and that

$$\sum_{n=1}^\infty \frac{(-1)^{n-1}}{2n - 1} = \frac{\pi}{4}.$$

5. Apply the technique of exercise 4 to the sequence (f_n), where for each natural number n and each $x \in [0, \frac{1}{2}]$, $f_n(x)$ is defined to be the nth partial sum of the series $\sum x^{n-1}$. Deduce that

$$\sum_{n=1}^\infty \frac{1}{n \, 2^n} = \log 2.$$

6. By analogy with Example 10.2(8), prove that if (a_n) is a sequence in the set $\{0, 1, 2\}$, then the "ternary decimal" $\sum_{n=1}^\infty a_n/3^n$ converges to a number in the interval $[0, 1]$.

7. The **Cantor set** C is defined to be the set of all those numbers that can be expressed as a ternary decimal $\sum_{n=1}^\infty a_n/3^n$, where (a_n) is any sequence in $\{0, 2\}$. For each of the following choices of the sequence (a_n), evaluate $\sum_{n=1}^\infty a_n/3^n$.
(a) $a_n = 0$ for every n.
(b) $a_n = 2$ for every n.
(c) $a_1 = 2$ and $a_n = 0$ for all $n \geq 2$.
(d) $a_1 = 0$ and $a_n = 2$ for all $n \geq 2$.
(e) $a_n = 0$ when n is odd and $a_n = 2$ when n is even.
(f) Show that no point of C lies in the interval $(\frac{1}{3}, \frac{2}{3})$.

10.4 SERIES REPRESENTATION OF THE EXPONENTIAL AND LOGARITHMIC FUNCTIONS

10.4.1 Series Representation of the Exponential Function. In this section we shall show that for every number x we have

$$e^x = \sum_{n=0}^{\infty} \frac{x^n}{n!}.$$

Suppose that x is any real number, and let n be any natural number. We are going to apply the Taylor mean value theorem (Theorem 7.13.6) to the exponential function exp on the interval I that lies between the numbers 0 and x:

Since $\exp^{(j)}(0) = 1$ for each j, the Taylor mean value theorem implies that for some number $c \in I$ we have

$$e^x = 1 + \frac{x^1}{1!} + \frac{x^2}{2!} + \frac{x^3}{3!} + \cdots + \frac{x^n}{n!} + \frac{x^{n+1}}{(n+1)!} e^c.$$

If $x < 0$, we have $c < 0$ and $e^c < 1$; and if $x > 0$, then $e^c < e^x$. So in both cases, $e^c < 1 + e^x$; and it follows that

$$\left| e^x - \sum_{j=0}^{n} \frac{x^j}{j!} \right| = \left| \frac{x^{n+1}}{(n+1)!} e^c \right| \le \frac{|x|^{n+1}}{(n+1)!} (1 + e^x).$$

So in order to prove that $\sum_{j=0}^{n} x^j/j! \to e^x$ as $n \to \infty$, all we have to show is that $|x|^{n+1}/(n+1)! \to 0$ as $n \to \infty$.

For each n, write $a_n = |x|^n/n!$. To show that $a_n \to 0$ as $n \to \infty$, choose a natural number N such that $N > |x|$. Then whenever $n \ge N$, we have

$$a_{n+1} = \frac{|x|^{n+1}}{(n+1)!} = \frac{|x|^{n+1}}{(n+1)n!} = \frac{|x|}{n+1} a_n \le a_n;$$

and so starting at $n = N$, the sequence (a_n) is decreasing. Define $\alpha = \lim_{n \to \infty} a_n$. Letting $n \to \infty$ in the identity $a_{n+1} = [|x|/(n+1)] a_n$ yields $\alpha = 0\alpha$, from which we deduce that $\alpha = 0$. This completes the proof. An alternative way of showing that $a_n \to 0$ will be provided in Example 10.8.5(2).

10.4.2 Series Representation of log(1 + x).

In this section we shall show that whenever $-1 < x \le 1$, we have

$$\log(1 + x) = \sum_{n=1}^{\infty} \frac{(-1)^{n-1} x^n}{n}.$$

We established this identity for the special case $x = 1$ in Example 10.2(7), and as a matter of fact, all we are going to do now is repeat the reasoning of that example for a general point x in the interval $(-1, 1]$. Let $x \in (-1, 1]$. For every $t \in (-1, 1)$ and every natural number n, we define $f_n(t)$ to be the nth partial sum of the series $\Sigma(-t)^{n-1}$; in other words, we define

$$f_n(t) = 1 - t + t^2 - \cdots + (-t)^{n-1} = \frac{1 - (-t)^n}{1 + t},$$

and for each n we define $f_n(1) = \frac{1}{2}$.

For each n the function f_n is continuous at every point of $(-1, 1]$, except at 1; and therefore if I is the closed interval whose endpoints are 0 and x, each f_n is Riemann-integrable on I:

It is easy to see that if $x \geq 0$, then $0 \leq f_n(t) \leq 1$ for each n and every $t \in I$. In the event that $x < 0$, it follows easily that $0 < f_n(t) \leq 1/(1 + x)$ for each n and every $t \in I$. Therefore, since $f_n(t) \to 1/(1 + t)$ as $n \to \infty$ for every $t \in I$, it follows from the bounded convergence theorem that

$$\int_0^x f_n \to \int_0^x \frac{1}{1 + t} \, dt = \log(1 + x).$$

But for every n we have

$$\int_0^x f_n = \int_0^x [1 - t + t^2 - \cdots + (-t)^{n-1}] \, dt = \sum_{j=1}^n \frac{(-1)^{j-1} x^j}{j},$$

and it follows that

$$\sum_{j=1}^n \frac{(-1)^{j-1} x^j}{j} \to \log(1 + x) \qquad \text{as} \qquad n \to \infty,$$

which is what we set out to prove.

10.5 ELEMENTARY PROPERTIES OF SERIES

We begin by showing that series have the same sort of properties of linearity, nonnegativity, and additivity that we saw in Chapter 8 for integrals.

10.5.1 Theorem: Linearity of Series. Suppose that (a_n) and (b_n) are given sequences, that c is a real number, and that both of the series $\Sigma \, a_n$ and $\Sigma \, b_n$ converge.

(1) The series $\Sigma(a_n + b_n)$ also converges, and we have

$$\sum_{n=1}^\infty (a_n + b_n) = \sum_{n=1}^\infty a_n + \sum_{n=1}^\infty b_n.$$

(2) The series $\Sigma \, ca_n$ converges, and we have $\sum_{n=1}^\infty ca_n = c \sum_{n=1}^\infty a_n$.

■ **Proof.** Part (1) can be proved by noting that for every n we have

$$\sum_{j=1}^{n} (a_j + b_j) = \sum_{j=1}^{n} a_j + \sum_{j=1}^{n} b_j$$

and then letting $n \to \infty$. The proof of (2) is similar. ■

10.5.2 Theorem: Nonnegativity of Series. Suppose that $a_n \leq b_n$ for every natural number n and that the two series $\Sigma\, a_n$ and $\Sigma\, b_n$ converge. Then $\Sigma_{n=1}^{\infty} a_n \leq \Sigma_{n=1}^{\infty} b_n$.

■ **Proof.** For every n we have $\Sigma_{j=1}^{n} a_j \leq \Sigma_{j=1}^{n} b_j$. Therefore, the result follows at once by letting $n \to \infty$. ■

10.5.3 Theorem: Additivity of Series. Given any sequence (a_n) and any natural number N, the convergence of $\Sigma\, a_n$ is independent of whether we start the series at $n = 1$ or at $n = N$; and in the event that the series converges, we have

$$\sum_{n=1}^{\infty} a_n = \sum_{n=1}^{N} a_n + \sum_{n=N+1}^{\infty} a_n.$$

■ **Proof.** Whenever $n > N$, we have

$$\sum_{j=1}^{n} a_j = \sum_{j=1}^{N} a_j + \sum_{j=N+1}^{n} a_j,$$

and the result follows at once by letting $n \to \infty$. ■

10.5.4 Theorem: A Necessary Condition for Convergence of a Series. If $\Sigma\, a_n$ converges, then $a_n \to 0$ as $n \to \infty$.

■ **Proof.** Suppose that $\Sigma\, a_n$ converges, and put $\alpha = \Sigma_{n=1}^{\infty} a_n$. This means that $\Sigma_{j=1}^{n} a_j \to \alpha$ as $n \to \infty$, and we also have $\Sigma_{j=1}^{n-1} a_j \to \alpha$ as $n \to \infty$. Therefore,

$$a_n = \sum_{j=1}^{n} a_j - \sum_{j=1}^{n-1} a_j \to \alpha - \alpha = 0 \qquad \text{as} \qquad n \to \infty.$$ ■

10.5.5 Failure of the Converse of Theorem 10.5.4. Theorem 10.5.4 tells us that *if* a given series $\Sigma\, a_n$ converges, *then* $a_n \to 0$. It is not true, however, that if $a_n \to 0$, then the series $\Sigma\, a_n$ must converge. We have seen, for example, that the series $\Sigma \log(1 + 1/n)$ diverges, even though $\log(1 + 1/n) \to 0$ as $n \to \infty$.

Roughly speaking, a series $\Sigma\, a_n$ will converge if and only if a_n tends to zero *quickly enough*, but it is no easy task to say just what these words mean. Much of this chapter will be devoted to answering the question of precisely what we should mean when we say "quickly enough."

10.6   THE COMPARISON PRINCIPLE FOR SERIES WHOSE TERMS ARE NONNEGATIVE

10.6.1   Introduction.   As we have seen, there are two ways in which a given series $\Sigma\, a_n$ might diverge: either $\Sigma_{j=1}^{n}\, a_j \to \pm\infty$ as $n \to \infty$, or $\Sigma_{j=1}^{n}\, a_j$ might not tend to a limit at all. If, however, we restrict our attention to series $\Sigma\, a_n$ for which $a_n \geq 0$ for every n, then the situation is considerably simpler. In this case, if we define $A_n = \Sigma_{j=1}^{n}\, a_j$ for every n, then the sequence (A_n) is *increasing*; and therefore, $\Sigma\, a_n$ will be convergent if and only if the sequence (A_n) of its nth partial sums is bounded. So if $a_n \geq 0$ for each n, then in order to show that the series $\Sigma\, a_n$ is convergent, it would be sufficient to find a bounded sequence (B_n) such that $A_n \leq B_n$ for all n. This, in a nutshell, is the comparison principle, which is one of the most important tools for determining the convergence or divergence of a given series. As you may have noticed, this principle is almost a carbon copy of the comparison principle we saw for integrals in Theorem 9.1.7 and Section 9.4.

Our first application of the comparison principle is the integral test.

10.6.2   The Integral Test.   Suppose f is a nonnegative decreasing function on $[1, \infty)$. Then the series $\Sigma\, f(n)$ converges if and only if the improper integral $\int_{1}^{\to\infty} f$ converges.

■ **Proof.**   We remark first that since f is a monotone function, its Riemann integrability on $[1, n]$ for every natural number n follows from Theorem 8.16.1. We now define $A_n = \Sigma_{j=1}^{n}\, f(j)$ and $B_n = \int_{1}^{n} f$ for each n; and to prove the theorem, we shall show that (A_n) is bounded if and only if (B_n) is bounded. See Figure 10.1.

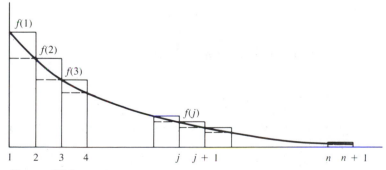

Figure 10.1

Now for every natural number j and every number $t \in [j, j + 1]$, we have $f(j + 1) \leq f(t) \leq f(j)$, and it follows that

$$\int_j^{j+1} f(j + 1) \leq \int_j^{j+1} f \leq \int_j^{j+1} f(j);$$

in other words,

$$f(j + 1) \leq \int_j^{j+1} f \leq f(j).$$

Summing from 1 to n, we obtain

$$\sum_{j=1}^{n} f(j + 1) \leq \int_1^{n+1} f \leq \sum_{j=1}^{n} f(j);$$

in other words, $A_{n+1} - f(1) \leq B_{n+1} \leq A_n$ for every n. From this inequality we deduce that (A_n) is bounded if and only if (B_n) is bounded, which is what we needed to prove. ∎

10.6.3 The p-Series: An Application of the Integral Test. Given any number p, the **p-series** is the series $\Sigma\, 1/n^p$. If $p \leq 0$, then $1/n^p$ fails to tend to 0 as $n \to \infty$, and the p-series diverges. Assume now that $p > 0$, and define $f(x) = 1/x^p$ for $x \in [1, \infty)$. The function f is positive and decreasing; and as we saw in Chapter 9, the integral $\int_1^{\to\infty} f$ converges when $p > 1$ and diverges when $p \leq 1$. We therefore deduce from the integral test that the p-series $\Sigma\, 1/n^p$ converges when $p > 1$ and diverges when $p \leq 1$.

10.6.4 An Alternative Approach to the p-Series. The purpose of this section is to show how the p-series defined in Section 10.6.3 may be tested for convergence without making use of the integral test. This will enable you to test these important series for convergence even if you have not read Chapters 8 and 9.

We begin by showing that $\Sigma\, 1/n$ diverges. The idea of the proof is suggested by the following array:

$$1 \quad \tfrac{1}{2} \quad \tfrac{1}{3} \quad \tfrac{1}{4} \quad \tfrac{1}{5} \quad \tfrac{1}{6} \quad \tfrac{1}{7} \quad \tfrac{1}{8} \quad \tfrac{1}{9} \quad \tfrac{1}{10} \quad \tfrac{1}{11} \quad \tfrac{1}{12} \quad \tfrac{1}{13} \quad \tfrac{1}{14} \quad \tfrac{1}{15} \quad \tfrac{1}{16} \quad \tfrac{1}{17} \cdots$$

$$1 \quad \tfrac{1}{2} \quad \tfrac{1}{4} \quad \tfrac{1}{4} \quad \tfrac{1}{8} \quad \tfrac{1}{8} \quad \tfrac{1}{8} \quad \tfrac{1}{8} \quad \tfrac{1}{16} \quad \tfrac{1}{16} \quad \tfrac{1}{16} \quad \tfrac{1}{16} \quad \tfrac{1}{16} \quad \tfrac{1}{16} \quad \tfrac{1}{16} \quad \tfrac{1}{16} \quad \tfrac{1}{32} \cdots$$

It is easy to see that for each natural number n we have

$$\sum_{j=1}^{2^n} \frac{1}{j} \geq 1 + \frac{1}{2} + \frac{2}{4} + \frac{4}{8} + \cdots + \frac{2^{n-1}}{2^n} \geq 1 + \frac{n}{2}.$$

This shows that the sequence of partial sums of the series $\Sigma\, 1/n$ is unbounded, and therefore, the series diverges.

If $p \leq 1$, then for each natural number n we have $\Sigma_{j=1}^{n}\, 1/j^p \geq \Sigma_{j=1}^{n}\, 1/j$; and since the latter expression tends to infinity as $n \to \infty$, it follows that the sequence of partial sums of $\Sigma\, 1/n^p$ is unbounded. Therefore, when $p \leq 1$, the series $\Sigma\, 1/n^p$

diverges. It remains to show that when $p > 1$, the series $\Sigma \ 1/n^p$ converges. Suppose $p > 1$. The idea of the proof is suggested by the following array:

$$1 \quad \frac{1}{2^p} \quad \frac{1}{3^p} \quad \frac{1}{4^p} \quad \frac{1}{5^p} \quad \frac{1}{6^p} \quad \frac{1}{7^p} \quad \frac{1}{8^p} \quad \frac{1}{9^p} \quad \frac{1}{10^p} \quad \frac{1}{11^p} \quad \frac{1}{12^p} \quad \frac{1}{13^p} \quad \frac{1}{14^p} \quad \frac{1}{15^p} \quad \frac{1}{16^p} \cdots$$

$$1 \quad \frac{1}{2^p} \quad \frac{1}{2^p} \quad \frac{1}{4^p} \quad \frac{1}{4^p} \quad \frac{1}{4^p} \quad \frac{1}{4^p} \quad \frac{1}{8^p} \quad \frac{1}{8^p} \quad \frac{1}{8^p} \quad \frac{1}{8^p} \quad \frac{1}{8^p} \quad \frac{1}{8^p} \quad \frac{1}{8^p} \quad \frac{1}{8^p} \quad \frac{1}{16^p} \cdots$$

Looking at this array, one may see that for each natural number n

$$\sum_{j=1}^{2^n-1} \frac{1}{j^p} \leq 1 + \frac{2}{2^p} + \frac{4}{4^p} + \frac{8}{8^p} + \cdots + \frac{2^{n-1}}{(2^{n-1})^p}$$

$$= 1 + \frac{1}{2^{p-1}} + \frac{1}{(2^{p-1})^2} + \frac{1}{(2^{p-1})^3} + \cdots + \frac{1}{(2^{p-1})^{n-1}}.$$

The latter expression is the nth partial sum of a geometric series and does not exceed $2^{p-1}/(2^{p-1} - 1)$. This shows that the sequence of partial sums of the series $\Sigma \ 1/n^p$ is bounded and, therefore, that this series converges.

10.6.5   The Comparison Tests

(1) Suppose (a_n) and (b_n) are sequences of nonnegative numbers and K is some positive number; and suppose that for all sufficiently large n we have $a_n \leq Kb_n$. Then if $\Sigma \ b_n$ is convergent, the series $\Sigma \ a_n$ is also convergent.

■ **Proof.**   Choose a natural number N such that $a_n \leq Kb_n$ for all $n \geq N$; and for each $n \geq N$ define $A_n = \Sigma_{j=N}^n a_j$ and $B_n = \Sigma_{j=N}^n b_j$. We observe that $A_n \leq KB_n$ for all $n \geq N$. Therefore, if the series $\Sigma \ b_n$ is convergent—in other words, if (B_n) is bounded—the sequence (A_n) will also be bounded; and the series $\Sigma \ a_n$ will converge. ■

(2) Suppose that (a_n) and (b_n) are sequences of positive numbers and that the sequence (a_n/b_n) is bounded. Then if $\Sigma \ b_n$ is convergent, the series $\Sigma \ a_n$ is also convergent.

■ **Proof.**   Using the fact that (a_n/b_n) is bounded, choose a number K such that for every n we have $a_n \leq Kb_n$. The result now follows from part (1). ■

(3) Suppose that (a_n) and (b_n) are sequences of positive numbers and that for some positive number δ we have $a_n/b_n \geq \delta$ for all n sufficiently large. Then if $\Sigma \ a_n$ is convergent, the series $\Sigma \ b_n$ is also convergent.

■ **Proof.**   Choose a number $\delta > 0$ such that $a_n/b_n \geq \delta$ for n sufficiently large. Then for n sufficiently large we have $b_n \leq (1/\delta)a_n$, and the result now follows at once from part (1). ■

(4) Suppose that (a_n) and (b_n) are sequences of nonnegative numbers and that the sequence (a_n/b_n) is convergent and has a nonzero limit. Then if either one of the series Σa_n and Σb_n is convergent, so is the other.

■ **Proof.** This follows at once from (2) and (3). ■

10.6.6 Some Examples. The examples in this section illustrate how the comparison tests may be used to determine whether a given series converges or diverges.

(1) In Section 10.2 we saw that the series $\Sigma \log(1 + 1/n)$ diverges. Now it is a simple matter to show that $[\log(1 + 1/n)]/(1/n) \to 1$ as $n \to \infty$ [see Exercise 7.16(3)], and so we can use the comparison test to obtain yet another proof that the series $\Sigma 1/n$ diverges.

(2) The series $\Sigma (\log n)/\sqrt{1 + n^3}$ converges. To see this, put $a_n = (\log n)/\sqrt{1 + n^3}$ and $b_n = 1/n^{1.4}$ for all n. The convergence of Σa_n follows at once from the fact that Σb_n is a convergent p-series and the fact that $a_n/b_n \to 0$ as $n \to \infty$.

(3) The series $\Sigma 1/(1 + 1/n)^{(n^2)}$ converges. Using the binomial theorem, one may show easily that for each n we have $(1 + 1/n)^n > 2$; and so for each n we have

$$\left(1 + \frac{1}{n}\right)^{(n^2)} = \left[\left(1 + \frac{1}{n}\right)^n\right]^n > 2^n.$$

The convergence of the series $\Sigma 1/(1 + 1/n)^{(n^2)}$ therefore follows from the convergence of the geometric series $\Sigma 1/2^n$.

(4) Even though the p-series $\Sigma 1/n^p$ converges whenever $p > 1$, the series $\Sigma 1/n^{(1+1/n)}$ diverges. To see this, we compare the series with $\Sigma 1/n$. To make this comparison, we use the fact that $x^{1/x} \to 1$ as $x \to \infty$. So if $a_n = 1/n^{(1+1/n)}$ and $b_n = 1/n$, then $a_n/b_n \to 1$ as $n \to \infty$.

10.6.7 A Comparison of the Series Σa_n and $\Sigma \log(1 - a_n)$. Suppose $0 < a_n < 1$ for every natural number n.

(1) $a_n \to 0$ as $n \to \infty$ if and only if $\log(1 - a_n) \to 0$ as $n \to \infty$.

(2) Σa_n is convergent if and only if $\Sigma \log(1 - a_n)$ is convergent.

(3) If we define $p_n = (1 - a_1)(1 - a_2)(1 - a_3) \cdots (1 - a_n)$ for each n, then Σa_n is convergent if and only if $\lim_{n \to \infty} p_n > 0$.

■ **Proof.** Part (1) is clear. To prove part (2), we note first that unless $a_n \to 0$ as $n \to \infty$, both of the series diverge. Now let us assume that $a_n \to 0$ as $n \to \infty$. From the fact that $x/\log(1 - x) \to -1$ as $x \to 0$, it is clear that $a_n/[-\log(1 - a_n)] \to 1$

as $n \to \infty$; and since $-\log(1 - a_n) > 0$ for each n, part (2) follows at once from Theorem 10.6.5(4).

Finally, to prove part (3), we observe that the sequence (p_n) as defined above is decreasing, and that for each n we have

$$\log p_n = \sum_{j=1}^{n} \log(1 - a_j).$$

It therefore follows from part (2) that $\Sigma\, a_n$ diverges if and only if $\log p_n \to -\infty$ as $n \to \infty$, and this is equivalent to the condition $p_n \to 0$. ∎

10.6.8 The Conditional Convergence of the Integral in Example 9.5.6.

Most of the time, the integral test is used when we know how a given integral behaves and we want to test a given series. Sometimes, however, the same comparison principle that gives us the integral test may be used to test a given integral by comparing it with a series whose behavior is known to us. In this section we shall use the divergence of the series $\Sigma\, 1/n$ to prove that the integral $\int_0^{\to\infty} |\sin x|/x\, dx$ diverges. As you may recall, we mentioned in Section 9.5.6 that this integral is divergent, and we promised to prove this in Chapter 10. For each natural number n we define $A_n = \int_0^{2n\pi} |\sin x|/x\, dx$; and in order to show that $\int_0^{\to\infty} |\sin x|/x\, dx$ diverges, we shall show that $A_n \to \infty$ as $n \to \infty$. We start with the observation that if n is any integer, and if $2n\pi + \pi/4 \le x \le 2n\pi + 3\pi/4$, then $\sin x \ge 1/\sqrt{2}$. Therefore, for every natural number n we have

$$\int_{2n\pi}^{2(n+1)\pi} \frac{|\sin x|}{x}\, dx \ge \frac{1}{\sqrt{2}} \int_{2n\pi+\pi/4}^{2n\pi+3\pi/4} \frac{1}{x}\, dx$$

$$\ge \frac{1}{\sqrt{2}} \int_{2n\pi+\pi/4}^{2n\pi+3\pi/4} \frac{1}{2(n+1)\pi}\, dx = \frac{1}{4(n+1)\sqrt{2}}.$$

From this we see that if n is any natural number, then

$$A_n = \sum_{j=1}^{n} \int_{2(j-1)\pi}^{2j\pi} \frac{|\sin x|}{x}\, dx \ge \frac{1}{4\sqrt{2}} \sum_{j=1}^{n} \frac{1}{j};$$

and so the divergence of the series $\Sigma\, 1/n$ implies that $A_n \to \infty$ as $n \to \infty$.

10.7 EXERCISES

In these exercises you should interpret each series as starting at a high enough value of n to ensure that all its terms are defined.

- **1.** Find the values of p for which the series $\Sigma\, 1/n(\log n)^p$ converges. *Hint:* Use the integral test.

- **2.** Determine whether the following series converge or diverge.

(a) $\sum \dfrac{1}{n^{3/2} + n}$

(b) $\sum \dfrac{1}{n^{3/2} - n}$

(c) $\sum \dfrac{n}{\sqrt{n^4 - n^2 + 2}}$

(d) $\sum \dfrac{n \log n}{\sqrt{n^5 - n^2 + 2}}$

▪ **3.** Determine whether the following series converge or diverge.

(a) $\sum \dfrac{1}{n^{[1 + (\log n)/n]}}$. *Hint:* Look at $\lim\limits_{x \to \infty} \dfrac{1}{x^{(\log x/x)}}$.

(b) $\sum \dfrac{1}{n^{[1 + (\log n)^2/n]}}$

(c) $\sum \dfrac{1}{(1 + 1/n)^{n \log n}}$

(d) $\sum \dfrac{1}{(1 + 1/n)^{n(\log n)^2}}$

(e) Now that you have seen that (c) diverges and (d) converges, state and prove a single theorem that contains both (c) and (d) as special cases.

(f) $\sum \dfrac{1}{(\log n)^3}$

(g) $\sum \dfrac{1}{(\log n)^n}$

(h) $\sum \dfrac{1}{(\log n)^{(\log n)}}$

(i) $\sum \dfrac{1}{(\log \log n)^{(\log n)}}$

(j) $\sum \dfrac{1}{(\log n)^{(\log\log n)}}$

(k) $\sum \dfrac{1}{(\log \log \log n)^{(\log n)}}$

(l) $\sum \dfrac{1}{(\log \log n)^{(\log\log n)}}$

(m) $\sum \dfrac{1}{[(\log \log n)^{(\log\log n)}]^{(\log\log n)}}$

(n) $\sum \dfrac{1}{(\log \log n)^{[(\log\log n)^{(\log\log n)}]}}$

▪ **4.** Given $x > 0$, determine for which numbers α the series $\sum [\sin(x/n)]^\alpha$ converges.

5. Prove that if (a_n) is a sequence of positive numbers and $\sum a_n$ converges, then $\sum a_n^2$ also converges.

6. Given $a_n > 0$ for every n, prove that the series $\sum a_n$ is convergent if and only if the series $\sum \log(1 + a_n)$ converges.

7. Given that $a_n > 0$ for every n and that for each n

$$p_n = (1 + a_1)(1 + a_2)(1 + a_3) \cdots (1 + a_n),$$

prove that $\sum a_n$ is convergent if and only if the sequence (p_n) is convergent.

8. Given that $0 < a_n < 1$ for every n and that for each n we have

$$p_n = (1 - a_1)(1 - a_2)(1 - a_3) \cdots (1 - a_n)$$

and

$$q_n = (1 + a_1)(1 + a_2)(1 + a_3) \cdots (1 + a_n),$$

prove that $p_n \to 0$ as $n \to \infty$ if and only if $q_n \to \infty$ as $n \to \infty$.

9. The purpose of this exercise is to develop a sharper form of the integral test. Suppose that f is a positive decreasing function on $[1, \infty)$ and that for each natural number n

$$a_n = \sum_{j=1}^{n} f(j) - \int_1^{n+1} f.$$

Prove that for each n we have $0 \leq a_n \leq f(1)$ and that the sequence (a_n) is increasing. You have therefore shown that the sequence (a_n) is convergent. Can you see that what you have proved implies that $\Sigma\, f(n)$ is convergent if and only if the integral $\int_1^{\to\infty} f$ converges?

10. Suppose that for each natural number n

$$a_n = \sum_{j=1}^{n} \frac{1}{j} - \log n.$$

Prove that the sequence (a_n) is convergent. The limit of this sequence is approximately 0.57721566 and is known as **Euler's constant.** It is traditional to denote Euler's constant as γ, and in the next few exercises you will find this notation useful.

11. Prove that $\Sigma_{j=n+1}^{2n} 1/j \to \log 2$ as $n \to \infty$. *Hint:* For each n

$$\sum_{j=n+1}^{2n} \frac{1}{j} = \sum_{j=1}^{2n} \frac{1}{j} - \log 2n - \sum_{j=1}^{n} \frac{1}{j} + \log n + \log 2.$$

■ 12. Find the sum of the series $\Sigma_{n=1}^{\infty} 1/n(2n-1)$. *Hint:* For each j we have

$$\frac{1}{j(2j - 1)} = \frac{2}{2j - 1} - \frac{1}{j},$$

and therefore,

$$\sum_{j=1}^{n} \frac{1}{j(2j - 1)} = 2\left(\frac{1}{1} + \frac{1}{3} + \frac{1}{5} + \cdots + \frac{1}{2n - 1}\right) - \sum_{j=1}^{n} \frac{1}{j}$$

$$= 2\sum_{j=1}^{2n} \frac{1}{j} - 2\left(\frac{1}{2} + \frac{1}{4} + \frac{1}{6} + \cdots + \frac{1}{2n}\right) - \sum_{j=1}^{n} \frac{1}{j}$$

$$= 2\sum_{j=1}^{2n} \frac{1}{j} - 2\sum_{j=1}^{n} \frac{1}{j}.$$

■ **13.** Find the sums of the series

$$\sum_{n=1}^{\infty} \frac{1}{n(2n+1)} \quad \text{and} \quad \sum_{n=1}^{\infty} \frac{1}{n(4n^2-1)}.$$

Using the first of these two series, deduce that if f is the function defined in Exercise 8.15(2), then $\int_0^1 f = 1 - \log 2$.

■ **14.** Find the limit $\lim_{n \to \infty} \sum_{j=n+1}^{n^p} 1/j \log j$, where p is any natural number.

10.8 THE RATIO TESTS

This section is really a continuation of our discussion of the comparison principle. The idea of the ratio test is to compare the behavior of two series $\Sigma\, a_n$ and $\Sigma\, b_n$ by comparing the two ratios a_{n+1}/a_n and b_{n+1}/b_n, which describe the rate of growth of the sequences (a_n) and (b_n). As we shall see in our first theorem, the smaller the ratio a_{n+1}/a_n, the more likely the series $\Sigma\, a_n$ is to converge.

10.8.1 The Fundamental Ratio Test. Suppose that (a_n) and (b_n) are sequences of positive numbers and that for n sufficiently large we have

$$\frac{a_{n+1}}{a_n} \le \frac{b_{n+1}}{b_n}.$$

Then if $\Sigma\, b_n$ is convergent, the series $\Sigma\, a_n$ is also convergent.

■ *Proof.* Choose a natural number N such that the inequality

$$\frac{a_{n+1}}{a_n} \le \frac{b_{n+1}}{b_n}$$

holds for all $n \ge N$. Then whenever $n \ge N$, we have $a_{n+1}/b_{n+1} \le a_n/b_n$; in other words, beginning at $n = N$, the sequence (a_n/b_n) is decreasing; and it follows that for all $n \ge N$ we have $a_n/b_n \le a_N/b_N$. So for all $n \ge N$ we have $a_n \le (a_N/b_N)b_n$; and since $\Sigma\, b_n$ converges, the convergence of $\Sigma\, a_n$ follows from the comparison test, Theorem 10.6.5(1). ■

10.8.2 A Simple Criterion for Divergence. Suppose that (a_n) is a sequence of positive numbers and that for all n sufficiently large we have $a_{n+1}/a_n \ge 1$. Then the series $\Sigma\, a_n$ diverges.

■ *Proof.* Choose a natural number N such that $a_{n+1}/a_n \ge 1$ for all $n \ge N$. Then since $a_n \ge a_N$ for all $n \ge N$, we see that the sequence (a_n) fails to converge to 0; and we conclude that $\Sigma\, a_n$ diverges. ■

Note that this criterion for divergence does not have a valid converse; in other words, even if $a_{n+1}/a_n < 1$ for n sufficiently large, there is no guarantee that $\Sigma\, a_n$ will converge. Look, for example, at $\Sigma\, 1/n$. If $a_n = 1/n$ for all n, then for each n we have $a_{n+1}/a_n = n/(n + 1) < 1$; but $\Sigma\, a_n$ diverges. More generally, for any positive number p if $a_n = 1/n^p$ for all n, then $a_{n+1}/a_n < 1$; and as we know, some p-series converge while others diverge. In some of the tests that follow, we shall see that if $a_{n+1}/a_n < 1$, then the series $\Sigma\, a_n$ will converge as long as a_{n+1}/a_n is not "too close" to 1.

10.8.3   Comparison with Geometric Series: d'Alembert's Test.

Suppose that (a_n) is a sequence of positive numbers and that for some number $\alpha < 1$ we have $a_{n+1}/a_n \le \alpha$ for all n sufficiently large. Then the series $\Sigma\, a_n$ converges.

■ **Proof.**   For all n sufficiently large we have $a_{n+1}/a_n \le \alpha^{n+1}/\alpha^n$, and since the geometric series $\Sigma\, \alpha^n$ converges, the convergence of $\Sigma\, a_n$ follows from Theorem 10.8.1.   ■

10.8.4   A Limit Form of d'Alembert's Test.

Suppose that (a_n) is a sequence of positive numbers, that $0 \le \alpha \le \infty$, and that $a_{n+1}/a_n \to \alpha$ as $n \to \infty$.

(1) If $\alpha < 1$, the series $\Sigma\, a_n$ converges.

(2) If $\alpha > 1$, the series $\Sigma\, a_n$ diverges.

■ **Proof.**   To prove part (1), assume that $\alpha < 1$. Choose a number β such that $\alpha < \beta < 1$:

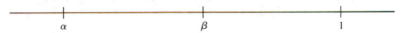

For n sufficiently large we have $a_{n+1}/a_n \le \beta$, and the convergence of $\Sigma\, a_n$ therefore follows from Theorem 10.8.3.

Now to prove part (2), assume that $\alpha > 1$. Then for n sufficiently large we have $a_{n+1}/a_n > 1$, and the divergence of $\Sigma\, a_n$ therefore follows from Theorem 10.8.2.   ■

Note that d'Alembert's test gives no information when $\alpha = 1$.

10.8.5   Some Illustrations of d'Alembert's Test

(1) Let $a_n = (n!)^2/(2n)!$ for all n. Then for each n we have

$$\frac{a_{n+1}}{a_n} = \frac{(n + 1)!(n + 1)!(2n)!}{(2n + 2)!n!n!} = \frac{(n + 1)^2}{(2n + 1)(2n + 1)}$$

and since the latter expression tends to $\frac{1}{4}$ as $n \to \infty$, we see that $\Sigma\, a_n$ converges.

(2) In Section 10.4.1 we showed that for any real number x we have $|x|^n/n! \to 0$ as $n \to \infty$. Now using d'Alembert's test, we can give a simple alternative proof of this. Put $a_n = |x|^n/n!$. Then

$$\frac{a_{n+1}}{a_n} = \frac{|x|}{n+1} \to 0 \qquad \text{as} \qquad n \to \infty,$$

and d'Alembert's test implies that $\Sigma\, a_n$ converges. It therefore follows from Theorem 10.5.4 that $a_n \to 0$ as $n \to \infty$.

(3) Let $a_n = 2^n n!/n^n$ for each n. It is not hard to show that $a_{n+1}/a_n \to 2/e$ as $n \to \infty$; and since $e > 2$, it follows from the limit form of d'Alembert's test that $\Sigma\, a_n$ converges.

(4) Let $a_n = 4^n(n!)^2/(2n)!$ for all n. In this case one can see easily that $a_{n+1}/a_n \to 1$ as $n \to \infty$, and so the limit form of d'Alembert's test is of no help. But since

$$\frac{a_{n+1}}{a_n} = \frac{2n+2}{2n+1} > 1$$

for each n, the divergence of the series $\Sigma\, a_n$ follows from Theorem 10.8.2.

(5) Let $a_n = n^n/e^n n!$ for each n. None of the previous ratio tests give any information about the convergence or divergence of $\Sigma\, a_n$, and it may not even be clear whether $a_n \to 0$ as $n \to \infty$. In the next few sections we shall study some more delicate ratio tests which will allow us to test a variety of series for which d'Alembert's test fails. We shall return to this series in Example 10.8.8(2), where we shall show that $\Sigma\, a_n$ diverges, and in Example 10.8.10(2), where we shall show that $a_n \to 0$.

10.8.6 Comparison with p-Series: Raabe's Test. Suppose (a_n) is a sequence of positive numbers.

(1) If for some number $p > 1$ we have $a_{n+1}/a_n \le 1 - p/n$ for all n sufficiently large, then $\Sigma\, a_n$ converges.

(2) If for some number $p \le 1$ we have $a_{n+1}/a_n \ge 1 - p/n$ for all n sufficiently large, then $\Sigma\, a_n$ diverges.

The proof of Raabe's test will make use of the inequalities contained in the following lemma.

Lemma

(1) Given $p > 1$ and $x \in (0, 1)$, we have $(1 - x)^p > 1 - px$.

(2) Given $0 < p < 1$ and $x \in (0, 1)$, we have $(1 - x)^p < 1 - px$.

■ *Proof.* We shall prove part (1). Part (2) follows from an almost identical argument. Assume now that $p > 1$; and for each $x \in [0, 1)$, define $f(x) = (1 - x)^p - 1 + px$. We need to show that $f(x) > 0$ for all $x \in (0, 1)$. Now $f(0) = 0$, and so this inequality will follow at once from the mean value theorems if we can show that $f'(x) > 0$ for all $x \in (0, 1)$. But for all $x \in (0, 1)$ we have

$$f'(x) = -p(1 - x)^{p-1} + p = p[1 - (1 - x)^{p-1}];$$

and since $0 < 1 - x < 1$ and $p - 1 > 0$, we see that $(1 - x)^{p-1} < 1$; and it follows that $f'(x) > 0$, as required. ■

■ *Proof of Raabe's Test.* We shall prove part (1). The proof of part (2) is similar and will be left as a simple exercise. Assume now that $p > 1$ and that for all sufficiently large n we have $a_{n+1}/a_n \leq 1 - p/n$. For all $n \geq 2$, define $b_n = 1/(n - 1)^p$. We deduce from the preceding lemma that whenever $n \geq 2$, we have

$$\frac{b_{n+1}}{b_n} = \left(1 - \frac{1}{n}\right)^p \geq 1 - \frac{p}{n};$$

and it follows that $a_{n+1}/a_n \leq b_{n+1}/b_n$ for n sufficiently large. The result now follows from Theorem 10.8.1. ■

10.8.7 A Limit Form of Raabe's Test.
Suppose that (a_n) is a sequence of positive numbers, that p is some real number, and that

$$n\left(1 - \frac{a_{n+1}}{a_n}\right) \to p \qquad \text{as} \qquad n \to \infty.$$

(1) If $p > 1$, then the series $\Sigma\, a_n$ converges.

(2) If $p < 1$, then the series $\Sigma\, a_n$ diverges.

■ *Proof.* We shall prove part (1). The proof of part (2) is similar and will be left as a simple exercise. Assume, then, that $p > 1$, and choose a number q such that $1 < q < p$. For n sufficiently large we have

$$n\left(1 - \frac{a_{n+1}}{a_n}\right) > q,$$

and for all such n we have $a_{n+1}/a_n < 1 - q/n$. Therefore, the convergence of $\Sigma\, a_n$ follows from the previous form of Raabe's test (Theorem 10.8.6). ■

10.8.8 Some Illustrations of Raabe's Test

(1) This example will be useful to us when we study the binomial series in Example 10.10.7(5) and Exercises 11.10(6) and 11.10(7). Suppose α is some real number; and for each n, take

$$a_n = \left| \frac{\alpha(\alpha - 1)(\alpha - 2) \cdots (\alpha - n + 1)}{n!} \right|.$$

In the event that α is a nonnegative integer, we see that $a_n = 0$ for all n sufficiently large, and $\Sigma\, a_n$ obviously converges. Otherwise, we see that for n sufficiently large we have

$$\frac{a_{n+1}}{a_n} = \left| \frac{\alpha - n}{n + 1} \right| = \frac{n - \alpha}{n + 1},$$

and from this it follows that

$$n \left(1 - \frac{a_{n+1}}{a_n} \right) \to 1 + \alpha \qquad \text{as} \qquad n \to \infty.$$

Therefore, $\Sigma\, a_n$ converges when $\alpha \geq 0$ and diverges when $\alpha < 0$.

(2) We now return to Example 10.8.5(5) and let $a_n = n^n/e^n n!$ for each n. In view of Exercise 7.16(8), it is not hard to see that

$$n \left(1 - \frac{a_{n+1}}{a_n} \right) \to \frac{1}{2} \qquad \text{as} \qquad n \to \infty,$$

and we conclude from Theorem 10.8.7 that $\Sigma\, a_n$ diverges.

10.8.9 A Criterion for the *n*th Term of a Series to Converge to Zero. The message of the ratio tests is that if (a_n) is a decreasing sequence of positive numbers, then the series $\Sigma\, a_n$ will converge as long as the ratio a_{n+1}/a_n is not "too close" to 1. In this section we shall see that even though the ratio a_{n+1}/a_n may be too close to 1 to allow convergence of $\Sigma\, a_n$, it may nevertheless satisfy a weaker criterion that allows us to deduce that $a_n \to 0$. This will be particularly useful to us when we come to Section 10.10.

Theorem. Suppose (a_n) is a decreasing sequence of positive numbers; and for each natural number n, define $b_n = 1 - a_{n+1}/a_n$. Then the sequence (a_n) converges to zero if and only if the series $\Sigma\, b_n$ diverges.

■ *Proof.* Since $1 - b_n = a_{n+1}/a_n$ for every n, it is clear that whenever $n \geq 2$, we have

$$a_n = a_1(1 - b_1)(1 - b_2)(1 - b_3) \cdots (1 - b_{n-1}).$$

The theorem therefore follows at once from Theorem 10.6.7.   ■

10.8.10 Some Examples. The examples in this subsection illustrate the power of Theorem 10.8.9.

(1) Suppose α is a real number; and for each n, define

$$a_n = \left| \frac{\alpha(\alpha - 1)(\alpha - 2) \cdot \cdot \cdot (\alpha - n + 1)}{n!} \right|.$$

In Example 10.8.8(1) we saw that $\Sigma \, a_n$ converges when $\alpha \geq 0$ and diverges when $\alpha < 0$. When we begin our study of binomial series in Example 10.10.7(5), however, it will be important to know for which negative numbers α the sequence (a_n) converges to zero. To answer this question, we note that for each n we have $a_{n+1}/a_n = (n - \alpha)/(n + 1)$. In the event that $\alpha \leq -1$, we see that $a_{n+1}/a_n \geq 1$ for each n, and (a_n) does not converge to zero. Suppose now that $-1 < \alpha < 0$. Then (a_n) is a decreasing sequence of positive numbers. Define $b_n = 1 - a_{n+1}/a_n$ for each n. Then $b_n = (1 + \alpha)/(n + 1)$ for each n, and it follows from the divergence of $\Sigma \, b_n$ that $a_n \to 0$ as $n \to \infty$.

(2) For each natural number n, define $a_n = n^n/e^n n!$. We saw in Example 10.8.8(2) that $\Sigma \, a_n$ diverges because

$$n \left(1 - \frac{a_{n+1}}{a_n} \right) \to \frac{1}{2} \quad \text{as} \quad n \to \infty.$$

Define $b_n = 1 - a_{n+1}/a_n$ for each n. Then for n sufficiently large we have $b_n > 1/3n$, and it follows from the divergence of $\Sigma \, b_n$ that $a_n \to 0$ as $n \to \infty$.

(3) We now generalize examples (1) and (2). Suppose that (a_n) is a sequence of positive numbers, that p is some real number, and that

$$n \left(1 - \frac{a_{n+1}}{a_n} \right) \to p \quad \text{as} \quad n \to \infty.$$

(a) If $p > 0$, then $a_n \to 0$ as $n \to \infty$.
(b) If $p < 0$, then (a_n) does not converge to zero.
Part (b) is clear because in this case (a_n) is increasing. To prove part (a), define $b_n = 1 - a_{n+1}/a_n$ for each n, and note that $\Sigma \, b_n$ diverges.

Although the limit form of Raabe's test (Theorem 10.8.7) gives no information when $p = 1$, the test can be sharpened in a way that allows us to test many of the series for which this happens. The following starred subsection can be omitted without loss of continuity.

10.8.11*   A Sharper Version of Raabe's Test.   Suppose that (a_n) is a sequence of positive numbers, that p is some real number, and that

$$n \log n \left(1 - \frac{1}{n} - \frac{a_{n+1}}{a_n} \right) \to p \quad \text{as} \quad n \to \infty.$$

(1) If $p > 1$, then the series $\Sigma \, a_n$ converges.

(2) If $p < 1$, then the series $\Sigma \, a_n$ diverges.

■ *Proof.* We shall prove part (1) and leave the proof of part (2) as an exercise. Assume now that $p > 1$, and choose a number q such that $1 < q < p$. For all $n \geq 3$, define

$$b_n = \frac{1}{(n-1)[\log(n-1)]^q},$$

and observe that it follows from the integral test that $\Sigma \, b_n$ converges. From Example 7.15.4(3) we see that

$$n \log n \left(1 - \frac{1}{n} - \frac{b_{n+1}}{b_n}\right) \to q \quad \text{as} \quad n \to \infty,$$

and it follows that for all sufficiently large n we have

$$n \log n \left(1 - \frac{1}{n} - \frac{a_{n+1}}{a_n}\right) > n \log n \left(1 - \frac{1}{n} - \frac{b_{n+1}}{b_n}\right).$$

For all such n we have $a_{n+1}/a_n < b_{n+1}/b_n$, and it follows from the fundamental ratio test that $\Sigma \, a_n$ converges. ■

10.9 EXERCISES

■ **1.** In each of the following cases, determine whether the series $\Sigma \, a_n$ converges or diverges.

(a) $a_n = \dfrac{[(2n)!]^3}{[(3n)!]^2}$ **(b)** $a_n = \dfrac{3^n}{n!}$

(c) $a_n = \dfrac{3^{(n^2)}}{n!}$ **(d)** $a_n = \dfrac{3^{(n \log n)}}{n!}$

(e) $a_n = \dfrac{(\log n)^n}{(\exp n)(\log 2)(\log 3) \cdots (\log n)}$

■ **2. (a)** Prove that the series $\Sigma \, n^\alpha/n!$ converges for all numbers α.
    (b) Find for which numbers α the series $\Sigma \, n^{\alpha n}/n!$ converges.
    (c) Determine whether the series $\Sigma n^{n - \log n}/n!$ converges.

3. Prove that the series $\Sigma \, (2n)!/4^n(n!)^2$ diverges, and then prove that the series $\Sigma \, [(2n)!/4^n(n!)^2] x^{2n}$ converges when $|x| < 1$ and diverges when $|x| \geq 1$.

■ **4.** Find for which positive numbers x the series

$$\sum \frac{n!}{x(x+1) \cdots (x+n-1)}$$

converges and for which positive numbers x the nth term of this series converges to zero.

5. Prove that if $a_n = e^n n!/n^n$, then the series Σa_n diverges. *Hint:* Use Theorem 10.8.7. Alternatively, take another look at Exercise 9.3(16), and show that $a_{n+1}/a_n > 1$ for each n.

6. Find for which positive numbers x the series $\Sigma n! x^n/n^n$ converges.

■ **7.** Prove that the series $\Sigma[(2n)!/4^n(n!)^2]^\alpha$ converges when $\alpha > 2$ and diverges when $\alpha < 2$. *Hint:* Use Exercise 7.16(12). When $\alpha = 2$, the series may be tested by using the sharper form of Raabe's test.

■ **8.** Determine for which positive numbers α and β the series

$$\Sigma \sqrt{\frac{\alpha(\alpha + 1)(\alpha + 2) \cdots (\alpha + n - 1)}{\beta(\beta + 1)(\beta + 2) \cdots (\beta + n - 1)}}$$

converges.

* **9. Gauss's test** says that if (a_n) is a sequence of positive numbers and if for some number $\alpha > 1$ we have

$$n^\alpha \left(1 - \frac{1}{n} - \frac{a_{n+1}}{a_n}\right) \to 0 \qquad \text{as} \qquad n \to \infty,$$

then Σa_n diverges. Show that this theorem follows at once from the sharper version of Raabe's test. Use Gauss's test to settle the case $\alpha = 2$ of exercise 7.

10. Cauchy's root test says that if (a_n) is a sequence of nonnegative numbers, and if for some number $\alpha < 1$ we have $(a_n)^{1/n} \leq \alpha$ for all n sufficiently large, then Σa_n converges; and if $(a_n)^{1/n} \geq 1$ for infinitely many natural numbers n, then Σa_n diverges. Prove this test.

11. Suppose that (a_n) is a sequence of positive numbers and that for some number $q > 0$ we have $n(1 - a_{n+1}/a_n) \geq q$ for n sufficiently large. Prove that the sequence (a_n) is eventually decreasing and that $a_n \to 0$ as $n \to \infty$.

12. Suppose that (a_n) is a sequence of positive numbers and that for some number $q > 0$ we have $n \log n (1 - a_{n+1}/a_n) \geq q$ for n sufficiently large. Prove that $a_n \to 0$ as $n \to \infty$.

***13.** Prove that the sharper version of Raabe's test deserves its name. In other words, prove that any series that can be tested for convergence by Raabe's test can also be tested by the sharper version of Raabe's test. Give an example of a series which can be tested by the sharper version of Raabe's test but not by Raabe's test itself.

***14.** The sharper version of Raabe's test was obtained by comparing a given series Σa_n with a series of the form $\Sigma 1/n(\log n)^p$. Can you develop an even sharper theorem which would involve a comparison with a series of the form $\Sigma 1/n(\log n)(\log \log n)^p$?

10.10 ABSOLUTE AND CONDITIONAL CONVERGENCE

We began our discussion of the comparison principle with the observation that series $\Sigma \, a_n$ for which $a_n \geq 0$ for all n are considerably easier to test for convergence than general series of real numbers. We pointed out that the reason for this is that if $a_n \geq 0$ for all n, then the sequence of nth partial sums $\Sigma_{j=1}^{n} \, a_j$ must be increasing. Throughout Sections 10.6–10.9 we relied heavily on this fact, and up to this point we have said very little about the convergence of a general series. Apart from what we learned about some special series by evaluating the limits of their nth partial sums, the only information we have right now about the convergence or divergence of a general series is Theorem 10.5.4: A series can converge only if its nth term tends to zero. We now begin our study of general series; and as you will see, this section is a close analogue of Section 9.5 in which we studied the convergence of general integrals.

10.10.1 Definition of Absolute Convergence. (Compare with Definition 9.5.1.) A series $\Sigma \, a_n$ is said to **converge absolutely** if the series $\Sigma \, |a_n|$ converges.

10.10.2 Theorem. (Compare with Theorem 9.5.2.) Every absolutely convergent series is convergent. Furthermore, if $\Sigma \, a_n$ is absolutely convergent, then

$$\left| \sum_{n=1}^{\infty} a_n \right| \leq \sum_{n=1}^{\infty} |a_n|.$$

■ *Proof.* Suppose $\Sigma \, a_n$ is absolutely convergent. In order to prove that $\Sigma \, a_n$ converges, we shall express each term a_n as the difference $b_n - c_n$ of two nonnegative numbers whose sum is $|a_n|$. The numbers b_n and c_n that satisfy these requirements are

$$b_n = \frac{|a_n| + a_n}{2} \quad \text{and} \quad c_n = \frac{|a_n| - a_n}{2}.$$

Since $0 \leq b_n \leq |a_n|$ and $0 \leq c_n \leq |a_n|$ for all n, it follows from the comparison test that the series $\Sigma \, b_n$ and $\Sigma \, c_n$ are convergent. The convergence of $\Sigma \, a_n$ therefore follows from the identity $a_n = b_n - c_n$ and Theorem 10.5.1. The final assertion of the theorem now follows from the fact that for every n we have

$$\left| \sum_{j=1}^{n} a_j \right| \leq \sum_{j=1}^{n} |a_j|. \qquad ■$$

10.10.3 Definition of Conditional Convergence. (Compare with Definition 9.5.3.) Not every convergent series is absolutely convergent. For example, we saw in Example 10.2.7 that

$$\sum_{n=1}^{\infty} \frac{(-1)^{n-1}}{n} = \log 2,$$

even though the series $\Sigma\ 1/n$ diverges.

A convergent series that is not absolutely convergent is said to be **conditionally convergent**.

Just as we saw when we studied the convergence of integrals, the comparison principle does *not* apply to conditional convergence, and our principal theorem on conditional convergence is Dirichlet's test.

10.10.4   Dirichlet's Test.   (Compare with Theorem 9.5.4.) Suppose that (a_n) is a decreasing sequence of positive numbers and that $a_n \to 0$ as $n \to \infty$. Suppose that (b_n) is a sequence of real numbers and that for some number K we have $\left|\Sigma_{j=1}^{n} b_j\right| \le K$ for every n. Then the series $\Sigma\ a_n b_n$ converges.

■ *Proof.*   For each n, define $B_n = \Sigma_{j=1}^{n} b_j$. Then whenever $n \ge 2$, we have $b_n = B_n - B_{n-1}$; and therefore,

$$\sum_{j=1}^{n} a_j b_j = a_1 B_1 + a_2(B_2 - B_1) + a_3(B_3 - B_2) + \cdots + a_n(B_n - B_{n-1})$$

$$= B_1(a_1 - a_2) + B_2(a_2 - a_3) + \cdots + B_{n-1}(a_{n-1} - a_n) + a_n B_n.$$

We therefore need to show that as $n \to \infty$, both $\Sigma_{j=1}^{n-1} B_j(a_j - a_{j+1})$ and $a_n B_n$ tend to finite limits. Now to see that $\Sigma_{j=1}^{n-1} B_j(a_j - a_{j+1})$ tends to a finite limit as $n \to \infty$, we need to see that the series $\Sigma\ B_n(a_n - a_{n+1})$ converges; and since $\left|B_n(a_n - a_{n+1})\right| \le K(a_n - a_{n+1})$ for each n, the absolute convergence of this series will follow from the comparison test when we have shown that the series $\Sigma(a_n - a_{n+1})$ converges. But this is easy:

$$\sum_{j=1}^{n} (a_j - a_{j+1}) = a_1 - a_{n+1} \to a_1 \qquad \text{as} \qquad n \to \infty.$$

Finally, we need to see that the sequence $(a_n B_n)$ converges. Now for each n we have $\left|a_n B_n\right| \le K a_n$, and it therefore follows from the sandwich theorem that $a_n B_n \to 0$ as $n \to \infty$. This completes the proof.   ■

Our next theorem is the series analogue of Theorem 9.5.5. This theorem tells us something about the magnitude of the sum of a series whose convergence has been established by Dirichlet's test and the size of the error involved when we replace this sum by an nth partial sum.

10.10.5   Theorem.   Suppose (a_n) is a decreasing sequence of positive numbers, (b_n) is a sequence of real numbers, N is some natural number, and $K \ge 0$; and suppose that $\left|\Sigma_{j=N}^{n} b_j\right| \le K$ for all $n \ge N$. Then whenever $n \ge N$, we have $\left|\Sigma_{j=N}^{n} a_j b_j\right| \le K a_N$; and in the event that $\Sigma\ a_n b_n$ converges, we have $\left|\Sigma_{j=N}^{\infty} a_j b_j\right| \le K a_N$.

■ **Proof.** For each $n \geq N$, define $B_n = \sum_{j=N}^{n} b_j$. Then for all $n > N$ we have

$$\sum_{j=N}^{n} a_j b_j = a_N B_N + a_{N+1}(B_{N+1} - B_N) + \cdots + a_n(B_n - B_{n-1})$$

$$= B_N(a_N - a_{N+1}) + B_{N+1}(a_{N+1} - a_{N+2}) + \cdots$$

$$+ B_{n-1}(a_{n-1} - a_n) + a_n B_n,$$

and therefore,

$$\left| \sum_{j=N}^{n} a_j b_j \right| \leq \sum_{j=N}^{n-1} |B_j(a_j - a_{j+1})| + |a_n B_n|$$

$$\leq \sum_{j=N}^{n-1} K(a_j - a_{j+1}) + K a_n = K(a_N - a_n) + K a_n = K a_N. \qquad ■$$

Our final theorem in this section is Abel's theorem, which is the analogue of Theorem 9.8.7 for series and which provides us with important information about series of the type $\Sigma \, a_n x^n$. As you may know, series of this type are known as **power series.** We have already seen a number of important power series, like the geometric series $\Sigma \, x^{n-1}$ and the series representations of the exponential and logarithmic functions that were obtained in Section 10.4. We shall have more to say about power series in Chapter 11.

10.10.6 Abel's Theorem for Series. (Compare with Theorem 9.8.7.) Suppose the series $\Sigma \, a_n$ converges. Then the series $\Sigma \, a_n x^n$ converges absolutely whenever $|x| < 1$, and

$$\sum_{n=1}^{\infty} a_n x^n \to \sum_{n=1}^{\infty} a_n \qquad \text{as} \qquad x \to 1 \quad \text{(from the left)}.$$

■ **Proof.** We note first that since $\Sigma \, a_n$ converges, we have $a_n \to 0$; and therefore, for n sufficiently large we have $|a_n| < 1$. Therefore, if $|x| < 1$, we have $|a_n x^n| \leq |x|^n$ for n sufficiently large; and the absolute convergence of $\Sigma \, a_n x^n$ follows from the comparison test.

Now let $\varepsilon > 0$; and using the fact that $\Sigma \, a_n$ converges, choose a natural number N such that whenever $n \geq N$, we have

$$\left| \sum_{j=1}^{\infty} a_j - \sum_{j=1}^{n-1} a_j \right| < \frac{\varepsilon}{6},$$

in other words, $|\sum_{j=n}^{\infty} a_j| < \varepsilon/6$. Then whenever $n \geq N$, we have

$$\left| \sum_{j=N}^{n} a_j \right| = \left| \sum_{j=N}^{\infty} a_j - \sum_{j=n+1}^{\infty} a_j \right| < \frac{\varepsilon}{3};$$

and it follows from Theorem 10.10.5 that for all $x \in [0, 1]$ we have

$$\left| \sum_{j=N}^{\infty} a_j x^j \right| \leq \frac{\varepsilon x^N}{3} \leq \frac{\varepsilon}{3}.$$

Now using the fact that the polynomial $\sum_{j=1}^{N-1} a_j x^j$ is continuous at 1, choose $\delta > 0$ such that whenever $1 - \delta < x < 1$, we have

$$\left| \sum_{j=1}^{N-1} a_j x^j - \sum_{j=1}^{N-1} a_j \right| < \frac{\varepsilon}{3}.$$

Then whenever $1 - \delta < x \leq 1$, we have

$$\left| \sum_{j=1}^{\infty} a_j x^j - \sum_{j=1}^{\infty} a_j \right| \leq \left| \sum_{j=1}^{N-1} a_j x^j - \sum_{j=1}^{N-1} a_j \right| + \left| \sum_{j=N}^{\infty} a_j x^j \right| + \left| \sum_{j=N}^{\infty} a_j \right|$$

$$< \frac{\varepsilon}{3} + \frac{\varepsilon}{3} + \frac{\varepsilon}{3} = \varepsilon. \qquad \blacksquare$$

10.10.7 Some Examples

(1) Suppose that (a_n) is a decreasing sequence of positive numbers and that $a_n \to 0$ as $n \to \infty$. Since $|\sum_{j=1}^{n}(-1)^{j-1}| \leq 1$ for all n, it follows from Dirichlet's test that the series $\Sigma(-1)^{n-1}a_n$ converges and that for every natural N we have $|\sum_{n=N}^{\infty}(-1)^{n-1}a_n| \leq a_N$. This result is sometimes known as the **alternating series test** or the **Leibniz test**.

(2) In this example we investigate the convergence of the series $\Sigma(n!)^2 x^n/(2n)!$, where x is any real number. For each n, put $a_n = |(n!)^2 x^n/(2n)!|$. By looking at a_{n+1}/a_n, we see easily that when $|x| \geq 4$, the sequence (a_n) does not converge to 0; and consequently, the given series diverges. When $|x| < 4$, it follows at once from d'Alembert's test that the series converges.

(3) In this example we investigate the convergence of $\Sigma (2n)! x^n/(n!)^2$. It is easy to see that this series converges when $|x| < \frac{1}{4}$ and diverges when $|x| > \frac{1}{4}$. When $x = \frac{1}{4}$, the series diverges by Raabe's test (Theorem 10.8.7). When $x = -\frac{1}{4}$, the series is $\Sigma(-1)^n a_n$, where $a_n = (2n)!/4^n(n!)^2$; and since the sequence (a_n) is decreasing, it follows from Example 10.8.10(3) that $a_n \to 0$; and we deduce that the given series converges.

(4) We now investigate the convergence of the series $\Sigma(\sin nx)/n$, where x is any real number. If $\sin(x/2) = 0$, then x is a multiple of 2π, and the series is the trivial series $\Sigma\, 0$. From now on we shall assume that $\sin(x/2) \neq 0$. Now given any natural number n, making use of the trigonometric identity

$$\cos(\alpha - \beta) - \cos(\alpha + \beta) = 2 \sin \alpha \sin \beta,$$

we have

$$\sum_{j=1}^{n} \sin jx = \sum_{j=1}^{n} \frac{2 \sin(x/2) \sin jx}{2 \sin(x/2)}$$

$$= \frac{1}{2 \sin(x/2)} \sum_{j=1}^{n} \left[\cos(2j - 1)\frac{x}{2} - \cos(2j + 1)\frac{x}{2} \right]$$

$$= \frac{\cos(x/2) - \cos(2n + 1)(x/2)}{2 \sin(x/2)},$$

and it follows that

$$\left| \sum_{j=1}^{n} \sin jx \right| \le \frac{1}{|\sin(x/2)|}.$$

We can therefore deduce from Dirichlet's test that the series $\Sigma (\sin nx)/n$ converges for every x. In Chapter 11 we shall go one step further and obtain the sum of this series.

(5) Given any real number α and any natural number n, the **binomial coefficient** $\binom{\alpha}{n}$ is defined to be $\alpha(\alpha - 1)(\alpha - 2) \cdots (\alpha - n + 1)/n!$. If $a_n = |\binom{\alpha}{n}|$, then as we saw in Example 10.8.8(1), the series $\Sigma\, a_n$ converges when $\alpha \ge 0$ and diverges when $\alpha < 0$; in other words, the series $\Sigma\binom{\alpha}{n}$ converges absolutely when $\alpha \ge 0$ but does not converge absolutely when $\alpha < 0$. We saw in Example 10.8.10(1) that (a_n) does not converge to zero when $\alpha \le -1$, and in this case the series $\Sigma\binom{\alpha}{n}$ diverges. Suppose, finally, that $-1 < \alpha < 0$. For each n we have

$$\binom{\alpha}{n} = \frac{(-1)^n (-\alpha)(1 - \alpha)(2 - \alpha) \cdots (n - 1 - \alpha)}{n!}$$

$$= (-1)^n \left| \binom{\alpha}{n} \right| = (-1)^n a_n;$$

and since (a_n) decreases to zero in this case, it follows from Dirichlet's test that $\Sigma\binom{\alpha}{n}$ converges. In summary, then, the series $\Sigma\binom{\alpha}{n}$ converges absolutely if $\alpha \ge 0$, converges conditionally if $-1 < \alpha < 0$, and diverges if $\alpha \le -1$.

10.11 EXERCISES

- **1.** Determine whether the following series converge or diverge.

  (a) $\sum \dfrac{(-1)^n \log n}{n}$ **(b)** $\sum \dfrac{\sin(n\pi/4)}{n}$

  (c) $\sum \left(\dfrac{1}{2} - 1\right) \left(\dfrac{1}{3} - 1\right) \cdots \left(\dfrac{1}{n} - 1\right)$

  (d) $\sum \left(\dfrac{1}{2^\delta} - 1\right) \left(\dfrac{1}{3^\delta} - 1\right) \cdots \left(\dfrac{1}{n^\delta} - 1\right)$, where $\delta > 0$.

  (e) $\sum \dfrac{(2 \log 2 - 1)(3 \log 3 - 1) \cdots (n \log n - 1)}{n!(\log 2)(\log 3) \cdots (\log n)} (-1)^n$

▪ **2.** For each of the following series, determine the values of x for which the series is absolutely convergent and the values of x for which the series is conditionally convergent.

(a) $\sum \dfrac{(3x - 2)^n}{n}$ **(b)** $\sum \dfrac{(\log x)^n}{n}$

(c) $\sum \dfrac{(-1)^n x^n}{(\log n)^x}$ **(d)** $\sum \dfrac{(3n)! x^n}{(2n)! n!}$

(e) $\sum \dfrac{n! x^n}{n^n}$ **(f)** $\sum \dfrac{(nx)^n}{n!}$

▪ **3.** Find the values of x for which the following series converge.

(a) $\sum \dfrac{\cos nx}{\log n}$ **(b)** $\sum \dfrac{(-1)^n \sin nx}{n}$

(c) $\sum \dfrac{\cos^2 nx}{\log n}$ **(d)** $\sum \left| \dfrac{\sin nx}{n} \right|$

▪ **4.** Find the values of x and α for which the **binomial series** $\sum \binom{\alpha}{n} x^n$ converges.

5. Given $\alpha > -1$, prove that there is a number K such that for all $x \in [0, 1]$ and all natural numbers n we have $\left| \sum_{j=1}^{n} \binom{\alpha}{j} x^j \right| \leq K$. *Hint:* Use the convergence of the series $\sum \binom{\alpha}{n}$ together with Theorem 10.10.5.

6. Prove that if $|x| \leq 1$ and N is any natural number, then

$$\left| \sum_{n=N}^{\infty} \frac{(-1)^{n-1} |x|^n}{n} \right| \leq \frac{1}{N}.$$

The following two exercises will be used in Section 11.2.

7. Given $|x| < 1$, prove that

$$(1 - 2x \cos \theta + x^2) \sum_{n=1}^{\infty} x^{n-1} \cos n\theta = \cos \theta - x$$

and

$$(1 - 2x \cos \theta + x^2) \sum_{n=1}^{\infty} x^{n-1} \sin n\theta = \sin \theta.$$

8. (a) Prove that if $|x| < 1$, then the series $\sum x^{n-1} \sin n\theta$ and $\sum x^{n-1} \cos n\theta$ converge absolutely for all θ.

(b) Prove that if θ is not an even multiple of π, then there exists a number K such that for all $x \in [0, 1)$ and all natural numbers n we have

$$\left| \sum_{j=1}^{n} x^{j-1} \cos j\theta \right| \leq K \quad \text{and} \quad \left| \sum_{j=1}^{n} x^{j-1} \sin j\theta \right| \leq K.$$

What can be said if θ is an even multiple of π?

(c) Prove that if θ is not an odd multiple of π, then there exists a number K such that for all $x \in (-1, 0]$ and all natural numbers n we have

$$\left| \sum_{j=1}^{n} x^{j-1} \cos j\theta \right| \leq K \quad \text{and} \quad \left| \sum_{j=1}^{n} x^{j-1} \sin j\theta \right| \leq K.$$

■ **9.** Prove **Abel's test** for convergence, which states that if (a_n) is a decreasing sequence of positive numbers, and if the series $\Sigma\, b_n$ converges, then the series $\Sigma\, a_n b_n$ is convergent. This result may be proved by the method of proof of Dirichlet's test, but can you find a much shorter proof?

10. Suppose (a_n) is an increasing sequence and $0 \leq a_n \leq \alpha$ for all n. Prove that if a sequence (b_n) satisfies $\left| \Sigma_{j=N}^{n} b_j \right| \leq K$ for all $n \geq N$, then whenever $n \geq N$, we have $\left| \Sigma_{j=N}^{n} a_j b_j \right| \leq 2K\alpha$.

■ **11.** Give an example of a sequence (a_n) of positive numbers and a sequence (b_n) of real numbers such that the following conditions hold.
   (a) $a_n \rightarrow 0$ as $n \rightarrow \infty$.
   (b) The sequence whose nth member is $\Sigma_{j=1}^{n} b_j$ is bounded.
   (c) The series $\Sigma\, a_n b_n$ diverges.

12. Give an example of sequences (a_n) and (b_n) such that the following conditions hold.
   (a) (a_n) is a decreasing sequence of positive numbers.
   (b) The sequence whose nth member is $\Sigma_{j=1}^{n} b_j$ is bounded.
   (c) The series $\Sigma\, a_n b_n$ diverges.

*13. (a) Suppose that a_1 and δ are rational numbers, $\delta > 0$, that α is any real number, and that $\alpha - \delta < a_1 < \alpha$. Prove that there exist rationals a_2 and b_1 such that $\alpha - \delta/2 < a_2 < \alpha < b_1$ and $(b_1 - a_1) + (b_1 - a_2) = \delta$. Deduce that it is possible to find two sequences (a_n) and (b_n) of rationals such that for every n we have $\alpha - \delta/2^n < a_{n+1} < \alpha < b_n$ and $(b_n - a_n) + (b_n - a_{n+1}) = \delta/2^{n-1}$.
   (b) Given any real number α, prove that there exists a sequence (c_n) of rational numbers such that $\Sigma\, |c_n|$ converges to a rational number and $\Sigma\, c_n$ converges to α.
   (c) The proof of Theorem 10.10.2 makes use of the comparison test, which in turn makes use of the fact that every increasing bounded sequence is convergent. The proof of the latter result uses the completeness axiom for $\boldsymbol{R}$. Is it really necessary to make use of the completeness axiom for $\boldsymbol{R}$ in order to prove that every absolutely convergent series is convergent?

10.12 REARRANGEMENTS OF SERIES

A series $\Sigma\, b_n$ is said to be a **rearrangement** of a series $\Sigma\, a_n$ if there is a one–one function p from $\boldsymbol{Z}^+$ onto $\boldsymbol{Z}^+$ such that for every natural number n we have $b_n = a_{p(n)}$.

For example, we might have $a_2 + a_1 + a_4 + a_3 + \cdots$, which is the series $\Sigma\, a_{p(n)}$ where $p(n) = n + 1$ for n odd and $p(n) = n - 1$ for n even.

In this section we shall show that rearrangement of the terms of an absolutely convergent series always yields an absolutely convergent series with the same sum. This is by no means true for conditionally convergent series. As a matter of fact, it can be shown that if a series $\Sigma\, a_n$ converges conditionally, then some of the rearrangements of $\Sigma\, a_n$ will diverge and convergent rearrangements can be found converging to any number at all. For details of this phenomenon, see Apostol [1], Section 8.18.

In order to prove the theorems of this section, we need a little notation. Given a sequence (a_n) and a finite set S of natural numbers, say $S = \{j_1, j_2, \ldots, j_n\}$, the sum $\Sigma_{j \in S}\, a_j$ is defined to be $\Sigma_{k=1}^{n} a_{j_k}$. As an example, note that when $S = \{1, 2, \ldots, n\}$, the sum $\Sigma_{j \in S}\, a_j$ is just the nth partial sum of the series $\Sigma\, a_n$. Our first theorem in this section shows that when the terms of a series are nonnegative, then the sum of the series can be expressed simply in terms of sums of the form $\Sigma_{j \in S}\, a_j$.

10.12.1 Theorem. Suppose (a_n) is a sequence of nonnegative numbers. Then

$$\sum_{n=1}^{\infty} a_n = \sup\left\{\sum_{j \in S} a_j \,\middle|\, S \text{ is a finite subset of } \mathbf{Z}^+\right\}.$$

■ **Proof.** Call the right side of the proposed identity α. For every natural N we have $\Sigma_{n=1}^{N} a_n \le \alpha$; and letting $N \to \infty$, we obtain $\Sigma_{n=1}^{\infty} a_n \le \alpha$. On the other hand, if S is any finite subset of $\mathbf{Z}^+$ and N is the largest member of S, then

$$\sum_{j \in S} a_j \le \sum_{j=1}^{N} a_j \le \sum_{n=1}^{\infty} a_n,$$

and it follows that $\alpha \le \Sigma_{n=1}^{\infty} a_n$. ■

10.12.2 The Rearrangement Theorem for Nonnegative Series. Suppose that (a_n) is a sequence of nonnegative numbers and that $\Sigma\, b_n$ is any rearrangement of $\Sigma\, a_n$. Then

$$\sum_{n=1}^{\infty} a_n = \sum_{n=1}^{\infty} b_n.$$

■ **Proof.** This follows at once from Theorem 10.12.1. ■

10.12.3 The Rearrangement Theorem for Absolutely Convergent Series. Suppose that $\Sigma\, a_n$ converges absolutely and that $\Sigma\, b_n$ is a rearrangement of $\Sigma\, a_n$. Then $\Sigma\, b_n$ also converges absolutely and has the same sum.

■ *Proof.* From Theorem 10.12.2, we see that

$$\sum_{n=1}^{\infty} |b_n| = \sum_{n=1}^{\infty} |a_n| \quad \text{and that} \quad \sum_{n=1}^{\infty} \left(|b_n| + b_n \right) = \sum_{n=1}^{\infty} \left(|a_n| + a_n \right);$$

and from this the result follows at once.   ■

10.13   ITERATED SERIES

What we are about to study are the series analogues of the results on iterated integrals that we saw in Section 9.8, and our principal result in this section will be the analogue for series of the improper integral form of Fichtenholz's theorem (Theorem 9.8.10). We begin by introducing the concept of **iterated series.** Suppose f is a function defined on $\mathbf{Z}^+ \times \mathbf{Z}^+$; suppose that for every natural number n the series $\sum_m f(m, n)$ is convergent; and for every natural n, define $\phi(n) = \sum_{m=1}^{\infty} f(m, n)$. If the series $\sum \phi(n)$ converges, then we say that $\sum_n \sum_m f(m, n)$ converges to the number $\sum_{n=1}^{\infty} \phi(n)$. In other words, the iterated sum $\sum_{n=1}^{\infty} \sum_{m=1}^{\infty} f(m, n)$ is defined to be $\sum_{n=1}^{\infty} [\sum_{m=1}^{\infty} f(m, n)]$. The sum $\sum_{m=1}^{\infty} \sum_{n=1}^{\infty} f(m, n)$ is defined similarly.

Now just as we saw for integrals in Section 9.8.9, the order of summation in an iterated series is important; and we cannot, in general, guarantee the equality of the sums $\sum_{n=1}^{\infty} \sum_{m=1}^{\infty} f(m, n)$ and $\sum_{m=1}^{\infty} \sum_{n=1}^{\infty} f(m, n)$, even when they both exist. We begin with an example that shows how this can happen.

10.13.1   Example.   In the array of numbers shown in Figure 10.2, let m count the rows starting at the top and let n count the columns starting at the left. For each pair $(m, n) \in \mathbf{Z}^+ \times \mathbf{Z}^+$ we define $f(m, n)$ to be the number in row m and column n.

$$
\begin{array}{cccccccc}
-1 & \dfrac{1}{2} & \dfrac{1}{4} & \dfrac{1}{8} & \dfrac{1}{16} & \cdots & \dfrac{1}{2^n} & \cdots \\[2ex]
0 & -1 & \dfrac{1}{2} & \dfrac{1}{4} & \dfrac{1}{8} & \cdots & \dfrac{1}{2^{n-1}} & \cdots \\[2ex]
0 & 0 & -1 & \dfrac{1}{2} & \dfrac{1}{4} & \cdots & \dfrac{1}{2^{n-2}} & \cdots \\[2ex]
0 & 0 & 0 & -1 & \dfrac{1}{2} & \cdots & \dfrac{1}{2^{n-3}} & \cdots \\
\cdot & \cdot & \cdot & \cdot & \cdot & \cdot & \cdot \\
\cdot & \cdot & \cdot & \cdot & \cdot & \cdot & \cdot \\
\cdot & \cdot & \cdot & \cdot & \cdot & \cdot & \cdot \\
\end{array}
$$

Figure 10.2

For every natural number m we have $\sum_{n=1}^{\infty} f(m, n) = 0$, and therefore, $\sum_{m=1}^{\infty} \sum_{n=1}^{\infty} f(m, n) = 0$. On the other hand, for every natural number n we have

$$\sum_{m=1}^{\infty} f(m, n) = \frac{-1}{2^{n-1}} \quad \text{and so} \quad \sum_{n=1}^{\infty} \sum_{m=1}^{\infty} f(m, n) = -2.$$

So the order of summation in these iterated series cannot be interchanged. Notice, however, that the function f changes sign and that the series $\sum_{n} \sum_{m} |f(m, n)|$ does not converge.

10.13.2   Theorem: Interchange of the Order of Summation for Nonnegative Series.   Suppose f is a nonnegative function defined on $\mathbf{Z}^+ \times \mathbf{Z}^+$. Then

$$\sum_{m=1}^{\infty} \sum_{n=1}^{\infty} f(m, n) = \sum_{n=1}^{\infty} \sum_{m=1}^{\infty} f(m, n).$$

Note that the statement of this theorem includes the assertion that if either side of the identity is ∞, then so is the other.

■ *Proof.*   The proof of this theorem will follow roughly the same lines as the proof of Theorem 9.8.10. To obtain a contradiction, let us assume that

$$\sum_{m=1}^{\infty} \sum_{n=1}^{\infty} f(m, n) < \sum_{n=1}^{\infty} \sum_{m=1}^{\infty} f(m, n).$$

Given any natural number n, define $\phi(n) = \sum_{m=1}^{\infty} f(m, n)$, and define $\phi_m(n)$ for each m by $\phi_m(n) = \sum_{j=1}^{m} f(j, n)$. Note that for each n we have $\phi_m(n) \to \phi(n)$ as $m \to \infty$. Now using our assumption that

$$\sum_{m=1}^{\infty} \sum_{n=1}^{\infty} f(m, n) < \sum_{n=1}^{\infty} \phi(n),$$

choose a natural number N such that

$$\sum_{m=1}^{\infty} \sum_{n=1}^{\infty} f(m, n) < \sum_{n=1}^{N} \phi(n).$$

Since $\phi_m(n) \to \phi(n)$ for every n, it follows from the arithmetical rules for limits of sequences (Theorem 4.5.2) that

$$\sum_{n=1}^{N} \phi_m(n) \to \sum_{n=1}^{N} \phi(n) \quad \text{as} \quad m \to \infty;$$

and we can therefore choose a natural number M such that

$$\sum_{m=1}^{\infty} \sum_{n=1}^{\infty} f(m, n) < \sum_{n=1}^{N} \phi_M(n).$$

But this gives us

$$\sum_{m=1}^{\infty} \sum_{n=1}^{\infty} f(m, n) < \sum_{n=1}^{N} \sum_{m=1}^{M} f(m, n) = \sum_{m=1}^{M} \sum_{n=1}^{N} f(m, n) \le \sum_{m=1}^{M} \sum_{n=1}^{\infty} f(m, n)$$

$$\le \sum_{m=1}^{\infty} \sum_{n=1}^{\infty} f(m, n),$$

which is the desired contradiction. ∎

10.13.3 Theorem: Interchange of the Order of Summation for Absolutely Convergent Series.

Suppose that f is a function defined on $\mathbf{Z}^+ \times \mathbf{Z}^+$ and that the series $\sum_m \sum_n |f(m, n)|$ converges. Then $\sum_n \sum_m |f(m, n)|$ also converges, and

$$\sum_{m=1}^{\infty} \sum_{n=1}^{\infty} f(m, n) = \sum_{n=1}^{\infty} \sum_{m=1}^{\infty} f(m, n).$$

∎ **Proof.** As usual when we want to extend a theorem about nonnegative functions to a situation in which we assume absolute convergence, we apply the nonnegative case to each of the functions $|f| + f$ and $|f|$ and then subtract. We leave the details as an exercise. ∎

10.14 MULTIPLICATION OF SERIES

For convenience, we shall start our series at $n = 0$ in this section. The question we have to consider is how a product $(\sum_{n=0}^{\infty} a_n)(\sum_{n=0}^{\infty} b_n)$ can be "multiplied out," where $\sum a_n$ and $\sum b_n$ are two given convergent series. There are a number of different approaches to this problem, leading to concepts called the *Cauchy product*, the *Laurent product*, the *Fourier product*, and the *Dirichlet product*; but we shall be looking at only one of these, the Cauchy product. To motivate the Cauchy product, we shall consider two *power series* $\sum a_n x^n$ and $\sum b_n x^n$. Under reasonable circumstances, we might expect to have

$$(a_0 + a_1 x + a_2 x^2 + a_3 x^3 + \cdots)(b_0 + b_1 x + b_2 x^2 + b_3 x^3 + \cdots)$$

$$= a_0 b_0 + (a_1 b_0 + a_0 b_1)x + (a_2 b_0 + a_1 b_1 + a_0 b_2)x^2$$

$$+ (a_3 b_0 + a_2 b_1 + a_1 b_2 + a_0 b_3)x^3 + \cdots$$

$$+ (a_n b_0 + \cdots + a_0 b_n)x^n + \cdots.$$

Setting $x = 1$ in this identity suggests the definition of a Cauchy product.

We define the **Cauchy product** of two given series $\sum a_n$ and $\sum b_n$ to be the series $\sum c_n$, where for each n we have

$$c_n = a_n b_0 + a_{n-1} b_1 + \cdots + a_0 b_n = \sum_{j=0}^{n} a_{n-j} b_j.$$

Ideally, one would like to say that if $\Sigma\ c_n$ is the Cauchy product of two convergent series $\Sigma\ a_n$ and $\Sigma\ b_n$, then $\Sigma\ c_n$ converges; and

$$\sum_{n=0}^{\infty} c_n = \left(\sum_{n=0}^{\infty} a_n\right)\left(\sum_{n=0}^{\infty} b_n\right);$$

but unfortunately, the series $\Sigma\ c_n$ can diverge, as we see in the following two examples.

(1) For each n, define

$$a_n = b_n = \frac{(-1)^n}{\log(n + 2)}.$$

The series $\Sigma\ a_n$ and $\Sigma\ b_n$ converge by Dirichlet's test. Now let $\Sigma\ c_n$ be the Cauchy product of $\Sigma\ a_n$ and $\Sigma\ b_n$. Then for each n we have

$$c_n = \frac{(-1)^n}{\log(n + 2)\ \log 2} + \frac{(-1)^n}{\log(n + 1)\ \log 3}$$

$$+ \cdots + \frac{(-1)^n}{\log 2\ \log(n + 2)};$$

and so for each n we have

$$|c_n| \ge \frac{n}{[\log(n + 2)]^2}.$$

Therefore, $|c_n| \to \infty$, and it follows that $\Sigma\ c_n$ is divergent.

(2) For each n, define

$$a_n = b_n = \frac{(-1)^n}{\sqrt{n + 1}}.$$

Once again, $\Sigma\ a_n$ and $\Sigma\ b_n$ converge; but if $\Sigma\ c_n$ is the Cauchy product of $\Sigma\ a_n$ and $\Sigma\ b_n$, then $|c_n| \ge 1$ for each n. We leave the details as an exercise.

So as we have seen, the Cauchy product of two convergent series can diverge, but this is the only pathology that can occur. In this section we shall prove that if $\Sigma\ c_n$ is the Cauchy product of two convergent series $\Sigma\ a_n$ and $\Sigma\ b_n$, and if $\Sigma\ c_n$ happens to converge, then the identity

$$\sum_{n=1}^{\infty} c_n = \left(\sum_{n=0}^{\infty} a_n\right)\left(\sum_{n=0}^{\infty} b_n\right)$$

is assured. We shall also prove that if at least one of the convergent series $\Sigma\ a_n$ and $\Sigma\ b_n$ happens to converge absolutely, then the series $\Sigma\ c_n$ *will* converge. Finally, we shall use an interesting theorem of Neder to show that a sufficient condition for the convergence of $\Sigma\ c_n$ is that the two series $\Sigma\ a_n$ and $\Sigma\ b_n$ should converge and that the sequences (na_n) and (nb_n) should be bounded. Neder's theorem can be omitted without loss of continuity.

We begin this section on multiplication of series with a lemma which might be reminiscent of the dominated convergence theorem in Section 9.7 and which contains the technical information that we need.

10.14.1 Lemma. Suppose that (f_n) is a sequence of functions defined on the set P of nonnegative integers, and suppose that (f_n) converges pointwise on P to a function f; in other words, suppose that for each $j \in P$ we have $f_n(j) \to f(j)$ as $n \to \infty$. Suppose that $\Sigma \alpha_j$ is a given convergent series and that for all j and n we have $|f_n(j)| \le \alpha_j$. Then

$$\sum_{j=0}^{n} f_n(j) \to \sum_{j=0}^{\infty} f(j) \qquad \text{as} \qquad n \to \infty.$$

■ **Proof.** We remark first that for every j we have $|f(j)| \le \alpha_j$; and therefore, the absolute convergence of $\Sigma f(j)$ follows from the comparison test. Let $\varepsilon > 0$, and choose a natural number N_1 such that $\Sigma_{j=N_1}^{\infty} \alpha_j < \varepsilon/3$. Now that N_1 has been chosen, choose a natural number N_2 such that whenever $n \ge N_2$, we have

$$\left| \sum_{j=0}^{N_1} f_n(j) - \sum_{j=0}^{N_1} f(j) \right| < \frac{\varepsilon}{3}.$$

Define N to be the larger of the two numbers N_1 and N_2. Then whenever $n \ge N$, we have

$$\left| \sum_{j=0}^{n} f_n(j) - \sum_{j=0}^{\infty} f(j) \right| = \left| \sum_{j=0}^{N_1} f_n(j) - \sum_{j=0}^{N_1} f(j) + \sum_{j=N_1+1}^{n} f_n(j) - \sum_{j=N_1+1}^{\infty} f(j) \right|$$

$$\le \left| \sum_{j=0}^{N_1} f_n(j) - \sum_{j=0}^{N_1} f(j) \right| + \sum_{j=N_1+1}^{n} \alpha_j + \sum_{j=N_1+1}^{\infty} \alpha_j$$

$$< \frac{\varepsilon}{3} + \frac{\varepsilon}{3} + \frac{\varepsilon}{3} = \varepsilon. \qquad ■$$

10.14.2 Mertens's Theorem. If Σa_n converges and Σb_n converges absolutely, and if Σc_n is the Cauchy product of Σa_n and Σb_n, then Σc_n converges; and we have

$$\sum_{n=0}^{\infty} c_n = \left(\sum_{n=0}^{\infty} a_n \right) \left(\sum_{n=0}^{\infty} b_n \right).$$

■ **Proof.** Using the fact that Σa_n converges, choose a number K such that for every n we have $|\Sigma_{m=0}^{n} a_m| \le K$. Now from the definition of c_m as being $a_m b_0$

$+ a_{m-1}b_1 + \cdots + a_0 b_m$ for each m, we see that $\sum_{m=0}^{n} c_m$ is the sum of all the products $a_i b_j$ for which $i + j \leq n$. Therefore,

$$\sum_{m=0}^{n} c_m = \sum_{j=0}^{n} \sum_{i=0}^{n-j} a_i b_j = \sum_{j=0}^{n} f_n(j),$$

where $f_n(j) = b_j \left(\sum_{i=0}^{n-j} a_i \right)$. Since $|f_n(j)| \leq K|b_j|$ for all j and n, and since for every j we have

$$f_n(j) \to b_j \left(\sum_{i=0}^{\infty} a_i \right) \qquad \text{as} \qquad n \to \infty,$$

it follows from Lemma 10.14.1 that

$$\sum_{m=0}^{n} c_m = \sum_{j=0}^{n} f_n(j) \to \sum_{j=0}^{\infty} b_j \left(\sum_{i=0}^{\infty} a_i \right) = \left(\sum_{i=0}^{\infty} a_i \right) \left(\sum_{j=0}^{\infty} b_j \right) \qquad \text{as} \qquad n \to \infty. \qquad \blacksquare$$

Our next theorem is the original theorem of Cauchy. This is the special case of Mertens's theorem that results when we assume that *both* of the series $\Sigma\, a_n$ and $\Sigma\, b_n$ converge absolutely. In the proof that follows, we make use of Mertens's theorem; but as an exercise, you might like to construct an alternative proof that uses the results of Section 10.13.

10.14.3 Cauchy's Theorem. If $\Sigma\, a_n$ and $\Sigma\, b_n$ both converge absolutely, and if $\Sigma\, c_n$ is the Cauchy product of $\Sigma\, a_n$ and $\Sigma\, b_n$, then $\Sigma\, c_n$ converges absolutely; and we have

$$\sum_{n=0}^{\infty} c_n = \left(\sum_{n=0}^{\infty} a_n \right) \left(\sum_{n=0}^{\infty} b_n \right).$$

■ *Proof.* The only thing that needs to be proved is the absolute convergence of the series $\Sigma\, c_n$. But this follows from the fact that if $\Sigma\, d_n$ is the Cauchy product of the series $\Sigma\, |a_n|$ and $\Sigma\, |b_n|$, then $\Sigma\, d_n$ converges by Mertens's theorem, and $|c_n| \leq d_n$ for each n. ■

10.14.4 Abel's Theorem on Multiplication of Series. If $\Sigma\, a_n$ and $\Sigma\, b_n$ converge, if $\Sigma\, c_n$ is the Cauchy product of $\Sigma\, a_n$ and $\Sigma\, b_n$, and if the series $\Sigma\, c_n$ is convergent, then we have

$$\sum_{n=0}^{\infty} c_n = \left(\sum_{n=0}^{\infty} a_n \right) \left(\sum_{n=0}^{\infty} b_n \right).$$

■ *Proof.* This theorem is a simple consequence of Mertens's theorem and Abel's theorem for series (Theorem 10.10.6). From Abel's theorem for series we deduce that if $0 \leq x < 1$, then all three of the series $\Sigma\, a_n x^n$, $\Sigma\, b_n x^n$, and $\Sigma\, c_n x^n$ converge absolutely; and it follows from Mertens's theorem that

$$\sum_{n=0}^{\infty} c_n x^n = \left(\sum_{n=0}^{\infty} a_n x^n\right)\left(\sum_{n=0}^{\infty} b_n x^n\right).$$

Now letting $x \to 1$ and applying Abel's theorem again, we obtain the desired result. ∎

The following theorem can be omitted without loss of continuity.

10.14.5* Neder's Theorem. Suppose $\Sigma\, c_n$ is the Cauchy product of two convergent series $\Sigma\, a_n$ and $\Sigma\, b_n$, and suppose that it is possible to find two positive functions ϕ and ψ defined on $\mathbf{Z}^+$ and a positive number K such that for every n we have $\phi(n) + \psi(n) = n$, and

$$\sum_{\phi(n) \leq j \leq n} |a_j| \leq K \qquad \text{and} \qquad \sum_{\psi(n) \leq j \leq n} |b_j| \leq K.$$

Then $\Sigma\, c_n$ converges, and (of course) we have

$$\sum_{n=0}^{\infty} c_n = \left(\sum_{n=0}^{\infty} a_n\right)\left(\sum_{n=0}^{\infty} b_n\right).$$

∎ *Proof.* Suppose first that $\phi(n)$ does not tend to infinity as $n \to \infty$. Choose a natural number N such that for arbitrarily large values of n we have $\phi(n) \leq N$. For all such values of n we have

$$\sum_{j=0}^{n} |a_j| = \sum_{0 \leq n < \phi(n)} |a_j| + \sum_{\phi(n) \leq j \leq n} |a_j| \leq \sum_{j=1}^{N} |a_j| + K;$$

and it follows that the series $\Sigma\, a_n$ is absolutely convergent. So in the event that $\phi(n)$ does not tend to infinity as $n \to \infty$, the result follows at once from Mertens's theorem; and the result follows similarly if $\psi(n)$ does not tend to infinity as $n \to \infty$.

We shall therefore assume from now on that as $n \to \infty$, we have $\phi(n) \to \infty$ and $\psi(n) \to \infty$. Now we need some notation: As in Lemma 10.14.1, we shall denote the set of nonnegative integers as P. For each natural number n, define

$$W(n) = \{(i, j) \in P \times P \mid i + j \leq n\},$$

$$A(n) = \{(i, j) \in P \times P \mid i \leq \phi(n) \quad \text{and} \quad j \leq \psi(n)\},$$

$$B(n) = \{(i, j) \in W(n) \mid i > \phi(n) \quad \text{and} \quad j \leq \psi(n)\},$$

$$C(n) = \{(i, j) \in W(n) \mid i \leq \phi(n) \quad \text{and} \quad j > \psi(n)\}.$$

Note that for every n, $W(n)$ is the union of the mutually disjoint sets $A(n)$, $B(n)$, and $C(n)$. Therefore, given any natural number n, we have

$$\sum_{m=0}^{n} c_m = \sum_{m=0}^{n} \sum_{j=0}^{m} a_{m-j} b_j = \sum_{(i,j) \in W(n)} a_i b_j$$

$$= \sum_{(i,j) \in A(n)} a_i b_j + \sum_{(i,j) \in B(n)} a_i b_j + \sum_{(i,j) \in C(n)} a_i b_j.$$

We shall look at these three sums one at a time. From the definition of $A(n)$ we see that

$$\sum_{(i,j) \in A(n)} a_i b_j = \left(\sum_{i \leq \phi(n)} a_i \right) \left(\sum_{j \leq \psi(n)} b_j \right);$$

and since $\phi(n) \to \infty$ and $\psi(n) \to \infty$ as $n \to \infty$, it follows that

$$\sum_{(i,j) \in A(n)} a_i b_j \to \left(\sum_{i=0}^{\infty} a_i \right) \left(\sum_{j=0}^{\infty} b_j \right) \qquad \text{as} \qquad n \to \infty.$$

To complete the proof, we must therefore show that each of the sums

$$\sum_{(i,j) \in B(n)} a_i b_j \qquad \text{and} \qquad \sum_{(i,j) \in C(n)} a_i b_j$$

tends to zero as $n \to \infty$.

We look first at $\sum_{(i,j) \in B(n)} a_i b_j$. Let $\varepsilon > 0$, and choose a natural number N such that whenever $n \geq N$, we have $|\sum_{j=n}^{\infty} b_j| < \varepsilon$. Notice that whenever $n \geq N$, we have $|\sum_{j=N}^{n} b_j| < 2\varepsilon$. We now partition each set $B(n)$ into two parts:

$$B_1(n) = \{(i, j) \in B(n) \mid j \leq N\},$$

$$B_2(n) = \{(i, j) \in B(n) \mid j > N\}.$$

Now as long as n is sufficiently large to make $\psi(n) > N$, we see that whenever $j \leq N$, we have $\phi(n) \leq n - j$. Therefore,

$$\sum_{(i,j) \in B_1(n)} a_i b_j = \sum_{j=0}^{N} b_j \left(\sum_{\phi(n) < i \leq n-j} a_i \right) = \sum_{j=0}^{N} b_j \left(\sum_{i=0}^{n-j} a_i - \sum_{0 \leq i \leq \phi(n)} a_i \right)$$

$$\to \sum_{j=0}^{N} b_j \left(\sum_{i=0}^{\infty} a_i - \sum_{i=0}^{\infty} a_i \right) = 0 \qquad \text{as} \qquad n \to \infty.$$

The sum $\sum_{(i,j) \in B_2(n)} a_i b_j$ may be written as

$$\sum_{(i,j) \in B_2(n)} a_i b_j = \sum_{\phi(n) < i \leq n} \sum_{j=N+1}^{n-i} a_i b_j,$$

where the second of these summations is understood to be zero for those values of i for which $n - i < N + 1$. Therefore,

$$\left| \sum_{(i,j) \in B_2(n)} a_i b_j \right| \leq \sum_{\phi(n) < i \leq n} |a_i| \left| \sum_{j=N+1}^{n-i} b_j \right| \leq 2\varepsilon K,$$

and we have therefore shown that the sum $\Sigma_{(i,j)\in B(n)}\, a_i b_j$ can be made arbitrarily small by making n large enough. In other words,

$$\sum_{(i,j)\in B(n)} a_i b_j \to 0 \quad \text{as} \quad n \to \infty;$$

and the proof that $\Sigma_{(i,j)\in C(n)}\, a_i b_j \to 0$ as $n \to \infty$ is similar. This completes the proof. ∎

10.14.6* **The Edmonds-Hardy Theorem.** Suppose that $\Sigma\, c_n$ is the Cauchy product of two convergent series $\Sigma\, a_n$ and $\Sigma\, b_n$, and suppose that the sequences (na_n) and (nb_n) are bounded. Then $\Sigma\, c_n$ converges, and (of course) we have

$$\sum_{n=0}^{\infty} c_n = \left(\sum_{n=0}^{\infty} a_n\right)\left(\sum_{n=0}^{\infty} b_n\right).$$

▪ **Proof.** We make use of Neder's theorem with $\phi(n) = \psi(n) = n/2$. Choose a number α such that $|na_n| \le \alpha$ and $|nb_n| \le \alpha$ for all n. Then given any $n \ge 2$, we have

$$\sum_{\phi(n)\le j\le n} |a_j| \le \frac{\alpha}{n/2}\left(\frac{n}{2}+1\right) \le 2\alpha;$$

and similarly, $\Sigma_{\phi(n)\le j\le n} |b_j| \le 2\alpha$, so that the requirements of Neder's theorem are satisfied. ∎

10.15 EXERCISES

1. Find a rearrangement $\Sigma\, b_n$ of the series $\Sigma(-1)^n/n$ whose sum is ∞.

▪ 2. Calculate the Cauchy product of the series $\Sigma(-1)^n x^n$ and $\Sigma\, x^n$. By looking at the sums of these series, verify that Mertens's theorem holds for this product when $|x| < 1$.

▪ 3. (In order to do this exercise, you will need to make use of the binomial theorem.) Calculate the Cauchy product of the series $\Sigma\, x^n/n!$ and $\Sigma\, y^n/n!$; and by looking at the sums of these series, verify that Mertens's theorem holds for this product.

4. Prove that if $\alpha > 0$ and $\beta > 1$, then the Cauchy product of the series

$$\sum \frac{(-1)^n}{(n+1)^\alpha} \quad \text{and} \quad \sum \frac{(-1)^n}{(n+1)^\beta}$$

converges.

▪ 5. Given $S_n = \Sigma_{j=0}^n 1/(j+1)$, prove that

$$\sum_{n=0}^{\infty} \frac{(-1)^n}{n + 2} S_n = \frac{1}{2} (\log 2)^2.$$

Hint: $(\log 2)^2$ is the square of the sum of the series in Example 10.2(7).

6. Prove that if α and β are positive numbers and $\alpha + \beta \leq 1$, then the Cauchy product of the series

$$\sum \frac{(-1)^n}{(n + 1)^{\alpha}} \quad \text{and} \quad \sum \frac{(-1)^n}{(n + 1)^{\beta}}$$

diverges.

* **7.** State and prove an analogue of Mertens's theorem for improper integrals of the form $\int_0^{\to\infty}$.

* **8.** State and prove an analogue of Abel's theorem (Theorem 10.14.4) for improper integrals of the form $\int_0^{\to\infty}$.

* **9.** State and prove analogues of Neder's theorem and of the Hardy-Edmonds theorem for improper integrals of the form $\int_0^{\to\infty}$.

***10.** Prove that if $\sum c_n$ is the Cauchy product of two convergent series $\sum a_n$ and $\sum b_n$, and if for some number $\delta > 0$ the sequences $(na_n \log n)$ and $(n^{\delta} b_n)$ are bounded, then $\sum c_n$ converges. *Hint:* Define $\phi(n) = n^{\delta}$.

Chapter *11*

Sequences and Series of Functions

The topic of sequences and series of functions is not new to us as we begin this chapter. We began this topic with our study of the bounded convergence theorem in Chapter 8 and with the important applications of this theorem in Chapter 9; and we continued it in Chapter 10, where we encountered some series of functions. Therefore, we should look upon the present chapter as a *continuation* of our study of sequences and series of functions rather than the beginning of a brand new topic. We shall begin with a fairly general application of the bounded convergence theorem to series. Then we shall study a type of convergence called **uniform convergence**, which is generally stronger than bounded convergence and pointwise convergence; and finally, we shall make an in-depth study of power series.

11.1 THE BOUNDED CONVERGENCE THEOREM FOR SERIES

11.1.1 Definition of Bounded Convergence of a Series. (Compare with Section 8.20.3.) Suppose (f_n) is a sequence of functions defined on a set S and that

f is a function defined on S. We say that the series Σf_n **converges pointwise** to f on S if $f(x) = \Sigma_{n=1}^{\infty} f_n(x)$ for every $x \in S$. If Σf_n converges pointwise to f on S, and if there exists a number K such that for all $x \in S$ and $n \in \mathbf{Z}^+$ we have $|\Sigma_{j=1}^{n} f_j(x)| \leq K$, then we say that Σf_n **converges boundedly** to f on S.

Another way of saying that Σf_n converges boundedly to f on S is simply to say that the sequence of partial sums $\Sigma_{j=1}^{n} f_j$ converges boundedly to f on S; and by looking at the definition in this way, we arrive at the series analogue of the bounded convergence theorem given in the following subsection.

11.1.2 Bounded Convergence Theorem for Series.

Suppose that (f_n) is a sequence of Riemann-integrable functions on an interval $[a, b]$ and that the series Σf_n converges boundedly on $[a, b]$. Then the series $\Sigma \int_a^b f_n(x)\, dx$ converges, and in the event that $\Sigma_{n=1}^{\infty} f_n$ is Riemann-integrable on $[a, b]$, we have

$$\int_a^b \sum_{n=1}^{\infty} f_n(x)\, dx = \sum_{n=1}^{\infty} \int_a^b f_n(x)\, dx.$$

We often refer to this theorem more briefly by saying that if the sum of a boundedly convergent series of Riemann-integrable functions is Riemann-integrable, then the integral of this sum may be obtained by "integrating the series **term by term**." A corresponding result for differentiating the sum of a series term by term is the following analogue of Theorem 8.20.6.

11.1.3 Theorem on Differentiation of the Sum of a Series Term by Term.

Suppose that (f_n) is a sequence of differentiable functions on an interval $[a, b]$, that f_n' is continuous on $[a, b]$ for each n, and that the series $\Sigma f_n'$ converges boundedly on $[a, b]$ and has a continuous sum. Suppose, finally, that there exists a number c in $[a, b]$ for which the series $\Sigma f_n(c)$ converges. Then the series Σf_n converges pointwise on $[a, b]$ and has a differentiable sum, and we have $(\Sigma_{n=1}^{\infty} f_n)' = \Sigma_{n=1}^{\infty} f_n'$.

In the following section we demonstrate how the bounded convergence theorem can be used to obtain the sums of some interesting series. While not especially difficult, the section requires you to "dirty your hands" a little, and it may be omitted without loss of continuity.

11.2* SUMMATION OF SOME SPECIAL SERIES

11.2.1* An Application of the Bounded Convergence Theorem for Series.

In Example 10.10.7(4) we used Dirichlet's test to show that the series $\Sigma (\sin n\theta)/n$ is convergent for all θ, and in a similar way it may be shown that $\Sigma [(-1)^{n-1} \sin n\theta]/n$ is convergent for all θ, $\Sigma (\cos n\theta)/n$ converges if θ is not an even multiple of π, and $\Sigma [(-1)^{n-1} \cos n\theta]/n$ converges if θ is not an odd multiple of π.

We shall now find the sums of these series. We shall be making use of Exercises 10.11(7) and 10.11(8) in this section, and so it would be a good idea to review (or start) these exercises before you go any further. From Exercise 10.11(7) we know that for $|x| < 1$

$$\sum_{n=1}^{\infty} x^{n-1} \cos n\theta = \frac{\cos \theta - x}{1 - 2x \cos \theta + x^2}.$$

Now from Exercise 10.11(8)(b), we see that as long as θ is not an even multiple of π, we may apply the bounded convergence theorem for series to this identity and obtain

$$\sum_{n=1}^{\infty} \int_0^1 x^{n-1} \cos n\theta \; dx = \int_0^1 \sum_{n=1}^{\infty} x^{n-1} \cos n\theta \; dx = \int_0^1 \frac{\cos \theta - x}{1 - 2x \cos \theta + x^2} \; dx$$

and therefore,

$$\sum_{n=1}^{\infty} \frac{\cos n\theta}{n} = -\frac{1}{2} \log (2 - 2 \cos \theta).$$

Similarly, we see from Exercise 10.11(7) that for $|x| < 1$

$$\sum_{n=1}^{\infty} x^{n-1} \sin n\theta = \frac{\sin \theta}{1 - 2x \cos \theta + x^2},$$

and so for any θ we have

$$\sum_{n=1}^{\infty} \int_0^1 x^{n-1} \sin n\theta \; dx = \int_0^1 \sum_{n=1}^{\infty} x^{n-1} \sin n\theta \; dx = \int_0^1 \frac{\sin \theta}{1 - 2x \cos \theta + x^2} \; dx.$$

Therefore, in order to evaluate $\sum_{n=1}^{\infty} (\sin n\theta)/n$, we need to evaluate the integral

$$\int_0^1 \frac{\sin \theta}{1 - 2x \cos \theta + x^2} \; dx.$$

We shall show that if $0 < \theta < 2\pi$, then this integral is equal to $(\pi - \theta)/2$. This is clear when $\theta = \pi$, and so we shall assume that $\theta \neq \pi$. In this case

$$\int_0^1 \frac{\sin \theta}{1 - 2x \cos \theta + x^2} \; dx = \int_0^1 \frac{\sin \theta}{(x - \cos \theta)^2 + \sin^2 \theta} \; dx$$

$$= \arctan\left(\frac{1 - \cos \theta}{\sin \theta}\right) + \arctan(\cot \theta)$$

$$= \arctan\left(\tan \frac{1}{2} \theta\right) + \arctan\left[\tan\left(\frac{\pi}{2} - \theta\right)\right],$$

and by checking each of the cases $0 < \theta < \pi$ and $\pi < \theta < 2\pi$ separately, one can see that the latter expression is just $(\pi - \theta)/2$. We have therefore shown that for $0 < \theta < 2\pi$ we have

$$\sum_{n=1}^{\infty} \frac{\sin n\theta}{n} = \frac{1}{2}(\pi - \theta).$$

11.2.2*   Exercises

* 1. Replace x by $-x$ in the identities of Exercise 10.11(7) and then use the bounded convergence theorem for series to show that if θ is not an odd multiple of π, we have

$$\sum_{n=1}^{\infty} \frac{(-1)^{n-1} \cos n\theta}{n} = \frac{1}{2}\log(2 + 2\cos\theta).$$

* 2. Prove if $-\pi < \theta < \pi$, then

$$\sum_{n=1}^{\infty} \frac{(-1)^{n-1} \sin n\theta}{n} = \frac{\theta}{2}.$$

* 3. Use the method of Example 10.10.7(4) to prove that there exists a number K such that whenever $0 \leq \theta \leq \pi$ and n is any natural number, we have

$$\left| \sum_{j=1}^{n} (-1)^{j-1}(\pi - \theta)\sin j\theta \right| \leq K.$$

Deduce that

$$\sum_{n=1}^{\infty} \frac{(-1)^{n-1}}{n} \int_0^{\pi} (\pi - \theta)\sin n\theta \, d\theta = \int_0^{\pi} \frac{\theta}{2}(\pi - \theta) \, d\theta,$$

and therefore that

$$\sum_{n=1}^{\infty} \frac{(-1)^{n-1}}{n^2} = \frac{\pi^2}{12}.$$

* 4. Apply the method of exercise 3 to the identity

$$\sum_{n=1}^{\infty} \frac{\theta \sin n\theta}{n} = \frac{\theta}{2}(\pi - \theta).$$

Your final outcome should be the same.

* 5. Prove that

$$\sum_{n=1}^{\infty} \frac{1}{(2n)^2} = \frac{1}{4}\sum_{n=1}^{\infty} \frac{1}{n^2} \quad \text{and} \quad \sum_{n=1}^{\infty} \frac{1}{(2n-1)^2} = \frac{3}{4}\sum_{n=1}^{\infty} \frac{1}{n^2},$$

and deduce that

$$\sum_{n=1}^{\infty} \frac{(-1)^{n-1}}{n^2} = \frac{1}{2} \sum_{n=1}^{\infty} \frac{1}{n^2}.$$

* **6.** Prove that

$$\sum_{n=1}^{\infty} \frac{1}{n^2} = \frac{\pi^2}{6}, \qquad \sum_{n=1}^{\infty} \frac{1}{(2n)^2} = \frac{\pi^2}{24}, \qquad \text{and} \qquad \sum_{n=1}^{\infty} \frac{1}{(2n-1)^2} = \frac{\pi^2}{8}.$$

* **7.** Prove that if $0 \le \theta < \pi$, then the series $\sum [(-1)^{n-1} \sin nx]/n$ can be integrated term by term with respect to x on the interval $[0, \theta]$, and deduce that if $0 \le \theta \le \pi$, then we have

$$\sum_{n=1}^{\infty} \frac{(-1)^{n-1} \cos n\theta}{n^2} = \frac{\pi^2}{12} - \frac{\theta^2}{4}.$$

* **8.** Apply the method of exercise 7 to the series $\sum (\sin nx)/n$ on the interval $[\theta, \pi]$ where $0 \le \theta < \pi$, and deduce that if $0 \le \theta \le \pi$, then we have

$$\sum_{n=1}^{\infty} \frac{\cos n\theta}{n^2} = \frac{\pi^2}{6} - \frac{\pi\theta}{2} + \frac{\theta^2}{4}.$$

* **9.** Integrate the series in exercises 7 and 8 term by term, and show that whenever $0 \le \theta \le \pi$, we have

$$\sum_{n=1}^{\infty} \frac{\sin n\theta}{n^3} = \frac{2\pi^2\theta - 3\pi\theta^2 + \theta^3}{12}$$

and

$$\sum_{n=1}^{\infty} \frac{(-1)^{n-1} \sin n\theta}{n^3} = \frac{\pi^2\theta - \theta^3}{12},$$

and deduce that

$$\sum_{n=1}^{\infty} \frac{(-1)^{n-1}}{(2n-1)^3} = \frac{\pi^3}{32}.$$

*10. Justify the integration term by term of $\sum_{n=0}^{\infty} (1 - x)x^{3n}$ on the interval $[0, 1]$, and deduce that

$$\sum_{n=0}^{\infty} \frac{1}{(3n+1)(3n+2)} = \frac{\pi}{3\sqrt{3}}.$$

*11. Evaluate $\sum_{n=1}^{\infty} (5n + 2)/2n^2(2n + 1)$.

11.3 POINTWISE AND UNIFORM CONVERGENCE

11.3.1 The Need for a Stronger Type of Convergence.
As we have seen, the concept of bounded convergence plays a fundamental role in analysis, but even bounded convergence is inadequate from some points of view. The problem is that a sequence of continuous functions on a set S can quite easily converge boundedly to a function which is not continuous on S, and a sequence of functions which are Riemann-integrable on an interval $[a, b]$ can quite easily converge boundedly to a function which is not Riemann-integrable on $[a, b]$. The following two examples illustrate this.

(1) For each natural number n and each point $x \in [0, 2]$, we define

$$f_n(x) = \frac{x^n}{1 + x^n}.$$

It is easy to see that if $f(x) = 0$ for $0 \le x < 1$, and $f(1) = \frac{1}{2}$ and $f(x) = 1$ for $1 < x \le 2$, then the sequence (f_n) converges boundedly to f on $[0, 2]$. Notice that although every function f_n is continuous on $[0, 2]$, the limit function f is not.

(2) In Example 8.20.2(3) we saw a sequence (f_n) of step functions on $[0, 1]$ which converges boundedly on $[0, 1]$ to the function f which is 1 at every rational point of $[0, 1]$ and 0 at every irrational. This shows that a sequence of Riemann-integrable functions can converge boundedly to a nonintegrable function.

With these examples in mind, we now introduce a stronger kind of convergence of a sequence of functions, one that will guarantee that if each function f_n is continuous (or Riemann-integrable), then so is the limit function.

11.3.2 Definition of Uniform Convergence.
If (f_n) is a sequence of functions defined on a set S, then we say that (f_n) **converges uniformly** to a function f on S if

$$\sup\{|f_n(x) - f(x)| \,|\, x \in S\} \to 0 \qquad \text{as} \qquad n \to \infty.$$

If the sequence (f_n) converges uniformly to f on S, we write $f_n \to f$ *uniformly* on S.

As we did with bounded convergence and pointwise convergence, we extend this definition of uniform convergence to series by saying that a series $\Sigma\, f_n$ converges uniformly on a set S if the sequence of nth partial sums of this series converges uniformly on S.

The simple theorem that follows provides an alternative way of stating the definition of uniform convergence.

11.3.3 Theorem.
Given a sequence (f_n) of functions defined on a set S and a function f defined on S, the following two conditions are equivalent:

(1) $f_n \to f$ uniformly on S.

(2) For every number $\varepsilon > 0$ there exists a number N such that whenever $n \geq N$ and $x \in S$, we have $|f_n(x) - f(x)| < \varepsilon$.

■ *Proof.* For each n define $\alpha_n = \sup\{|f_n(x) - f(x)| \mid x \in S\}$. To prove that (1) $\Rightarrow$ (2), assume that (1) holds—in other words, that $\alpha_n \to 0$ as $n \to \infty$. Let $\varepsilon > 0$, and choose a natural number N such that $\alpha_n < \varepsilon$ whenever $n \geq N$. Then whenever $n \geq N$ and $x \in S$, we have $|f_n(x) - f(x)| \leq \alpha_n < \varepsilon$.

To prove that (2) $\Rightarrow$ (1), assume that (2) holds. We must now show that $\alpha_n \to 0$. Let $\varepsilon > 0$. Choose a natural number N such that whenever $n \geq N$ and $x \in S$, we have $|f_n(x) - f(x)| < \varepsilon/2$. It is now clear that whenever $n \geq N$, we have $\alpha_n < \varepsilon$.

■

11.3.4 A Comparison of the Definitions of Pointwise and Uniform Convergence.

Suppose that (f_n) is a sequence of functions defined on a set S and that f is a function defined on S. The condition $f_n \to f$ pointwise on S says that for every point $x \in S$ we have $f_n(x) \to f(x)$ as $n \to \infty$, and this may be stated in the following form:

> For every point $x \in S$ and every number $\varepsilon > 0$ there exists a number N such that whenever $n \geq N$, we have $|f_n(x) - f(x)| < \varepsilon$.

This is the same as saying the following:

> For every number $\varepsilon > 0$ and every point $x \in S$ there exists a number N such that whenever $n \geq N$, we have $|f_n(x) - f(x)| < \varepsilon$.

At first sight this definition of pointwise convergence may look bewilderingly similar to condition (2) in Theorem 11.3.3. You should notice, however, that in the definition of pointwise convergence we have to specify *both ε and x* before we can commit ourselves to a number N such that whenever $n \geq N$, we have $|f_n(x) - f(x)| < \varepsilon$. On the other hand, in the definition of uniform convergence no point x need be mentioned until *after* a value of N is specified; and therefore, in order to specify a value of N, we need only know the value of ε. The conditions of pointwise convergence and uniform convergence are therefore quite different. If a sequence (f_n) converges uniformly to a function f, then (f_n) will certainly converge pointwise to f; but as we see from the following example, the converse of this statement is not true.

> We define $f_n(x) = x^n$ for all $x \in [0, 1)$ and all $n \in \mathbf{Z}^+$. Then even though (f_n) converges pointwise (in fact, boundedly) to the constant function 0 on $[0, 1)$, the sequence (f_n) does not converge uniformly to 0 on $[0, 1)$, because for every n we have $\sup\{|f_n(x) - 0| \mid x \in [0, 1)\} = 1$.

Finally, it is worth noticing that as long as the limit function of a sequence is bounded, then uniform convergence is a stronger kind of convergence than bounded convergence. To see that it is, suppose that the sequence (f_n) converges uniformly to a bounded function f on a set S, and suppose that $|f(x)| \leq K$ for all $x \in S$. Using the fact that $f_n \to f$ uniformly on S, choose a natural number N such that whenever $n \geq N$ and $x \in S$, we have $|f_n(x) - f(x)| < 1$. Then whenever $n \geq N$ and $x \in S$, we have

$$|f_n(x)| = |f_n(x) - f(x) + f(x)| \leq |f_n(x) - f(x)| + |f(x)| < 1 + K,$$

and so starting at $n = N$, the sequence (f_n) converges boundedly to f.

11.3.5 Some Further Examples

(1) Let us look once again at Example 8.20.2(1). In that example we defined

$$f_n(x) = \frac{2n^2 x}{(1 + n^2 x^2)^2}$$

for all $n \in \mathbf{Z}^+$ and all points $x \in [0, 1]$. It is clear that the sequence (f_n) converges pointwise to the constant function 0; but since $f_n(1/n) = n/2$ for every n, we see that (f_n) does not converge boundedly to 0. Of course, as we saw in that example, $\int_0^1 f_n \to 1$ as $n \to \infty$, and so the fact that (f_n) does not converge boundedly follows also from the bounded convergence theorem.

(2) In this example we make a slight modification of example (1). For each point $x \in [0, 1]$ and each $n \in \mathbf{Z}^+$ we define

$$f_n(x) = \frac{2n x}{(1 + n^2 x^2)^2}.$$

These functions are slightly smaller than the functions in the preceding example, and of course, we have $f_n(x) \to 0$ as $n \to \infty$ for each $x \in [0, 1]$. This time, however, the sequence (f_n) converges boundedly. To see this, we observe that for each n

$$f_n'(x) = \frac{2n(1 - 3n^2 x^2)}{(1 + n^2 x^2)^3}.$$

It is therefore easy to see that the maximum value of f_n occurs at $1/n\sqrt{3}$, and that this maximum value is $9/8\sqrt{3}$, which is independent of n. So $f_n \to 0$ boundedly on $[0, 1]$. This convergence is not uniform, however, since for each n, we have $\sup f_n = 9/8\sqrt{3}$.

(3) Define $f_n(x) = x^n$ for all $x \in [0, 1)$ and all $n \in \mathbf{Z}^+$. We have seen that although $f_n \to 0$ boundedly on $[0, 1)$, this convergence is not uniform. Suppose now that $0 \leq \alpha < 1$. Then $f_n \to 0$ uniformly on $[0, \alpha]$. To see this, let $\varepsilon > 0$, and choose a number N such that whenever $n \geq N$, we have $\alpha^n < \varepsilon$. Then for all $n \geq N$ and $x \in [0, \alpha]$ we have $|f_n(x)| < \varepsilon$.

11.4 EXERCISES

■ **1.** In each of the following cases (f_n) is a sequence of functions defined on $[0, 1]$ by the given equation. Prove that $f_n \to 0$ pointwise on $[0, 1]$, and in each case, determine whether $f_n \to 0$ boundedly on $[0, 1]$ and whether $f_n \to 0$ uniformly on $[0, 1]$.

   (a) $f_n(x) = nx \exp(-nx)$ for all $x \in [0, 1]$.
   (b) $f_n(x) = n^2 x \exp(-nx)$ for all $x \in [0, 1]$.
   (c) $f_n(x) = nx \exp(-n^2 x^2)$ for all $x \in [0, 1]$.
   (d) $f_n(x) = nx \exp(-nx^2)$ for all $x \in [0, 1]$.
   (e) $f_n(x) = nx \exp(-n^2 x)$ for all $x \in [0, 1]$.

2. Given that $g_n \to 0$ uniformly on a set S and that $|f_n(x)| \leq g_n(x)$ whenever $n \in \mathbf{Z}^+$ and $x \in S$, prove that $f_n \to 0$ uniformly on S.

3. Given that $f_n \to f$ and $g_n \to g$ uniformly on S, where $S \subseteq \mathbf{R}$, prove that $f_n + g_n \to f + g$ uniformly on S. Prove that in the event that the functions f and g are bounded on S, then $f_n g_n \to fg$ uniformly on S.

■ **4.** Given that f is an unbounded function on a set S, prove that there exists a sequence (f_n) such that $f_n \to f$ uniformly on S but the sequence (f_n^2) does not converge uniformly to f^2 on S.

5. Prove that if a sequence (f_n) converges uniformly to 0 on a set S, then for every sequence (x_n) in S which converges to a point $x \in S$, we have $f_n(x_n) \to 0$ as $n \to \infty$.

■ **6.** Give an example to show that the converse of the result proved in exercise 5 is false.

*7. Prove that the converse of the result proved in exercise 5 is true if the set S is closed and bounded. *Note:* In order to do this exercise, you will probably need to make use of the notion of a subsequence which is defined in Chapter 4.

■ **8.** Prove that if (f_n) is a decreasing sequence of nonnegative continuous functions on a closed bounded set S, and if $f_n \to 0$ pointwise on S, then $f_n \to 0$ uniformly on S. Having done this, give an example of a decreasing sequence (f_n) of nonnegative continuous functions on $[0, 1)$ such that $f_n \to 0$ pointwise but not uniformly on $[0, 1)$.

11.5 THE IMPORTANT PROPERTIES OF UNIFORM CONVERGENCE

As we saw in Section 11.3.1, the limit of a boundedly convergent sequence of continuous functions need not be continuous, and the limit of a boundedly convergent sequence of Riemann-integrable functions need not be Riemann-integrable. As a matter of fact, you may recall that at the end of Chapter 8 the possibility that a boundedly convergent sequence of Riemann-integrable functions might have a nonintegrable limit was depicted as being the principal defect of the theory of Riemann integration. The important property of uniform convergence that sets it apart from the weaker mode of bounded convergence is that with uniform convergence this sort of pathology does not occur. In this section we shall see that the limit of a uniformly convergent sequence of continuous functions must be continuous, and that the limit of a uniformly convergent sequence of Riemann-integrable functions must be Riemann-integrable.

11.5.1   Theorem: Uniform Convergence and Continuity.

Suppose that a sequence (f_n) converges uniformly on a set S to a function f, that $x \in S$, and that for each n the function f_n is continuous at x. Then f is also continuous at x.

■ **Proof.**   Let $\varepsilon > 0$; and using the fact that $f_n \to f$ uniformly on S, choose a natural number N such that whenever $n \geq N$ and $t \in S$, we have $|f_n(t) - f(t)| < \varepsilon/3$. Now using the fact that f_N is continuous at x, choose a neighborhood U of x such that whenever $t \in S \cap U$, we have $|f_N(x) - f_N(t)| < \varepsilon/3$. Then whenever $t \in S \cap U$, we have

$$|f(x) - f(t)| = |f(x) - f_N(x) + f_N(x) - f_N(t) + f_N(t) - f(t)|$$

$$\leq |f(x) - f_N(x)| + |f_N(x) - f_N(t)| + |f_N(t) - f(t)|$$

$$< \frac{\varepsilon}{3} + \frac{\varepsilon}{3} + \frac{\varepsilon}{3} = \varepsilon. \qquad ■$$

11.5.2   Corollary.

Suppose that (f_n) is a sequence of continuous functions on a set S and that this sequence converges uniformly on S to a function f. Then f is continuous on S.

11.5.3   Theorem: Uniform Convergence and Riemann Integrability.

Suppose that (f_n) is a sequence of Riemann-integrable functions on an interval $[a, b]$ and that this sequence converges uniformly on $[a, b]$ to a function f. Then f is Riemann-integrable on $[a, b]$.

■ **Proof.**   We shall show that f satisfies condition (2) of Theorem 8.14.1. Let $\varepsilon > 0$, and choose a natural number N such that whenever $n \geq N$ and $x \in [a, b]$, we have

$$|f_n(x) - f(x)| < \frac{\varepsilon}{4(b - a)}.$$

Now using the fact that f_N is Riemann-integrable on $[a, b]$, and therefore that f_N satisfies condition (2) of Theorem 8.14.1, choose two step functions s and S on $[a, b]$ such that $s \leq f_N \leq S$ and $\int_a^b (S - s) < \varepsilon/2$. Now for every point $x \in [a, b]$ we have

$$s(x) - \frac{\varepsilon}{4(b - a)} \le f_N(x) - \frac{\varepsilon}{4(b - a)} < f(x)$$

$$< f_N(x) + \frac{\varepsilon}{4(b - a)} \le S(x) + \frac{\varepsilon}{4(b - a)};$$

and therefore, since

$$\int_a^b \left\{ \left[S + \frac{\varepsilon}{4(b - a)} \right] - \left[s - \frac{\varepsilon}{4(b - a)} \right] \right\} < \varepsilon,$$

it follows that f satisfies condition (2) of Theorem 8.14.1. ∎

11.5.4   Corollary.   Suppose that (f_n) is a sequence of Riemann-integrable functions on an interval $[a, b]$ and that this sequence converges uniformly on $[a, b]$ to a function f. Then $\int_a^b f_n \to \int_a^b f$ as $n \to \infty$.

∎ *Proof.*   Of course, this result follows at once from the bounded convergence theorem, but it is so easy to give a direct proof that it would be a shame not to do so. By Theorem 11.5.3, the function f is Riemann-integrable on $[a, b]$. Therefore,

$$\int_a^b |f_n - f| \le (b - a) \sup\{|f_n(x) - f(x)| \, | a \le x \le b\} \to 0 \qquad \text{as} \qquad n \to \infty. \quad ∎$$

11.6   SOME IMPORTANT TESTS FOR UNIFORM CONVERGENCE

11.6.1   The Cauchy Criterion for Uniform Convergence.   In Section 4.11 we saw that a necessary and sufficient condition for a given sequence (x_n) to converge is that (x_n) should be a Cauchy sequence. Now we shall see that a necessary and sufficient condition for a sequence (f_n) of functions to converge uniformly on a set S is that (f_n) should be "uniformly Cauchy" on S. The precise statement of this result is given in the following theorem.

Theorem.   Suppose (f_n) is a sequence of functions defined on a set S. Then the following two conditions are equivalent:

(1) The sequence (f_n) converges uniformly on S.

(2) For every $\varepsilon > 0$ there exists a natural number N such that whenever $m \ge N$ and $n \ge N$ and $x \in S$, we have $|f_m(x) - f_n(x)| < \varepsilon$.

∎ *Proof.*   To prove that (1) ⇒ (2), assume that (1) holds; and write the limit function of (f_n) as f. Let $\varepsilon > 0$, and choose a natural number N such that whenever

$n \geq N$ and $x \in S$, we have $|f_n(x) - f(x)| < \varepsilon/2$. Then whenever $m \geq N$ and $n \geq N$ and $x \in S$, we have

$$|f_m(x) - f_n(x)| = |f_m(x) - f(x) + f(x) - f_n(x)|$$

$$\leq |f_m(x) - f(x)| + |f(x) - f_n(x)| < \frac{\varepsilon}{2} + \frac{\varepsilon}{2} = \varepsilon.$$

This shows that $(1) \Rightarrow (2)$.

Now to show that $(2) \Rightarrow (1)$, assume that (2) holds. It is clear that for every point $x \in S$ the sequence $(f_n(x))$ is a Cauchy sequence of numbers, and by Theorem 4.11.3, this sequence must converge. Therefore, the sequence (f_n) converges pointwise on S. Denote its limit function as f. In order to complete the proof, we need to show that $f_n \to f$ uniformly on S. Let $\varepsilon > 0$, and choose a natural number N such that whenever $m \geq N$ and $n \geq N$ and $x \in S$, we have $|f_m(x) - f_n(x)| < \varepsilon/2$. We shall now show that whenever $n \geq N$ and $x \in S$, we have $|f_n(x) - f(x)| < \varepsilon$. For this purpose, let $n \geq N$ and $x \in S$. Since $f_m(x) \to f(x)$ as $m \to \infty$, we have

$$|f_n(x) - f_m(x)| \to |f_n(x) - f(x)| \qquad \text{as} \qquad m \to \infty;$$

and therefore, $|f_n(x) - f(x)| \leq \varepsilon/2 < \varepsilon$. ∎

The following analogue of the comparison test for uniform convergence is sometimes known as the **Weierstrass M-test.**

11.6.2 The Weierstrass Comparison Test for Uniform Convergence of a Series.
Suppose that (f_n) and (g_n) are sequences of functions defined on a set S, that $|f_n| \leq g_n$ for each n, and that the series $\Sigma\, g_n$ converges uniformly on S. Then the series $\Sigma\, f_n$ also converges uniformly on S.

∎ **Proof.** For each n define $F_n = \Sigma_{j=1}^{n} f_j$ and $G_n = \Sigma_{j=1}^{n} g_j$. We shall show first that if m and n are any two natural numbers and $x \in S$, then we have

$$|F_m(x) - F_n(x)| \leq |G_m(x) - G_n(x)|.$$

This is obvious if $m = n$. Assume that $m > n$. Then

$$|F_m(x) - F_n(x)| = \left| \sum_{j=n+1}^{m} f_j(x) \right| \leq \sum_{j=n+1}^{m} |f_j(x)|$$

$$\leq \sum_{j=n+1}^{m} g_j(x) = |G_m(x) - G_n(x)|,$$

and the case $m < n$ follows similarly.

We can now use the Cauchy criterion to show that the sequence (F_n) converges uniformly on S. Let $\varepsilon > 0$, and choose a natural number N such that whenever $m \geq N$ and $n \geq N$ and $x \in S$, we have $|G_m(x) - G_n(x)| < \varepsilon$. Then for $m \geq N$ and $n \geq N$ and $x \in S$, we have $|F_m(x) - F_n(x)| < \varepsilon$. ∎

Corollary. Suppose that (f_n) is a sequence of functions defined on a set S, that (α_n) is a sequence of numbers, $|f_n| \leq \alpha_n$ for each n, and that the series $\Sigma \, \alpha_n$ converges. Then the series $\Sigma \, f_n$ converges uniformly on S.

For series that fail to converge absolutely we need tests which are more delicate than the Weierstrass test, and for this purpose we use the following two theorems which depend heavily on Theorem 10.10.5.

11.6.3 Abel's Test for Uniform Convergence. Suppose that (f_n) and (g_n) are sequences of functions defined on a set S, that the function f_1 is bounded, that for every point $x \in S$, $(f_n(x))$ is a decreasing sequence of positive numbers, and that the series $\Sigma \, g_n$ converges uniformly on S. Then the series $\Sigma \, f_n g_n$ converges uniformly on S.

■ **Proof.** Choose a number $K > 0$ such that $f_1(x) \leq K$ for every $x \in S$; and for each n, define $H_n = \Sigma_{j=1}^{n} f_j g_j$ and $G_n = \Sigma_{j=1}^{n} g_j$. We shall use the Cauchy criterion to show that the sequence (H_n) converges uniformly on S. Let $\varepsilon > 0$, and choose N such that whenever $m \geq N$ and $n \geq N$ and $x \in S$, we have $|G_m(x) - G_n(x)| < \varepsilon/K$. The proof will be complete when we have shown that whenever $m \geq N$ and $n \geq N$ and $x \in S$, we have $|H_m(x) - H_n(x)| < \varepsilon$. Suppose, then, that $m \geq N$ and $n \geq N$ and $x \in S$. The inequality $|H_m(x) - H_n(x)| < \varepsilon$ is trivial if $m = n$, and the cases $m > n$ and $m < n$ are similar. We may therefore assume without loss of generality that $m > n$, and from Theorem 10.10.5 we obtain

$$|H_m(x) - H_n(x)| = \left| \sum_{j=n+1}^{m} f_j(x) g_j(x) \right| < f_{n+1}(x) \frac{\varepsilon}{K} \leq K \frac{\varepsilon}{K} = \varepsilon. \qquad ■$$

11.6.4 Dirichlet's Test for Uniform Convergence. Suppose that (f_n) and (g_n) are sequences of functions defined on a set S, that $f_n \rightarrow 0$ uniformly on S, that for every point $x \in S$, $(f_n(x))$ is a decreasing sequence of positive numbers, and that for some number K we have $|\Sigma_{j=1}^{n} g_j(x)| \leq K$ for all $n \in \mathbf{Z}^+$ and all $x \in S$. Then the series $\Sigma \, f_n g_n$ converges uniformly on S.

■ **Proof.** As before, define $H_n = \Sigma_{j=1}^{n} f_j g_j$. We need to see that the sequence (H_n) converges uniformly on S. Now whenever m and n are natural numbers and $m > n$ and $x \in S$, we have

$$\left| \sum_{j=n+1}^{m} g_j(x) \right| = \left| \sum_{j=1}^{m} g_j(x) - \sum_{j=1}^{n} g_j(x) \right| \leq \left| \sum_{j=1}^{m} g_j(x) \right| + \left| \sum_{j=1}^{n} g_j(x) \right| \leq 2K;$$

and it therefore follows from Theorem 10.10.5 that for all $x \in S$

$$|H_m(x) - H_n(x)| = \left| \sum_{j=n+1}^{m} f_j(x) g_j(x) \right| \leq 2K \, \sup\{f_{n+1}(t) \mid t \in S\}.$$

From this we conclude that the sequence (H_n) satisfies the Cauchy criterion for uniform convergence on S. ■

11.6.5 Some Examples

(1) For each $n \in \mathbf{Z}^+$ and $x \in [-1, 1]$, define $f_n(x) = x^n/n^2$. Since $|f_n| \leq 1/n^2$ for each n, it follows from the Weierstrass test that Σf_n converges uniformly on $[-1, 1]$. We often express this sort of statement more simply by saying that the series $\Sigma x^n/n^2$ is *uniformly convergent in x* on the interval $[-1, 1]$.

(2) From the identity

$$\sum_{j=1}^{n} (-1)^{j-1} x^j = \frac{x[1 - (-x)^n]}{1 + x},$$

we deduce that $|\Sigma_{j=1}^{n} (-1)^{j-1} x^j| \leq 2$ for all $n \in \mathbf{Z}^+$ and all $x \in [0, 1]$; and it therefore follows from Dirichlet's test that the series $\Sigma[(-1)^{n-1} x^n]/n$ converges uniformly in x on $[0, 1]$. As we have seen (Section 10.4.2), the sum of this series is $\log(1 + x)$.

(3) As we saw in Example 10.10.7(4), the series $\Sigma(\sin nx)/n$ converges pointwise in x on $\mathbf{R}$. We shall see now that whenever $0 < \delta \leq \pi$, this series converges uniformly in x on the interval $[\delta, 2\pi - \delta]$ but does not converge uniformly on $[0, 2\pi]$. Suppose $0 < \delta \leq \pi$. Given $x \in [\delta, 2\pi - \delta]$ and any natural number n, we have

$$\left| \sum_{j=1}^{n} \sin jx \right| \leq \frac{1}{|\sin(\delta/2)|},$$

and the uniform convergence of the series in $[\delta, 2\pi - \delta]$ follows from Dirichlet's test. To show that the series is not uniformly convergent on $[0, 2\pi]$, we shall observe that it does not satisfy condition (2) of Theorem 11.6.1. For each n, define $F_n(x) = \Sigma_{j=1}^{n}(\sin jx)/j$. Now for each n we have

$$\left| F_{2n} \left(\frac{1}{2n} \right) - F_n \left(\frac{1}{2n} \right) \right| = \sum_{j=n+1}^{2n} \frac{\sin(j/2n)}{j} \geq \sum_{j=n+1}^{2n} \frac{\sin(1/2)}{2n} = \frac{1}{2} \sin \frac{1}{2}.$$

11.7 EXERCISES

1. Prove that the series $\Sigma x^n/n!$ converges uniformly in x on every bounded interval, but that this series does not converge uniformly in x on $\mathbf{R}$.

■ **2.** Prove that the series $\Sigma[(2n)!/4^n(n!)^2]x^n$ does not converge uniformly in x on $(-1, 1)$, but that this series converges uniformly on $[-\delta, \delta]$ whenever $0 \leq \delta < 1$.

3. Prove that the series $\Sigma[(-1)^{n-1}/\sqrt{n}] \sin[1 + (x/n)]$ converges uniformly in x on $[-1, 1]$. Prove that the series converges pointwise on $\mathbf{R}$.

4. Prove that if

$$f(x) = \sum_{n=1}^{\infty} \frac{(-1)^{n-1}}{\sqrt{n}} \sin\left(1 + \frac{x}{n}\right)$$

for all $x \in \mathbf{R}$, then f is differentiable on $\mathbf{R}$ and for each x we have

$$f'(x) = \sum_{n=1}^{\infty} \frac{(-1)^{n-1}}{n^{3/2}} \cos\left(1 + \frac{x}{n}\right).$$

5. Prove that the series $\Sigma (x \log x)^n$ converges uniformly in x on the interval $(0, 1]$.

■ 6. Give an example of a sequence (f_n) of nonnegative, improper, Riemann-integrable functions on $[0, \infty)$ such that $f_n \to 0$ uniformly on $[0, \infty)$ but $\int_0^{\to\infty} f_n = 1$ for every n. Can this still happen if we require $f_n \leq f_1$ for every n?

■ 7. Determine whether the following statement is true or false: If f_n is uniformly continuous on a given set S for each $n \in \mathbf{Z}^+$, and if $f_n \to f$ uniformly on S, then f is uniformly continuous on S.

■ 8. Determine whether the following statement is true or false: If f_n is uniformly continuous on a given set S for infinitely many natural numbers n, and if $f_n \to f$ uniformly on S, then f is uniformly continuous on S.

9. A family $\mathcal{F}$ of functions is said to be **equicontinuous** on a set S if for every $\varepsilon > 0$ and every point $x \in S$ there exists a neighborhood U of x such that whenever $f \in \mathcal{F}$ and $t \in U \cap S$, we have $|f(t) - f(x)| < \varepsilon$. Prove that if a sequence (f_n) converges uniformly on S and each function f_n is continuous on S, then the family $\{f_n \mid n \in \mathbf{Z}^+\}$ is equicontinuous on S.

10. Invent a meaning for *equi-uniform continuity* of a family $\mathcal{F}$ on a set S, and decide whether (according to your choice of definition) exercise 9 has an analogue for equi-uniform continuity.

■ 11. Prove that if $\alpha \geq 0$, then the **binomial series** $\Sigma \binom{\alpha}{n} x^n$ converges uniformly in x on the interval $[-1, 1]$. Having done this, discuss the cases $-1 < \alpha < 0$ and $\alpha \leq -1$.

12. Suppose that (f_n) is a sequence of functions defined on a set S and that (f_n) converges uniformly on S to a function f. Suppose that x is a limit point of S and that for each n we have $f_n(t) \to \alpha_n$ as $t \to x$. Prove that the sequence (α_n) converges, and that if α is the limit of this sequence, then $f(t) \to \alpha$ as $t \to x$.

11.8 POWER SERIES

Up to this point we have seen only one important result about power series: Abel's theorem for series, which we saw as Theorem 10.10.6. On the other hand, we have already seen a number of important examples of power series. The simplest power series is the geometric series Σx^n, which we had already begun to develop in Example 4.10.2(3). Other important power series we have studied are $\Sigma\ x^n/n!$ and $\Sigma[(-1)^{n-1}x^n]/n$, and as we saw in Section 10.4, the first of these converges everywhere to e^x and the second converges at each point x in the interval $(-1, 1]$ to $\log(1 + x)$. In Example 10.10.7(5) we began our study of the binomial series $\Sigma\binom{\alpha}{n}x^n$, which we continued in Exercises 10.11(4) and 10.11(5), and which we shall complete in exercises 5, 6, and 7 of Section 11.10. In this section we shall make a systematic study of power series, and we shall show, among other things, that in their regions of convergence, power series can be differentiated and integrated term by term.

We begin with a theorem that will make it possible for us to define the important notion of *radius of convergence* of a power series.

11.8.1 Theorem.   Given any sequence (a_n), there exists a number $r \in [0, \infty]$ such that the power series $\Sigma\ a_nx^n$ converges absolutely when $|x| < r$ and diverges when $|x| > r$.

■ ***Proof.***   We observe first that if t and x are any two real numbers and $|t| \le |x|$, and if $a_nx^n \to 0$ as $n \to \infty$, then $a_nt^n \to 0$ as $n \to \infty$. Now define $r = \sup\{x\,|\,a_nx^n \to 0$ as $n \to \infty\}$. We see at once that the sequence (a_nx^n) converges to 0 whenever $|x| < r$ and fails to converge to 0 whenever $|x| > r$. The series $\Sigma\ a_nx^n$ therefore diverges whenever $|x| > r$. To complete the proof, we shall show that the series converges absolutely whenever $|x| < r$. Suppose that $|x| < r$, and choose a number δ such that $|x| < \delta < r$. Using the fact that $a_n\delta^n \to 0$ as $n \to \infty$, choose a natural number N such that whenever $n \ge N$, we have $|a_n\delta^n| \le 1$. Then whenever $n \ge N$, we have

$$\left|a_nx^n\right| \le \left|a_n\delta^n\left(\frac{x}{\delta}\right)^n\right| \le \left|\frac{x}{\delta}\right|^n,$$

and the absolute convergence of the series $\Sigma\ a_nx^n$ therefore follows from the comparison test and the convergence of the geometric series $\Sigma\ |x/\delta|^n$.   ■

11.8.2 Radius of Convergence of a Power Series.   Given a power series $\Sigma\ a_nx^n$, the number r such that $\Sigma\ a_nx^n$ converges for $|x| < r$ and diverges for $|x| > r$ is called the **radius of convergence** of the series $\Sigma\ a_nx^n$; and the set $\{x\,|\,\Sigma\ a_nx^n$ converges$\}$ is called the **interval of convergence** of the given series. The interval of convergence is therefore one of the four intervals $(-r, r), [-r, r), (-r, r], [-r, r]$. Note that if $r = 0$, then the series $\Sigma\ a_nx^n$ converges only when $x = 0$; and if $r = \infty$, then $\Sigma\ a_nx^n$ converges absolutely for all real numbers x. For practical purposes, when we want to find the radius of convergence of a given series, we often find it easiest to use d'Alembert's test, and this should not be surprising considering that the proof of Theorem 11.8.1 involved comparison with a geometric series. Using d'Alembert's test, we can easily check that the radii of convergence of the series

$$\Sigma\ x^n, \quad \Sigma\frac{x^n}{n!}, \quad \Sigma\frac{(-1)^{n-1}x^n}{n}, \quad \Sigma\frac{x^n}{2^nn^2}, \quad \Sigma\ n!x^n, \quad \text{and} \quad \Sigma\frac{n!x^n}{n^n}$$

are 1, ∞, 1, 2, 0, and e, respectively. The intervals of convergence of these series are $(-1, 1)$, $\mathbf{R}$, $(-1, 1]$, $[-2, 2]$, $\{0\}$, and $(-e, e)$.

11.8.3 Theorem: The Uniform Convergence of a Power Series. Suppose r is the radius of convergence of the series $\Sigma\ a_n x^n$.

(1) Whenever $0 \leq \delta < r$, the series $\Sigma\ a_n x^n$ converges uniformly in x on the interval $[-\delta, \delta]$.

(2) If the series $\Sigma\ a_n r^n$ converges, and $0 \leq \delta < r$, then $\Sigma\ a_n x^n$ converges uniformly in x on the interval $[-\delta, r]$.

(3) If the series $\Sigma\ a_n(-r)^n$ converges, and $0 \leq \delta < r$, then $\Sigma\ a_n x^n$ converges uniformly in x on the interval $[-r, \delta]$.

(4) If $\Sigma\ a_n x^n$ converges at both endpoints of its interval of convergence, then the series converges uniformly in x on $[-r, r]$.

■ *Proof.* Assume $0 \leq \delta < r$. Since $|a_n x^n| \leq |a_n \delta^n|$ for all $x \in [-\delta, \delta]$ and all n, part (1) follows at once from the Weierstrass test for uniform convergence and the convergence of the series $\Sigma\ |a_n \delta^n|$.

Now to prove part (2), assume that $\Sigma\ a_n r^n$ converges. In order to show that $\Sigma\ a_n x^n$ converges uniformly on $[-\delta, r]$, it is sufficient to show that the series converges uniformly on $[0, r]$. To do this, we shall use Abel's test for uniform convergence (Theorem 11.6.3). For each $n \in \mathbf{Z}^+$, let g_n be the constant function $a_n r^n$, and define $f_n(x) = (x/r)^n$ for $0 \leq x \leq r$. Then since $\Sigma\ g_n$ converges uniformly on $[0, r]$ and (f_n) is a decreasing sequence of nonnegative, bounded functions, and since $a_n x^n = f_n(x)g_n(x)$ for each x, we may conclude from Abel's test that $\Sigma\ a_n x^n$ converges uniformly in x on $[0, r]$.

Part (3) follows similarly; this time we take g_n to be the constant function $a_n(-r)^n$ and $f_n(x) = (-x/r)^n$ for $-r \leq x \leq 0$. Finally, part (4) follows by combining parts (2) and (3). ■

11.8.4 Theorem: Continuity of the Sum of a Power Series. Suppose I is the interval of convergence of a power series $\Sigma\ a_n x^n$; and for each $x \in I$, define $f(x) = \Sigma_{n=0}^{\infty} a_n x^n$. Then f is continuous on I.

■ *Proof.* Suppose the radius of convergence of the series $\Sigma\ a_n x^n$ is r. Now let $c \in I$. To prove that f is continuous at c, we shall assume first that $|c| < r$. Choose δ such that $|c| < \delta < r$. The continuity of f at c now follows at once from Theorem 11.5.1 and the fact that the series $\Sigma\ a_n x^n$ converges uniformly in x on $[-\delta, \delta]$. In the event that $c = r$, the continuity of f at c follows similarly in view of the uniform convergence of $\Sigma\ a_n x^n$ on $[0, r]$. The case $c = -r$ is analogous. ■

11.8.5 Theorem: Term-by-Term Integration of a Power Series. Suppose I is the interval of convergence of a power series $\Sigma\ a_n x^n$; and for each $x \in I$, define $f(x) = \Sigma_{n=0}^{\infty} a_n x^n$. Then given any point $c \in I$, we have

$$\int_0^c f = \sum_{n=0}^{\infty} a_n \frac{c^{n+1}}{n+1}.$$

■ **_Proof._** The result is obvious if $c = 0$, so we shall assume that $c \neq 0$. Define $J = [0, c]$ in the case $c > 0$ and $J = [c, 0]$ in the case $c < 0$. By Theorem 11.8.4, f is continuous on J; and therefore, the result will follow from the bounded convergence theorem for series (Theorem 11.1.2), when we have shown that $\Sigma\ a_n x^n$ converges boundedly in x on J. But this follows from Theorem 10.10.5 because the series $\Sigma\ a_n c^n$ converges and because $a_n x^n = a_n c^n (x/c)^n$ for all x and n. ■

11.8.6 Theorem: Term-by-Term Differentiation of a Power Series. Suppose that the radius of convergence r of the power series $\Sigma\ a_n x^n$ is positive and that I is the interval of convergence of the "derived" series $\Sigma\ n a_n x^{n-1}$.

(1) r is also the radius of convergence of the series $\Sigma\ n a_n x^{n-1}$.

(2) The series $\Sigma\ a_n x^n$ converges at each point $x \in I$; and if for each $x \in I$ we define $f(x) = \Sigma_{n=0}^{\infty} a_n x^n$, then whenever $x \in I$, we have

$$f'(x) = \sum_{n=1}^{\infty} n a_n x^{n-1}.$$

■ **_Proof._** We shall show first that the series $\Sigma\ a_n x^n$ converges at every point $x \in I$. Let $x \in I$, and assume without loss of generality that $x > 0$. The case $x < 0$ is analogous. Since the series $\Sigma\ n a_n t^{n-1}$ converges boundedly in t on the interval $[0, x]$ and has a continuous sum, and since $\Sigma\ a_n 0^n$ converges, it follows from Theorem 11.1.3 that $\Sigma\ a_n x^n$ converges. Furthermore, if we define $f(x) = \Sigma_{n=0}^{\infty} a_n x^n$ for each $x \in I$, then the same argument shows that for each such x, we have $f'(x) = \Sigma_{n=1}^{\infty} n a_n x^{n-1}$. This proves part (2) of the theorem. To complete the proof of part (1), we need to show that whenever $|x| < r$, the series $\Sigma\ n a_n x^{n-1}$ converges. Suppose $|x| < r$, and choose a number δ such that $|x| < \delta < r$. Now for each n

$$|n a_n x^{n-1}| = a_n \delta^n \left[\frac{n}{|x|} \left(\frac{|x|}{\delta} \right)^n \right]$$

and therefore, since $n(|x|/\delta)^n \to 0$ as $n \to \infty$, and since the series $\Sigma\ a_n \delta^n$ converges, the convergence of $\Sigma\ n a_n x^{n-1}$ follows from the comparison test. ■

Note that although the series $\Sigma\ a_n x^n$ and $\Sigma\ n a_n x^{n-1}$ have the same radius of convergence, these series need not behave in the same way at the endpoints of their intervals of convergence. For example, the interval of convergence of the series $\Sigma\ x^n/n$ is $[-1, 1)$, but the interval of its differentiated series $\Sigma\ x^{n-1}$ is $(-1, 1)$. Now because a power series can be differentiated term by term to yield another power series with the same radius of convergence, we can repeat this differentiation process as often as we like. In our next theorem we shall state this precisely, and we shall observe that if a function can be expressed as the sum of a power series, then the coefficients in this power series are uniquely determined by the behavior of the function near 0.

11.8.7 Theorem. Suppose that the power series $\Sigma\, a_n x^n$ has a positive radius of convergence r; and whenever $|x| < r$, define $f(x) = \Sigma_{m=0}^{\infty} a_m x^m$.

(1) The nth derivative $f^{(n)}(x)$ exists for every natural number n and every point $x \in (-r, r)$, and we have

$$f^{(n)}(x) = \sum_{m=n}^{\infty} m(m - 1)(m - 2) \cdots (m - n + 1)a_m x^{m-n}.$$

(2) For each natural number n we have $a_n = f^{(n)}(0)/n!$.

■ **Proof.** From Theorem 11.8.6 we obtain $f'(x) = \Sigma_{m=1}^{\infty} m a_m x^{m-1}$ whenever $|x| < r$, and since the latter power series also converges in $(-r, r)$, we may apply Theorem 11.8.6 again and obtain

$$f''(x) = \sum_{m=2}^{\infty} m(m - 1)a_m x^{m-2}$$

for $|x| < r$. This method can obviously be repeated as often as we like; and on repeating it n times, we obtain the identity in part (1). To prove part (2), all we have to do is substitute $x = 0$ in this identity. Since all but the first term in the expression

$$\sum_{m=n}^{\infty} m(m - 1)(m - 2) \cdots (m - n + 1)a_m x^{m-n}$$

contain a factor x, and therefore drop out when $x = 0$, the substitution $x = 0$ yields $f^{(n)}(0) = n!a_n$, as required. ■

11.8.8 Maclaurin Series. If a function f has derivatives of all orders at 0, then the **Maclaurin series** of f is defined to be the series $\Sigma\, [f^{(n)}(0)/n!]x^n$.

What we saw in Theorem 11.8.7 is that if a function f can be written as the sum of a power series in an interval $(-r, r)$, then this power series has to be the Maclaurin series of f. Putting it another way, either the Maclaurin series of a given function f will converge to f in a given interval $(-r, r)$, or there is no power series at all that can converge to f in $(-r, r)$. We can also conclude from Theorem 11.8.7 that if two functions f and g can both be expressed as the sums of power series in a given interval $(-r, r)$, and if $f(x) = g(x)$ for all points x in some neighborhood of 0, then the functions f and g are identical everywhere in $(-r, r)$. A much stronger result will be proved in Section 11.11.10.

It is worth mentioning that there are functions which have derivatives of all orders everywhere but which cannot be expressed as sums of power series. An oft-quoted example of this type of pathology is the function f defined in Example 11.8.9 that follows.

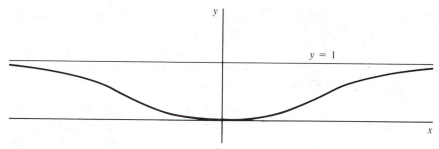

Figure 11.1

11.8.9   Example.   We define

$$f(x) = \exp(-x^{-2}) \quad \text{for} \quad x \neq 0, \quad \text{and} \quad f(0) = 0.$$

See Figure 11.1. As we shall see in a moment, this function has the interesting property that $f^{(n)}(0) = 0$ for every n, and because of this, the Maclaurin series of f is the series $\Sigma\, 0x^n$. Therefore, even though the Maclaurin series of f converges everywhere, it does not converge to the number $f(x)$ unless $x = 0$.

As a first step toward showing that $f^{(n)}(0) = 0$ for every n, we remark that if g is any rational function, then $g(u)e^{-u} \to 0$ as $u \to \infty$. Next, by substituting $u = 1/x^2$, we observe that if h is any rational function, then $h(x) \cdot \exp(-x^{-2}) \to 0$ as $x \to 0$. It is now easy to evaluate $f'(0)$:

$$f'(0) = \lim_{x \to 0} \frac{f(x) - f(0)}{x - 0} = \lim_{x \to 0} \frac{1}{x} \exp(-x^{-2}) = 0.$$

Given any number $x \neq 0$, we have $f'(x) = 2x^{-3} \exp(-x^{-2})$, and from this it follows that

$$f''(0) = \lim_{x \to 0} \frac{f'(x) - f'(0)}{x - 0} = \lim_{x \to 0} 2x^{-4} \exp(-x^{-2}) = 0.$$

Continuing in this way, we may see that $f^{(n)}(0) = 0$ for every n.

11.8.10   Power Series Centered Around an Arbitrary Point.   A *power series centered around a number c* is a series of the form $\Sigma\, a_n(x - c)^n$.

Note that if r is the radius of convergence of the power series $\Sigma\, a_n u^n$, then the series $\Sigma\, a_n(x - c)^n$ converges absolutely when $|x - c| < r$ and diverges when $|x - c| > r$. Furthermore, if $f(x) = \Sigma_{n=0}^{\infty} a_n(x - c)^n$ for $|x - c| < r$, then defining $g(u) = \Sigma_{n=0}^{\infty} a_n u^n$ for $|u| < r$, we see that $g(u) = f(u + c)$ whenever $|u| < r$; and therefore for each n we have

$$a_n = \frac{g^{(n)}(0)}{n!} = \frac{f^{(n)}(c)}{n!}.$$

So the only power series centered around c that can converge to a given function f in a neighborhood of c is the series $\Sigma[f^{(n)}(c)/n!]\,(x - c)^n$. This series is called the

Taylor series *of f centered around c.* Of course, when $c = 0$, then the Taylor series of f centered around c is just the Maclaurin series of f.

11.9   THE TRIGONOMETRIC FUNCTIONS

In Section 9.2 we gave an official definition of the exponential and logarithmic functions, and we promised that definitions of the trigonometric functions would be given after we had studied power series. We are now ready to keep this promise. In elementary mathematics the trigonometric functions are defined by appealing to the idea of an angle, which in turn refers to the area of a circular sector, or alternatively, to the length of a circular arc. We could use this sort of method here too; and as a matter of fact, if we had wanted to define the functions this way, we could have done so any time after Chapter 8. However, the angle approach is not the most convenient method of defining the trigonometric functions rigorously. Instead, we shall define the functions cos and sin to be the sums of their Maclaurin series.

To motivate the definition, we should bear in mind that we want cos and sin to satisfy the following conditions:

$$\cos' = -\sin, \quad \sin' = \cos, \quad \sin 0 = 0, \quad \text{and} \quad \cos 0 = 1.$$

Using these conditions, we see easily that the Maclaurin series of cos and sin are the series

$$\sum \frac{(-1)^n x^{2n}}{(2n)!} \quad \text{and} \quad \sum \frac{(-1)^n x^{2n+1}}{(2n + 1)!},$$

respectively; and we should note that both of these series have an infinite radius of convergence.

With this in mind, we *define*

$$\cos x = \sum_{n=0}^{\infty} \frac{(-1)^n x^{2n}}{(2n)!} \quad \text{and} \quad \sin x = \sum_{n=0}^{\infty} \frac{(-1)^n x^{2n+1}}{(2n + 1)!}$$

for all real numbers x.

The conditions $\cos' = -\sin$, $\sin' = \cos$, $\sin 0 = 0$, and $\cos 0 = 1$ are clearly satisfied, and one may also see that for every number x we have $\cos(-x) = \cos x$ and $\sin(-x) = -\sin x$.

11.9.1   Theorem.   For all numbers a and b we have the following:

(1) $\cos(a - b) = \cos a \cos b + \sin a \sin b$.

(2) $\cos(a + b) = \cos a \cos b - \sin a \sin b$.

(3) $\sin(a - b) = \sin a \cos b - \sin b \cos a$.

(4) $\sin(a + b) = \sin a \cos b + \sin b \cos a$.

■ *Proof.*   We shall prove the identities (1) and (3). The others will then follow by the usual methods. Let a and b be any real numbers, and for every number x, define

$$f(x) = \cos(a - x) - \cos a \cos x - \sin a \sin x$$

and

$$g(x) = \sin(a - x) - \sin a \cos x + \sin x \cos a,$$

and define $h = f^2 + g^2$. We see easily that $f' = g$ and $g' = -f$; and therefore, $h' = 2ff' + 2gg' = 0$; from which it follows that the function h is constant. But it is clear that $h(0) = 0$, and we therefore have $[f(x)]^2 + [g(x)]^2 = 0$ for every number x. Therefore, $f(x) = g(x) = 0$ for every number x, and the theorem follows on substituting $x = b$.   ■

Using Theorem 11.9.1, one may deduce all the usual identities of trigonometry, including the identity $\cos^2 + \sin^2 = 1$, from which it follows that neither $|\cos|$ nor $|\sin|$ can exceed 1.

11.9.2 Lemma.   There exists a least positive number x such that $\cos x = 0$.

■ *Proof.*   Define $E = \{x > 0 \mid \cos x = 0\}$. We need to show that the set E has a least member. Now since $\cos 0 = 1$, we have $E = \{x \geq 0 \mid \cos x = 0\}$, and it follows from the continuity of the function cos that E is closed. Therefore, in order to show that E has a least member, all we need to show is that E is not empty. For this purpose we shall use Bolzano's intermediate value theorem (Theorem 6.8.1). We know that $\cos 0 > 0$; and in order to show that E is nonempty, we shall show tht $\cos 2 < 0$. Using the fourth mean value theorem, choose a number $x \in (0, 2)$ such that

$$\cos 2 = \cos 0 + \frac{\cos^{(1)}(0)}{1!} 2 + \frac{\cos^{(2)}(0)}{2!} 2^2 + \frac{\cos^{(3)}(0)}{3!} 2^3 + \frac{\cos^{(4)}(x)}{4!} 2^4.$$

This gives us $\cos 2 = 1 - 2 + \frac{2}{3} \cos x$; and since $\cos x \leq 1$, it follows that $\cos 2 < 0$, as promised.   ■

11.9.3 The Number π.   We define the number π to be twice the least positive number x for which $\cos x = 0$. In other words, the least positive number at which the cosine function is zero is $\pi/2$.

11.9.4 The Periodic Behavior of the Functions cos and sin.   We begin this section by discussing the behavior of cos and sin on the interval $[0, \pi/2]$. Since $\cos 0 = 1$, it follows from Bolzano's intermediate value theorem that $\cos x > 0$ for all $x \in [0, \pi/2]$. From the fact that $\sin' = \cos$, we deduce that sin is strictly increasing on the interval $[0, \pi/2]$ and that $\sin x > 0$ whenever $x \in (0, \pi/2]$. Therefore, since $\sin^2 (\pi/2) = 1 - \cos^2(\pi/2) = 1$, we have $\sin(\pi/2) = 1$. Finally, since $\cos' = -\sin$, the function cos is strictly decreasing on $[0, \pi/2]$.

We now discuss the behavior of cos and sin on the interval $[\pi/2, \pi]$. For this purpose we shall use the identities $\cos(\pi/2 + x) = -\sin x$ and $\sin(\pi/2 + x) = \cos x$, which now follow directly from Theorem 11.9.1. Now as we saw above, sin increases from 0 to 1 on the interval $[0, \pi/2]$. It therefore follows from the identity $\cos(\pi/2 + x) = -\sin x$ that cos decreases from 0 to -1 on the interval $[\pi/2, \pi]$. Furthermore, since cos decreases from 1 to 0 on $[0, \pi/2]$, it follows from the identity $\sin(\pi/2 + x) = \cos x$ that sin decreases from 1 to 0 on $[\pi/2, \pi]$.

We can now turn our attention to the interval $[\pi, 2\pi]$, using the identities $\sin(\pi + x) = -\sin x$ and $\cos(\pi + x) = -\cos x$; and continuing in this way, one may verify all the usual relationships between the trigonometric functions sin and cos and the number π. In particular, one may see that sin and cos are periodic functions with period 2π. It would be a good exercise for you to write a careful and complete argument showing how all the familiar properties of cos and sin can be made to follow from their definitions.

11.10 EXERCISES

1. Given any number x, prove that $\sin x = 0$ if and only if x is an integer multiple of π. Prove that $\cos x = 0$ if and only if x is an odd multiple of $\pi/2$. Prove that for any integer n we have $\cos n\pi = (-1)^n$.

2. Prove that if n is any integer and x is any real number, then $\sin(x + 2n\pi) = \sin x$ and $\cos(x + 2n\pi) = \cos x$. Prove that if α is a given number, and if for every number x we have $\sin(x + \alpha) = \sin x$, then α must be an integer multiple of 2π.

3. Prove that the restriction of the function sin to the interval $[-\pi/2, \pi/2]$ is a strictly increasing function from $[-\pi/2, \pi/2]$ onto $[-1, 1]$. Prove that if $\arcsin u$ is defined for every point u in $[-1, 1]$ to be the unique number $x \in [-\pi/2, \pi/2]$ such that $u = \sin x$, then for every $u \in [-1, 1]$ we have

$$\arcsin u = \int_0^u \frac{1}{\sqrt{1 - t^2}}\, dt.$$

4. By analogy with exercise 3, give a definition of the function arctan, and deduce some of its properties.

5. The purpose of this exercise is to get you ready for exercise 6. Prove the following identity about binomial coefficients for all real numbers α and all integers $n \geq 0$:

$$(n + 1) \binom{\alpha}{n + 1} + n \binom{\alpha}{n} = \alpha \binom{\alpha}{n}.$$

6. Prove that if we define $f(x) = \sum_{n=0}^{\infty} \binom{\alpha}{n} x^n$ whenever $|x| < 1$, where α is any given real number, then the function f satisfies the differential equation $(1 + x)f'(x) = \alpha f(x)$ whenever $|x| < 1$. Recall that the interval of convergence of this series was found in Example 10.10.7(5) and Exercise 10.11(4).

7. Prove that if f is the function defined in exercise 6 and if $g(x) = f(x)(1 + x)^{-\alpha}$ for $|x| < 1$, then g is the constant function 1. Deduce that at every point x in the interval of convergence of the binomial series $\Sigma\binom{\alpha}{n}x^n$, we have

$$(1 + x)^\alpha = \sum_{n=0}^{\infty}\binom{\alpha}{n}x^n.$$

8. Prove that if $f(x) = \sqrt{1 - x}$ for $0 \le x \le 1$, then there exists a sequence of polynomials which converges uniformly to f on $[0, 1]$.

9. Prove that if $a > 0$ and if $g(x) = |x|$ for $-a \le x \le a$, then there exists a sequence of polynomials that converges uniformly to g on $[-a, a]$. *Hint:* If f is the function defined in exercise 8, then for every point $x \in [-a, a]$ we have $g(x) = af(1 - x^2/a^2)$.

Exercise 9 is of considerable importance because it may be used as the starting point for a proof of an important result known as the **Stone-Weierstrass theorem.** If you would like to read this interesting theorem, you can do so now by starting at Corollary 7.27 of Rudin [13].

11.11 ANALYTIC FUNCTIONS

As we have seen, many of the functions which play a role in analysis can be expressed as sums of power series. The ability to represent a function as the sum of a power series is a valuable tool, and it is therefore quite natural that we should give some special attention to the functions that can be represented in this way. These are the functions that we call *analytic functions*. Note, however, that in order to be able to represent a function as the sum of a power series, we do not necessarily have to be able to express it in its entire domain as the sum of a single power series of the form $\Sigma\, a_n x^n$. It may be that a series of this type would be capable of representing the function only near the origin 0, and that as we move further away from 0, we might need to turn to power series which are centered around some other point. Look, for example, at the function f defined by $f(x) = 1/(1 - x)$ for all $x \ne 1$. The power series $\Sigma\, x^n$ can represent f only in the interval $(-1, 1)$; but in spite of this, if c is any number except 1, we have

$$f(x) = \frac{1}{1 - c - (x - c)} = \left(\frac{1}{1 - c}\right)\frac{1}{1 - \left(\frac{x - c}{1 - c}\right)} = \frac{1}{1 - c}\sum_{n=0}^{\infty}\left(\frac{x - c}{1 - c}\right)^n$$

provided that $|x - c| < |1 - c|$. We see, then, that if c is any number unequal to 1, then f can be represented near c by a power series centered around c. This suggests the following definition of an analytic function.

11.11.1 Definition of an Analytic Function. A function f defined on an open set U is said to be **analytic** *on* U if for every point $c \in U$ there exists a number $\delta >$

0 and a sequence (a_n) of real numbers such that for every point $x \in (c - \delta, c + \delta)$ we have $f(x) = \sum_{n=0}^{\infty} a_n(x - c)^n$.

In view of Section 11.8.10, we can restate this definition as follows: A function f defined on an open set U is said to be *analytic on U* if for every point $c \in U$ there exists a number $\delta > 0$ such that for every point $x \in (c - \delta, c + \delta)$ we have

$$f(x) = \sum_{n=0}^{\infty} \frac{f^{(n)}(c)}{n!} (x - c)^n.$$

11.11.2 Some Examples of Analytic Functions

(1) The exponential function exp is analytic on $\boldsymbol{R}$ because if c is any point in $\boldsymbol{R}$, then for every number x we have

$$e^x = e^c e^{x-c} = e^c \sum_{n=0}^{\infty} \frac{(x - c)^n}{n!}.$$

(2) The function sin is analytic on $\boldsymbol{R}$ because if c is any point in $\boldsymbol{R}$, then for every number x we have

$$\sin x = \sin(x - c + c) = \sin(x - c) \cos c + \sin c \cos(x - c)$$

$$= \cos c \sum_{n=0}^{\infty} \frac{(-1)^n(x - c)^{2n+1}}{(2n + 1)!} + \sin c \sum_{n=0}^{\infty} \frac{(-1)^n(x - c)^{2n}}{(2n)!}.$$

(3) For each $x > -1$, define $f(x) = \log(1 + x)$. We shall show that f is analytic on $(-1, \infty)$. To see this, let $c \in (-1, \infty)$.

Choose $\delta > 0$ such that $(c - \delta, c + \delta) \subseteq (-1, \infty)$; in other words, $0 < \delta \leq c + 1$. Then whenever $x \in (c - \delta, c + \delta)$, we have $|(x - c)/(1 + c)| < 1$; and therefore,

$$f(x) = \log(1 + c + x - c) = \log(1 + c) + \log\left(1 + \frac{x - c}{1 + c}\right)$$

$$= \log(1 + c) + \sum_{n=0}^{\infty} \frac{(-1)^n(x - c)^{n+1}}{(n + 1)(1 + c)^{n+1}}.$$

(4) Define $f(x) = 1/x$ for all $x \neq 0$. To see that f is analytic on $\boldsymbol{R} \backslash \{0\}$, let $c \in \boldsymbol{R} \backslash \{0\}$. Then whenever $|x - c| < |c|$, we have

$$f(x) = \frac{1}{c + x - c} = \frac{1}{c[1 + (x - c)/c]} = \frac{1}{c} \sum_{n=0}^{\infty} \frac{(-1)^n(x - c)^n}{c^n}.$$

(5) Define $f(x) = 1/(1 + x^2)$ for all $x \in \mathbf{R}$. The function f is analytic on $\mathbf{R}$, but this would be quite hard to show at present. The fact that f is analytic will follow at once from Theorem 11.11.5.

11.11.3 Theorem: Some Simple Properties of Analytic Functions. Suppose that f and g are analytic on an open set U.

(1) The sum, difference, and product of f and g are all analytic on U.

(2) The function f is differentiable on U, and its derivative f' is also analytic on U. Consequently, f has derivatives of all orders on U.

■ **Proof.**   Let $c \in U$, and choose $\delta > 0$ such that in $(c - \delta, c + \delta)$ both f and g can be written as power series centered around c. For $|x - c| < \delta$, suppose

$$f(x) = \sum_{n=0}^{\infty} a_n(x - c)^n \quad \text{and} \quad g(x) = \sum_{n=0}^{\infty} b_n(x - c)^n.$$

Then for each such x we have

$$f(x) + g(x) = \sum_{n=0}^{\infty} (a_n + b_n)(x - c)^n,$$

and so the function $f + g$ is analytic on U. Similarly, $f - g$ is analytic on U. Now using Mertens's theorem (Theorem 10.14.2), or even Cauchy's theorem (Theorem 10.14.3), we deduce that whenever $|x - c| < \delta$, we have

$$f(x)g(x) = \sum_{n=0}^{\infty} \sum_{m=0}^{n} a_{n-m}b_m(x - c)^n;$$

and from this it follows that the function fg is also analytic on U. This proves part (1) of the theorem, and part (2) is immediate.   ■

In view of Theorem 11.11.3, a natural question to ask is whether the quotient of two analytic functions is always analytic as long as the denominator is not zero. The answer is yes; but before we can prove this, we need to deal with the composition of two analytic functions. The lemma that follows will help us over the technicalities of the composition theorem.

11.11.4 Lemma. Suppose that the power series $\Sigma\, b_n x^n$ converges whenever $|x| < r$, and suppose that for every natural number n and every point $x \in (-r, r)$ we have

$$\left(\sum_{m=0}^{\infty} b_m x^m \right)^n = \sum_{m=0}^{\infty} b_m(n)x^m \quad \text{and} \quad \left(\sum_{m=0}^{\infty} |b_m x^m| \right)^n = \sum_{m=0}^{\infty} c_m(n)|x|^m.$$

Then for all m and n we have $|b_m(n)| \leq c_m(n)$.

■ *Proof.* We prove this lemma by induction on n. Given any natural number n, let p_n be the statement that for all integers $m \geq 0$ we have $|b_m(n)| \leq c_m(n)$. Since $c_m(1) = |b_m| = |b_m(1)|$ for all m, the statement p_1 is clearly true. Now given any n for which statement p_n is true, since the series $\Sigma_m b_m(n + 1)x^m$ is the Cauchy product of $\Sigma_m b_m(n)x^m$ and $\Sigma_m b_m(1)x^m$, and $\Sigma_m c_m(n + 1)|x|^m$ is the Cauchy product of $\Sigma_m c_m(n)|x|^m$ and $\Sigma_m c_m(1)|x|^m$, we see that for every integer $m \geq 0$ we have

$$|b_m(n + 1)| = \left| \sum_{j=0}^{m} b_{m-j}(n)b_j(1) \right| \leq \sum_{j=0}^{m} |b_{m-j}(n)b_j(1)|$$

$$\leq \sum_{j=0}^{m} c_{m-j}(n)c_j(1) = c_m(n + 1).$$

The result therefore follows by induction. ■

11.11.5 The Composition Theorem.

Suppose that g is analytic on an open set U, that V is an open set which includes the range of g, and that f is analytic on V. Then the composition $f \circ g$ of f and g is analytic on U.

■ *Proof.* Let $\alpha \in U$. We need to show that in some neighborhood of α, $f \circ g$ can be written as a power series centered around α. Define $\beta = g(\alpha)$, and using the fact that f is analytic on V, choose a number $\varepsilon > 0$ such that in the interval $(\beta - \varepsilon, \beta + \varepsilon)$, f can be written as a power series centered around β. For $|y - \beta| < \varepsilon$, let us write

$$f(y) = \sum_{n=0}^{\infty} a_n(y - \beta)^n.$$

Now using the fact that g is analytic on U, choose a number $\delta_1 > 0$ such that in the interval $(\alpha - \delta_1, \alpha + \delta_1)$, g can be written as a power series centered around α. Suppose that whenever $|x - \alpha| < \delta_1$,

$$g(x) - \beta = \sum_{m=0}^{\infty} b_m(x - \alpha)^m.$$

We note that $b_0 = 0$ and that for all $x \in (\alpha - \delta_1, \alpha + \delta_1)$ we have

$$g(x) = \beta + \sum_{m=0}^{\infty} b_m(x - \alpha)^m.$$

Now using the fact that $b_0 = 0$, choose a number $\delta > 0$ such that $\delta \leq \delta_1$ and such that whenever $|x - \alpha| < \delta$, we have $\Sigma_{m=0}^{\infty} |b_m(x - \alpha)^m| < \varepsilon/2$.

By analogy with the notation in Lemma 11.11.4, write

$$\left[\sum_{m=0}^{\infty} b_m(x - \alpha)^m \right]^n = \sum_{m=0}^{\infty} b_m(n)(x - \alpha)^m$$

and

$$\left[\sum_{m=0}^{\infty} |b_m(x - \alpha)^m|\right]^n = \sum_{m=0}^{\infty} c_m(n)|x - \alpha|^m$$

for all $n \in \mathbf{Z}^+$ and all $x \in (\alpha - \delta, \alpha + \delta)$. From Lemma 11.11.4 it follows that $|b_m(n)| \le c_m(n)$ for all m and n; and so it follows that whenever $|x - \alpha| < \delta$, we have

$$\sum_{n=0}^{\infty} \sum_{m=0}^{\infty} |a_n| \, |b_m(n)| \, |x - \alpha|^m \le \sum_{n=0}^{\infty} \sum_{m=0}^{\infty} |a_n| \, |c_m(n)| \, |x - \alpha|^m$$

$$= \sum_{n=0}^{\infty} |a_n| \left[\sum_{m=0}^{\infty} |b_m(x - \alpha)^m|\right]^n \le \sum_{n=0}^{\infty} |a_n| \left(\frac{\varepsilon}{2}\right)^n < \infty.$$

We may therefore deduce from Theorem 10.13.3 that whenever $|x - \alpha| < \delta$,

$$f(g(x)) = \sum_{n=0}^{\infty} a_n[g(x) - \beta]^n = \sum_{n=0}^{\infty} a_n \left[\sum_{m=0}^{\infty} b_m(x - \alpha)^m\right]^n$$

$$= \sum_{n=0}^{\infty} a_n \sum_{m=0}^{\infty} b_m(n)(x - \alpha)^m = \sum_{m=0}^{\infty} \sum_{n=0}^{\infty} a_n b_m(n)(x - \alpha)^m$$

$$= \sum_{m=0}^{\infty} \left[\sum_{n=0}^{\infty} a_n b_m(n)\right](x - \alpha)^m,$$

and since the latter expression is a power series centered around α, the proof is complete. ∎

11.11.6 Theorem. Suppose that f and g are analytic on an open set U and that $g(x) \ne 0$ for all $x \in U$. Then the function f/g is analytic on U.

∎ **Proof.** The function $1/g$ is the composition of g and the reciprocal function; and since we saw in Example 11.11.2(4) that the reciprocal function is analytic on $\mathbf{R}\backslash\{0\}$, it follows from the composition theorem that $1/g$ is analytic on U. Therefore, f/g, being the product of two analytic functions, is analytic on U. ∎

The next theorem should come as no surprise; the sum of a power series is analytic.

11.11.7 Theorem. Suppose that the power series $\Sigma \, a_n x^n$ has radius of convergence $r > 0$; and for each point $x \in (-r, r)$, define $f(x) = \Sigma_{n=0}^{\infty} a_n x^n$. Then the function f is analytic on $(-r, r)$.

■ **Proof.** Suppose $c \in (-r, r)$. Choose $\delta > 0$ such that $(c - \delta, c + \delta) \subseteq (-r, r)$. Note that $\delta \leq r - |c|$:

Therefore, whenever $x \in (c - \delta, c + \delta)$, we have $|c| + |x - c| < r$; and it follows that

$$\sum_{n=0}^{\infty} \sum_{m=0}^{\infty} |a_n| \binom{n}{m} |c|^{n-m} |x - c|^m = \sum_{n=0}^{\infty} \sum_{m=0}^{n} |a_n| \binom{n}{m} |c|^{n-m} |x - c|^m$$

$$= \sum_{n=0}^{\infty} |a_n| (|c| + |x - c|)^n < \infty.$$

It therefore follows from Theorem 10.13.3 that whenever $x \in (c - \delta, c + \delta)$,

$$f(x) = \sum_{n=0}^{\infty} a_n x^n = \sum_{n=0}^{\infty} a_n [c + (x - c)]^n = \sum_{n=0}^{\infty} \sum_{m=0}^{n} a_n \binom{n}{m} c^{n-m} (x - c)^m$$

$$= \sum_{n=0}^{\infty} \sum_{m=0}^{\infty} a_n \binom{n}{m} c^{n-m} (x - c)^m = \sum_{m=0}^{\infty} \sum_{n=0}^{\infty} a_n \binom{n}{m} c^{n-m} (x - c)^m$$

$$= \sum_{m=0}^{\infty} \left[\sum_{n=0}^{\infty} a_n \binom{n}{m} c^{n-m} \right] (x - c)^m,$$

and this expresses the function in the desired form. ■

In the theorem that follows we see an important property of the set of points at which an analytic function is zero.

11.11.8 Theorem. Suppose that f is analytic on an open set U, that $c \in U$, and that $f(c) = 0$. Then there are two possibilities: Either

(1) $f(x) = 0$ for all points x lying sufficiently close to c.

or

(2) $f(x) \neq 0$ for all points x lying sufficiently close to c, other than c itself.

■ **Proof.** Choose a number $\delta > 0$ and a sequence (a_n) such that whenever $|x - c| < \delta$, we have $f(x) = \sum_{n=0}^{\infty} a_n (x - c)^n$. In the event that $a_n = 0$ for every n, we see at once that $f(x) = 0$ for every point $x \in (c - \delta, c + \delta)$. Suppose now that not all the numbers a_n are zero, and define N to be the least integer n for which $a_n \neq 0$. Then for every point $x \in (c - \delta, c + \delta)$ we have

$$f(x) = \sum_{n=N}^{\infty} a_n(x - c)^n = (x - c)^N g(x),$$

where for each point $x \in (c - \delta, c + \delta)$, $g(x) = \sum_{n=N}^{\infty} a_n(x - c)^{n-N}$. Since $g(c) = a_N$ and g is continuous at c, we deduce that $g(x) \neq 0$ for all points x lying sufficiently close to c; and it follows that as long as $x \neq c$ and x lies sufficiently close to c, we have $f(x) \neq 0$. ∎

11.11.9 Theorem. Suppose that f is analytic on an open interval I, that $E = \{x \in I | f(x) = 0\}$, and that E has a limit point that lies in I. Then $E = I$; in other words, f is the constant function zero.

■ **Proof.** From Theorem 11.11.8 we observe that whenever $x \in I$ and x is a limit point of E, the function f must be zero in some neighborhood of x. Now to obtain a contradiction, assume that f is not the constant zero, and choose a point $c \in I$ such that $f(c) \neq 0$. We shall now assume that E has a limit point less than c. The case in which E has a limit point greater than c is analogous. Define $\alpha = \sup\{x \in \mathscr{L}E \mid x < c\}$. Then $\alpha \in I$, and α is a limit point of E; and from the above observation we conclude that f is zero in some neighborhood of α. It follows that there must be an interval of points of $\mathscr{L}E$ lying between α and c, contradicting the choice of α. ∎

Finally, as promised, we give a sharper version of the result stated in Section 11.8.8.

11.11.10 Corollary. Suppose that $f(x) = \sum_{n=0}^{\infty} a_n x^n$ and $g(x) = \sum_{n=0}^{\infty} b_n x^n$ whenever $|x| < r$. Suppose that $E = \{x \mid f(x) = g(x)\}$ and that E has a limit point α that lies in $(-r, r)$. Then $f(x) = g(x)$ for every $x \in (-r, r)$, and for every integer $n \geq 0$ we have $a_n = b_n$.

■ **Proof.** By Theorem 11.11.7, the functions f and g are analytic on $(-r, r)$; and therefore, $f - g$ is an analytic function on $(-r, r)$ whose set of zeros has a limit point in $(-r, r)$. It follows from Theorem 11.11.9 that $f - g$ is the constant function zero, and it now follows from Sections 11.8.7 and 11.8.8 that $a_n = b_n$ for each n. ∎

11.12 EXERCISES

1. Prove that the function arctan is analytic on $\mathbf{R}$.

2. Given $f(x) = x^{1/3}$ for all x, prove that f is analytic on $\mathbf{R}\backslash\{0\}$. Determine whether f is analytic on $\mathbf{R}$.

3. Prove that the function tan is the sum of its Maclaurin series in some neighborhood of 0.

4. Given $f(0) = 0$ and $f(x) = \exp(-x^{-2})$ for all $x \neq 0$, prove that f is analytic on $R \backslash \{0\}$. Is f analytic on R? (Take another look at Example 11.8.9.)

5. Prove that a rational function is analytic on any open set in which its denominator does not vanish.

6. Given that f is analytic and nonconstant on an open interval I and that $a \in I$, prove that there exists a natural number n such that $f^{(n)}(a) \neq 0$.

Hints and Solutions to Selected Exercises

Exercises 1.4

Do not read these answers until you have written out your own version of the denials of the given sentences. The denial of a sentence can often be written in many different ways, and you should not feel that in order to be correct, your version has to be exactly the same as the one you will find below. On the other hand, you should check very carefully to see that the *meaning* of your version is the same.

1. What you said yesterday is correct and Jim does not have red hair.
2. You do not take me for a fool and you are not a fool yourself.
3. Either he did not walk into my office this morning, or he did not tell me a pack of lies, or he did not punch me on the nose.
4. Not all that glitters is gold.
5. Either you are wrong or I am right.
6. Either we are both right or we are both wrong.
7. Every cat scratches and someone in this room is not a liar.
8. I sometimes sleep without dreaming.

9. There is at least one person who cannot remember the battle of Waterloo but who is worth listening to on military subjects.
10. None of us is out of breath or at least one of us is not fat.
11. No one in this room is smoking.
12. The percentage of people in this room who are smoking is not fifty.
13. It is without regret that I inform you that someone in this room is smoking.
14. For every number x there is a number $u > x$ such that $f(u) \leq g(u)$.
15. It is possible to find numbers u and v such that $u > 50$, and $v > 50$, and $|f(u) - f(v)| \geq 2$.
16. For every number p it is possible to find numbers u and v such that $u > p$, and $v > p$, and $|f(u) - f(v)| \geq 2$.
17. There exists a positive number ε and there exist numbers x and t such that $x > 7$, and $t > 7$, and $|f(x) - f(t)| \geq \varepsilon$.
18. There exists a positive number ε such that for every number p it is possible to find numbers x and t such that $x > p$, and $t > p$, and $|f(x) - f(t)| \geq \varepsilon$.
19. For every number p it is possible to find a positive number ε and it is possible to find numbers x and t such that $x > p$, and $t > p$, and $|f(x) - f(t)| \geq \varepsilon$.
20. There exists a function h which is continuous on $(0, 1)$ and which has the property that for every number x in $(0, 1)$, h fails to be differentiable at x.
21. For every number w there exists a member x of A such that $x \geq w$.
22. It is possible to find a member x of A and a member y of B such that $x \geq y$.
23. It is possible to find members x and y of A such that $x \neq y$ and $|x - y| < 1$.
24. There exists a positive number ε such that whenever x and y are members of A, either $x = y$ or $|x - y| \geq \varepsilon$.
25. It is possible to find two sets P and Q of numbers which have the following two properties:
    (i) Whenever $x \in P$ and $y \in Q$, we have $x < y$.
    (ii) Given any number w, if $x \leq w$ for every member x of P, then there must exist a member y of Q such that $y < w$.

Exercises 2.3

4. x is not an upper bound of A.
5. $x \notin A$ and x is an upper bound of A.
8. Let α be any number. Since $\varnothing$ is empty, $\varnothing$ has no member greater than α. Therefore, α is an upper bound of $\varnothing$.
9. Choose a lower bound α of A and an upper bound β. Define p to be the larger of the two numbers $|\alpha|$ and $|\beta|$. Then show that p is an upper bound of B.

Exercises 2.5

5. Let $y \in B$. Choose $x \in A$ such that $x > y$. From this we deduce that $y < x \leq \sup A$, and it follows that $\sup A$ is an upper bound of B. Therefore, since $\sup B$ is the *least* upper bound of B, we deduce that $\sup B \leq \sup A$. One may show similarly that $\sup A \leq \sup B$.
6. The given information tells us that for every point $y \in B$, y is an upper bound

of A; and therefore, $y \geq \sup A$. Therefore, $\sup A$ is a lower bound of B. Continue from here.

7. Let $\delta > 0$. Since $\delta + \sup A > \inf B$, the number $\delta + \sup A$ is not a lower bound of B. Choose a point $y \in B$ such that $\delta + \sup A > y$. We now have $y - \delta < \sup A$, and so $y - \delta$ is not an upper bound of A. Continue from here.

12. Let $z \in C$, and choose $x \in A$ and $y \in B$ such that $z = x + y$. Since $z = x + y \leq \sup A + \sup B$, we deduce that $\sup A + \sup B$ is an upper bound of C. Therefore, $\sup C \leq \sup A + \sup B$. Now to obtain a contradiction, assume that $\sup C < \sup A + \sup B$. Using the fact that $\sup C - \sup B < \sup A$, choose $x \in A$ such that $\sup C - \sup B < x$, and then using the fact that $\sup C - x < \sup B$, choose $y \in B$ such that $\sup C - x < y$. But this yields $\sup C < x + y$, which is impossible since $x + y \in C$.

13. Read Lemma 8.14.3.

17. See Section 6.8 for a complete solution.

Exercises A.4

1. The equation $y = (3x - 2)/(x + 1)$ holds when $x = (2 + y)/(3 - y)$, and it follows from this that f is one–one and that the range of f is $\mathbf{R}\backslash\{3\}$.

2. The range of f is $(-2, \frac{1}{2})$.

5. f is one–one but g might not be. In the event that f maps A onto B, g is also one–one. Prove these assertions.

7. Suppose that f is a one–one function from A into B. Choose an element $a \in A$. Now for each point $y \in B$, if y lies in the range of f, define $g(y) = x$, where x is the unique point of A such that $y = f(x)$; and define $g(y) = a$, otherwise. Show that g maps B onto A.

8. Suppose g is a function from B onto A. For each point $x \in A$, choose a point $y \in B$ such that $x = g(y)$; and denote the chosen point y as $f(x)$. This defines a one–one function f from A into B.

Exercises A.8

1. Choose one–one functions f and g such that $f : A \to \mathbf{Z}^+$ and $g : B \to \mathbf{Z}^+$. Define $h : A \times B \to \mathbf{Z}^+ \times \mathbf{Z}^+$ by $h(x, y) = (f(x), g(y))$ for each point (x, y) of $A \times B$. Now show that h is one–one, and deduce that $A \times B \sim\subseteq \mathbf{Z}^+$.

4. Each point on the perpendicular bisector of AB is the center of a circle that passes through the points A and B. Since two circles can intersect at only two points, it follows that these circles have no points of intersection in $\mathbf{R}^2\backslash\{A, B\}$, and since the set of these circles is uncountable, not all of them can contain points in the countable set $(\mathbf{Q} \times \mathbf{Q})\backslash\{A, B\}$.

8. The set $\mathbf{R}$ of all real numbers is uncountable. Therefore, since the set of algebraic numbers is countable (see exercise 6), there must be uncountably many transcendental numbers.

9. The result follows simply from the fact that $S = \bigcup_{n=1}^{\infty} \{x \in S \mid f(x) > 1/n\}$; and therefore, for some natural n the set $\{x \in S \mid f(x) > 1/n\}$ is uncountable. As a matter of fact, this set is uncountable for *all* sufficiently large n.

10. The function f defined by $f(x, y) = x + y$ for all $(x, y) \in A \times B$ maps the countable set $A \times B$ onto $A + B$. Therefore, $A + B$ is countable.

13. Suppose S is an uncountable set of real numbers. For each integer n, define $S_n = S \cap [n, n + 1]$. From Theorem A.7.8 we know that for some n the set S_n is uncountable, and by the Bolzano-Weierstrass theorem this set S_n must have a limit point.

14. A nonempty open set must include an interval of positive length. Therefore, every nonempty open set is equivalent to $\mathbf{R}$.

15. Note that since H is nonempty and every point of H is a limit point of H, the set H must be infinite. Now to obtain a contradiction, assume that H is countable, and write $H = \{x_n | n \in \mathbf{Z}^+\}$. Using the fact that H is infinite, choose a point $y_1 \in H \backslash \{x_1\}$. Choose numbers a_1 and b_1 such that $a_1 < y_1 < b_1$ and such that the interval $[a_1, b_1]$ does not contain x_1. Define $H_1 = [a_1, b_1] \cap H$. We now repeat this process. Using the fact that H_1 is infinite, choose numbers a_2 and b_2 such that the interval $[a_2, b_2]$ does not contain x_2 and such that the set $H_2 = [a_2, b_2] \cap H$ is infinite. Continue this process. This yields a contracting sequence (H_n) of nonempty, closed, bounded sets with an empty intersection.

Exercises A.10

1. For each point f in $\{0, 1\}^A$, define $\phi(f) = \{x \in A | f(x) = 1\}$ and note that $f = \chi_{\phi(f)}$. This defines a function ϕ from $\{0, 1\}^A$ onto $\pi(A)$. It is easy to see that ϕ is one–one and that ϕ maps $\{0, 1\}^A$ onto $\pi(A)$.

2. A member of $(A^B)^C$ is a function $f: C \rightarrow A^B$. If f is such a function, then for each point $y \in C$, $f(y)$ is a function from B into A; in other words, $f(y)(x) \in A$ whenever $x \in B$. Now given $f \in (A^B)^C$, define $\phi(f)$ to be the function from $B \times C$ into A defined by $\phi(f)(x, y) = f(y)(x)$ for each point $(x, y) \in B \times C$. Show that ϕ is a one–one function from $(A^B)^C$ onto $A^{B \times C}$.

3. $\mathbf{R} \sim \pi(\mathbf{Z}^+) \sim \{0, 1\}^{\mathbf{Z}^+}$, and therefore, by exercise 2, $\mathbf{R}^{\mathbf{Z}^+} \sim (\{0, 1\}^{\mathbf{Z}^+})^{\mathbf{Z}^+} \sim \{0, 1\}^{\mathbf{Z}^+ \times \mathbf{Z}^+} \sim \{0, 1\}^{\mathbf{Z}^+} \sim \mathbf{R}$.

4. Suppose n is any natural. Since $\mathbf{R}^n \sim \subseteq \mathbf{R}^{\mathbf{Z}^+}$, it follows from exercise 3 that $\mathbf{R}^n \sim \subseteq \mathbf{R}$. But it is clear that $\mathbf{R} \sim \subseteq \mathbf{R}^n$, and therefore, $\mathbf{R}^n \sim \mathbf{R}$. Finally, if I is any interval of positive length, then $I \sim \mathbf{R} \sim \mathbf{R}^3$, and this is true even if the length of the interval is 0.0001.

Exercises 3.4

4. Suppose that $x \in A$ and that U is a neighborhood of y. Choose $\delta > 0$ such that $(y - \delta, y + \delta) \subseteq U$. Observe that $(x + y - \delta, x + y + \delta) \subseteq A + U$.

Exercises 3.6

6. Suppose A is nonempty, closed, and bounded above. Define $\alpha = \sup A$. To obtain a contradiction, assume that $\alpha \in \mathbf{R} \backslash A$. Using the fact that $\mathbf{R} \backslash A$ is open, choose $\delta > 0$ such that $(\alpha - \delta, \alpha + \delta) \subseteq \mathbf{R} \backslash A$, and observe that $\alpha - \delta$ is an upper bound of A.

9. Suppose $z \in A + U$. Choose $x \in A$ and $y \in U$ such that $z = x + y$. Now use Exercise 3.4(4).

10. See Theorem 4.13.3.

Exercises 3.9

1. Suppose $x \in \bar{A}$. Then since every neighborhood of x intersects with A, we see that $\boldsymbol{R}\backslash A$ is not a neighborhood of x. Now suppose that $\boldsymbol{R}\backslash A$ is not a neighborhood of A. Then nor is any subset of $\boldsymbol{R}\backslash A$; and therefore, every neighborhood of A must contain some points of A.

4. Given any point $x \in \boldsymbol{R}\backslash H$, we see that $\boldsymbol{R}\backslash H$ is a neighborhood of x which does not intersect with A. Therefore, no point of $\boldsymbol{R}\backslash H$ can lie in $\bar{A}$.

6. Let $x \in \boldsymbol{R}$. To prove that $x \in \overline{U \cap V}$, let $\delta > 0$. Since $x \in \bar{U}$, the set $(x - \delta, x + \delta) \cap U$ must be nonempty. Choose a point $y \in (x - \delta, x + \delta) \cap U$. From the fact that $y \in \bar{V}$ and that $(x - \delta, x + \delta) \cap U$ is a neighborhood of y, we conclude that $(x - \delta, x + \delta) \cap U \cap V \neq \emptyset$.

Exercises 3.11

6. We need to show that $\bar{U} \subseteq \mathscr{L}(U)$. Let $x \in \bar{U}$, and let $\delta > 0$. Since the set $(x - \delta, x + \delta) \cap U$ is nonempty and open, it includes an interval of positive length and is therefore infinite.

7. Define $\alpha = \sup A$. We note that $\alpha \notin A$. To show that $\alpha \in \mathscr{L}(A)$, let $\delta > 0$. Since $\alpha - \delta < \alpha$, we see that $\alpha - \delta$ is not an upper bound of A. Choose $x \in A$ such that $\alpha - \delta < x$, and note that $\alpha - \delta < x < \alpha$. This shows that $(\alpha - \delta, \alpha + \delta) \cap A\backslash\{\alpha\} \neq \emptyset$.

8. Use exercise 7.

Exercises 3.14

2. This exercise was done in the proof of the Bolzano-Weierstrass theorem, where we showed that if A is bounded and infinite, and if every nonempty subset of A has a least member, then there must be a nonempty subset with no greatest member.

Exercises 4.4

3. Define $x_n = m/k$ if n can be written in the form $2^m 3^k$, where m and k are naturals and $2k \leq m \leq 3k$; and define $x_n = 1$, otherwise. Note that $x_n = 1$ for infinitely many naturals n and that the range of (x_n) is $(\boldsymbol{Q} \cap [2, 3]) \cup \{1\}$.

4. $(0, \infty)$ is a neighborhood of x. Therefore, (x_n) is eventually in $(0, \infty)$.

5. If we had $x < 0$, then $(-\infty, 0)$ would be a neighborhood of x. But (x_n) is not frequently in $(-\infty, 0)$.

8. Use the fact that $x_n \leq 1/n$ for each n.

9. There are many ways to do this. One way is to prove by induction that if $n \geq 6$, then $2^n/n! < 1/n$.

11. Let $\varepsilon > 0$. Choose a natural k such that $1/k < \varepsilon$. Define $N = 2^{k-1}$.

12. Suppose $x_n \to x$. Let $\varepsilon > 0$. Since $(x - 5\varepsilon, x + 5\varepsilon)$ is a neighborhood of x, we know that (x_n) is eventually in $(x - 5\varepsilon, x + 5\varepsilon)$. Now suppose that for every $\varepsilon > 0$ there exists a natural N such that whenever $n \geq N$, we have $|x_n - x| < 5\varepsilon$. To prove that $x_n \to x$, let $\varepsilon > 0$. We now apply the given condition to the positive number $\varepsilon/5$. Choose a natural N such that whenever $n \geq N$, we have $|x_n - x| < 5\varepsilon/5$.

17. Use the fact that $(n^3 - n^2 + 1)/(n^2 + 5) = n - 1 + 1/(n^2 + 5)$.

20. **(a)** Suppose that (x_n) and (y_n) are eventually close and that $x_n \to x$, where $x \in \mathbf{R}$. To show that $y_n \to x$, let $\varepsilon > 0$. Choose a natural N_1 such that whenever $n \geq N_1$, we have $|x_n - y_n| < \varepsilon/2$. Choose a natural N_2 such that whenever $n \geq N_2$, we have $|x_n - x| < \varepsilon/2$. Define N to be the larger of the two numbers N_1 and N_2. Then whenever $n \geq N$, we have $|y_n - x| = |y_n - x_n + x_n - x| \leq |y_n - x_n| + |x_n - x| < \varepsilon/2 + \varepsilon/2 = \varepsilon$.

Exercises 4.6

8. Let $\varepsilon > 0$. Choose a natural M such that whenever $n \geq M$, we have $|x_n| < \varepsilon/2$. Now using the fact that $(x_1 + x_2 + x_3 + \cdots + x_M)/n \to 0$ as $n \to \infty$, choose a natural N_1 such that whenever $n \geq N_1$, we have $|x_1 + x_2 + x_3 + \cdots + x_M|/n < \varepsilon/2$. Define N to be the larger of the two numbers M and N_1. Then whenever $n \geq N$, we have

$$\left| \frac{x_1 + x_2 + x_3 + \cdots + x_n}{n} \right| \leq \left| \frac{x_1 + x_2 + x_3 + \cdots + x_M}{n} \right|$$

$$+ \left| \frac{x_{M+1} + x_{M+2} + x_{M+3} + \cdots + x_n}{n} \right|$$

$$< \frac{\varepsilon}{2} + \frac{(n - M)\varepsilon}{2n} < \varepsilon.$$

9. Suppose $x_n \to x$ as $n \to \infty$, and for each n, define $y_n = x_n - x$. Note that for each n

$$\left| \frac{x_1 + x_2 + x_3 + \cdots + x_n}{n} - x \right| = \left| \frac{y_1 + y_2 + y_3 + \cdots + y_n}{n} \right|$$

and make use of exercise 8.

15. Let x be a real partial limit of (x_n), and to show that x is a partial limit of (z_n), let $\varepsilon > 0$. Note that for each n we have $|z_n - x| \leq |x_n y_n - x y_n| + |x y_n - x|$. Choose a natural N_1 such that $|y_n - 1| < 1$ whenever $n \geq N_1$. Then whenever $n \geq N_1$, we have $|z_n - x| \leq 2|x_n - x| + |x| |y_n - 1|$. Choose a natural N_2 such that whenever $n \geq N_2$, we have $|x| |y_n - 1| < \varepsilon/2$. Define N to be the larger of the two numbers N_1 and N_2. There are infinitely many naturals $n \geq N$ such that $|x_n - x| < \varepsilon/4$, and for all such n we have $|z_n - x| < \varepsilon$. One may show similarly that if either $-\infty$ or ∞ is a partial limit of (x_n), then it is also a partial limit of (z_n). Therefore, every partial limit of (x_n) is a partial limit of (z_n). Now since $x_n = z_n/y_n$ for all sufficiently large n, and since $1/y_n \to 1$, the same argument shows that every partial limit of (z_n) is a partial limit of (x_n).

Exercises 4.8

1. Suppose that A is unbounded below. For every natural n, using the fact that the number $-n$ is not a lower bound of A, choose a member (which we shall call x_n) of A such that $x_n \leq -n$. It is easy to show that $x_n \to -\infty$.

2. Suppose $\alpha = \sup A$. Then for each natural n the number $\alpha - 1/n$ is not an upper bound of A. For each natural n, choose a member (which we shall call x_n) such that $x_n > \alpha - 1/n$. Now show that $x_n \to \alpha$.

Exercises 4.14

3. (e) From part (d) it follows that (x_n) is convergent. Suppose $x_n \to x$ as $n \to \infty$. Since $8x_{n+1}^3 = 6x_n + 1$ for each n, it follows that $8x^3 = 6x + 1$. We now observe that the latter equation has three solutions of the form $x = \cos \theta$. The equation becomes $8 \cos^3 \theta - 6 \cos \theta = 1$, and this may be written as $\cos 3\theta = \frac{1}{2}$. This implies that θ must be of the form $\pi/9 + 2k\pi/3$ for some integer k, and it is easy to deduce that the three solutions of the equation $8x^3 = 6x + 1$ are $x = \cos \pi/9$, $x = -\cos 2\pi/9$, and $x = -\cos 4\pi/9$. Since the limit of (x_n) cannot be negative, it therefore follows that $x = \cos \pi/9$.

9. To obtain a contradiction, suppose that for some natural N we have $x \notin H_N$. Since $\mathbf{R} \backslash H_N$ is a neighborhood of x and x is a partial limit of (x_n), there must be infinitely many naturals n such that $x_n \notin H_N$. Choose $n \geq N$ such that $x_n \notin H_N$. But this is impossible since $x_n \in H_n \subseteq H_N$.

*10. Yes. Suppose H is a closed, bounded set. First establish the following result: For every number $\delta > 0$ there exists a finite subset A of H such that for every point $x \in H$ we have $(x - \delta, x + \delta) \cap A \neq \emptyset$. To do this for a given $\delta > 0$, observe that a set A with the required properties can be found by selecting one point of H between $n\delta$ and $(n + 1)\delta$ (whenever possible) for each integer n. Having done this, choose a finite subset A_n of H for each natural n such that whenever $x \in H$, we have $(x - 1/n, x + 1/n) \cap H \neq \emptyset$. Using the fact that $\bigcup_{n=1}^{\infty} A_n$ is countable, choose a sequence (x_n) whose range is this set. Now show that H is the set of partial limits of (x_n).

Exercises 5.4

1. (c) Whenever $x \neq 2$, we have $(x^3 - 8)/(x^2 + x - 6) = (x^2 + 2x + 4)/(x + 3)$; and we deduce that

$$\left| \frac{x^3 - 8}{x^2 + x - 6} - \frac{12}{5} \right| = \left| \frac{(x - 2)(5x + 8)}{5(x + 3)} \right|;$$

and as long as $|x - 2| < 1$, the latter expression does not exceed $23|x - 2|/20$. Let $\varepsilon > 0$. Define δ to be the smaller of the two numbers 1 and $20\varepsilon/23$. Then whenever $x \neq 2$ and $|x - 2| < \delta$, we have

$$\left| \frac{x^3 - 8}{x^2 + x - 6} - \frac{12}{5} \right| < \varepsilon.$$

6. Define $g(x) = f(x)$ for all $x \in (-\infty, a) \cap S$. To prove that (a) $\Rightarrow$ (b), assume that (a) holds. Let V be a neighborhood of α. Choose $\delta > 0$ such that whenever $x \in (-\infty, a) \cap S$ and $|x - a| < \delta$, we have $g(x) \in V$. Now given $x \in S$ such that $0 < a - x < \delta$, we certainly have $x \in (-\infty, \delta)$ and $|x - a| < \delta$; and therefore, $f(x) = g(x) \in V$. The proof that (b) $\Rightarrow$ (a) is similar.

8. It is clear that if $f(x) \to \alpha$ as $x \to a$, then $f(x) \to \alpha$ as $x \to a+$. Suppose now that $f(x) \to \alpha$ as $x \to a+$. Let V be a neighborhood of α. Choose $\delta_1 > 0$ such that for all points $x \in S \cap (a, a + \delta_1)$ we have $f(x) \in V$. Choose $\delta_2 > 0$ such that $(a - \delta_2, a) \cap S = \emptyset$. Define δ to be the smaller of the two numbers δ_1 and δ_2. Then whenever $x \in S \setminus \{a\}$ and $|x - a| < \delta$, we have $f(x) \in V$. This shows that $f(x) \to \alpha$ as $x \to a$.

Exercises 6.5

7. Because S is closed, the limit of the sequence (x_n) must be a point of S. Therefore, if $x = \lim_{n \to \infty} x_n$, then f is continuous at x; and it follows that $f(x_n) \to f(x)$ as $n \to \infty$.

8. Let $\varepsilon > 0$. Choose δ_1 such that whenever $x \in S$ and $|x - a| < \delta_1$, we have $|f(x) - f(a)| < \varepsilon$. Using the fact that S is open, choose $\delta_2 > 0$ such that the interval $(a - \delta_2, a + \delta_2) \subseteq S$. Now define δ to be the smaller of the two numbers δ_1 and δ_2.

12. **(i)** Yes. Prove this.
   (ii) No. Construct an example by making one of the functions zero at a.

20. Use the fact that $R \setminus \{x | f(x) \in H\} = \{x | f(x) \in R \setminus H\}$, and apply exercise 19.

Exercises 6.9

7. The answer is yes. To show that S is closed, use the method of exercise 1. To show that S is bounded, define $f(x) = 1/(1 + x^2)$.

Exercises 6.13

3. Suppose that f is an increasing function on an interval S, and define E to be the set of those points of S at which f is discontinuous. For each point $x \in E$ we have $\lim_{x-} f < \lim_{x+} f$; and using this inequality, choose a rational number which we shall call $g(x)$ such that $\lim_{x-} f < g(x) < \lim_{x+} f$. This defines a one–one function g from E into Q.

8. To show that f is continuous on $[0, 1]$, look separately at the point 0 and the points $a \in (0, 1]$. The uniform continuity follows at once from the continuity. Now to obtain a contradiction, assume that f is lipschitzian on $[0, 1]$, and choose a number k such that $|f(x) - f(t)| \leq k|x - t|$ for all points x and t in $[0, 1]$. In particular, we have $\sqrt{x} \leq kx$ for all $x \in [0, 1]$, but this is impossible when $0 < x < 1/k^2$.

9. **(b)** To obtain a contradiction, assume that f is not bounded, and choose a sequence (x_n) as in part (a). Choose a limit point x of the set $E = \{x_n | n \in$

Z^+}. Now for every number $\delta > 0$ the interval $(x - \delta/2, x + \delta/2)$ contains infinitely many points of E; and by choosing two such points x_m and x_n of E, we obtain $|x_m - x_n| < \delta$ even though $|f(x_m) - f(x_n)| \geq 1$. This contradicts the uniform continuity of f.

Exercises 7.11

5. Look at Section 7.12.1.

Exercises 7.14

1. Let x and t be any two points of S, and suppose that $x < t$. Using the (first) mean value theorem, choose $c \in (x, t)$ such that $f(t) = f(x) + (t - x)f'(c)$. Since $f'(c) = 0$, it follows that $f(x) = f(t)$, and this shows that f is constant.

8. Let V be a neighborhood of w. Choose $\delta > 0$ such that whenever $a < x < a + \delta$, we have $f'(x) \in V$. We now show that whenever $a < x < a + \delta$, we have $[f(x) - f(a)]/(x - a) \in V$. Let $x \in (a, a + \delta)$. Choose $c \in (a, x)$ such that $[f(x) - f(a)]/(x - a) = f'(c)$. The desired result follows at once from the observation that $f'(c) \in V$.

9. The equation of the tangent line to the graph of f at $(a, f(a))$ is $y = f(a) + (x - a)f'(a)$. The precise statement of the fact that the point $(b, f(b))$ lies under this tangent line is the statement that $f(a) + (b - a)f'(a) - f(b) > 0$. For each $x \in [a, b]$, define $g(x) = f(a) + (x - a)f'(a) - f(x)$. We observe that for each $x \in (a, b)$, we have $g''(x) = -f''(x) > 0$. Therefore, applying the second mean value theorem to g on $[a, b]$, we can choose $c \in (a, b)$ such that

$$g(b) = g(a) + (b - a)g'(a) + \tfrac{1}{2}(b - a)^2 g''(c) = \tfrac{1}{2}(b - a)^2 g''(c) > 0.$$

10. Define $g(x) = f(x) - f(a) - \{[f(b) - f(a)]/(b - a)\}(x - a)$ for all $x \in [a, b]$. We need to show that $g(x) > 0$ for all $x \in (a, b)$. Using the fact that $g(a) = g(b) = 0$, choose $c \in (a, b)$ such that $g'(c) = 0$. Since $g''(x) = f''(x) < 0$ for all $x \in (a, b)$, we know that g' is strictly decreasing on (a, b). Therefore, $g'(x) > 0$ for $a < x < c$ and $g'(x) < 0$ for $c < x < b$. Since g is strictly increasing on $[a, c]$ and $g(a) = 0$, it follows that $g(x) > 0$ for $x \in (a, c]$; and since g is strictly decreasing on $[c, b]$ and $g(b) = 0$, it follows that $g(x) > 0$ for $x \in [c, b)$.

14. Write an almost exact copy of the proof of Theorem 7.13.9.

19. In order to show that $g'(x) \geq 0$ for all $x \in (0, \infty)$, we need to show that whenever $x > 0$, we have $xf'(x) - f(x) \geq 0$. Let $x > 0$. Choose $c \in (0, x)$ such that $f'(c) = [f(x) - f(0)]/(x - 0)$. Then $f(x)/x = f'(c) \leq f'(x)$, and the required inequality follows.

Exercises 7.16

1. $\frac{1}{6}$ **2.** $\frac{1}{2}$ **3.** 1   **4.** Put $x = e^u$. The limit is 0.

6. $-\frac{5}{4}$ **7.** e **8.** $e/2$ **9.** 1   **10.** 1   **11.** $\frac{1}{2}$

12. $\alpha/2$ **13.** 1   **14.** 0   **15.** 1   **16.** e

Exercises 8.4

7. Write $\{x \in \mathbf{R} \,|\, f(x) \neq 0\} = \{x_0, x_1, x_2, \ldots, x_n\}$, where $x_0 < x_1 < \cdots < x_n$. Define $a = x_0$ and $b = x_n$, and define $\mathcal{P} = (x_0, x_1, \ldots, x_n)$. We observe that $\mathcal{P}$ is a partition of $[a, b]$, f steps within $\mathcal{P}$, and $\Sigma(\mathcal{P}, f) = 0$.

11. Choose an interval $[a, b]$ such that both f and g are zero outside $[a, b]$. Then replace each integral of the type $\int_{-\infty}^{\infty}$ by an integral from a to b.

Exercises 8.6

7. Suppose E is elementary. Choose an interval $[a, b]$ such that $E \subseteq [a, b]$, and choose a partition $\mathcal{P} = (x_0, x_1, \ldots, x_n)$ of $[a, b]$ such that χ_E steps within $\mathcal{P}$. The set E is the union of some of the intervals (x_{i-1}, x_i) and some of the points of $\mathcal{P}$.

Exercises 8.8

1. Use the fact that $\chi_{A \cup B} = \chi_A + \chi_B - \chi_{A \cap B}$.

4. The elementary set $A \setminus H$ is not finite, and therefore, by exercise 3 we deduce that $m(A \setminus H) \neq 0$.

6. Given an elementary subset H of A, since every interval of positive length contains some irrationals, we deduce that H includes no interval of positive length.

7. It is easy to see that A is not elementary. (Prove this!) Now let $\varepsilon > 0$. We shall find a closed elementary set H and an open elementary set U such that $H \subseteq A \subseteq U$ and $m(U) - m(H) < \varepsilon$. Choose a natural N such that $x_{2N} < \varepsilon/2$. Define $H = \cup_{n=1}^{N-1} [x_{2n+1}, x_{2n}]$. Using Theorem 8.7.5, choose an open elementary set V including H such that $m(V) < m(H) + \varepsilon/2$, and define $U = (0, \varepsilon/2) \cup V$.

Exercises 8.15

4. Let $\varepsilon > 0$. Define $E = \{x \in [a, b] \,|\, f(x) \geq \varepsilon\}$; and using the fact that E is finite, express E in the form $\{x_1, x_2, \ldots, x_n\}$, where $x_1 < x_2 < \cdots < x_n$. Define $\mathcal{P}$ to be the partition $(a, x_1, x_2, \ldots, x_n, b)$ of $[a, b]$. Since $\mathfrak{u}(\mathcal{P}, f)(x) \leq \varepsilon$ for every point x in $[a, b]$ unless x is a point of $\mathcal{P}$, it is clear that $\int_a^b \mathfrak{u}(\mathcal{P}, f) \leq \varepsilon (b - a)$.

5. In view of Lemma 5.8.1, this exercise follows from exercise 4.

7. Suppose that $a \leq b \leq c$ and that f is Riemann-integrable on each of the intervals $[a, b]$ and $[b, c]$. We need to show that f is Riemann-integrable on $[a, c]$. Choose a pair of sequences (s_n) and (S_n) of step functions on $[a, b]$ which squeezes f on $[a, b]$, and choose a pair of sequences (t_n) and (T_n) of step functions on $[b, c]$ which squeezes f on $[b, c]$. For each n, define the functions g_n and G_n on $[a, c]$ as follows:

$$ g_n(x) = \begin{cases} s_n(x), & \text{for } a \leq x \leq b, \\ t_n(x), & \text{for } b < x \leq c, \end{cases} $$

and

$$G_n(x) = \begin{cases} S_n(x), & \text{for } a \le x \le b, \\ \\ T_n(x), & \text{for } b < x \le c. \end{cases}$$

Then g_n and G_n are step functions on $[a, c]$ for each n; and since

$$\int_a^c (G_n - g_n) = \int_a^b (S_n - s_n) + \int_b^c (T_n - t_n) \to 0 \qquad \text{as} \qquad n \to \infty,$$

we see that the pair (G_n) and (g_n) squeezes f on $[a, c]$.

Exercises 8.18

1. In the integral $\int_0^{\pi/2} f(\sin x)\, dx$, make the substitution $t = \pi - x$.

Exercises 8.21

4. Yes, the same thing is true for bounded convergence. Prove this.
5. Notice that $f_n' \to 0$ boundedly on $[0, 1]$. Now for each n, define $\alpha_n = f_n(a)$; and observe that for each point $x \in [a, b]$ and each natural n we have $f_n(x) - \alpha_n = \int_a^x f_n'$.

Exercises 8.23

1. Suppose that for some point $c \in [a, b]$ the sequence $(f_n(c))$ converges. Observe that for every $x \in [a, b]$ we have $f_n(x) = f_n(c) + \int_c^x f_n'$. Now use Theorem 8.22.1.
2. Assume Theorem 8.22.1, and suppose that a sequence (f_n) converges boundedly on an interval $[a, b]$ to an integrable function f. In order to show that $\int_a^b f_n \to \int_a^b f$, apply Theorem 8.22.1 to the sequence $f_1, f, f_2, f, f_3, f, \ldots$.
3. Write $E = \{x_n \mid n \in \mathbf{Z}^+\}$. For each n, define $E_n = \{x_i \mid 1 \le i \le n\}$, and define $f_n = \chi_{E_n}$. We see that for each n the function f_n is a step function and $\int_a^b f_n = 0$. Therefore, since $f_n \to \chi_E$ boundedly on $[a, b]$, in any theory of integration which admits a bounded convergence theorem and in which the pointwise limit of a sequence of step functions is integrable, we must have $\int_a^b \chi_E = 0$.

Exercises 9.3

18. To obtain a contradiction, assume that e is rational, and choose a natural $n > 2$ such that ne is an integer. We now obtain

$$0 < n!\left(e - \sum_{j=0}^n \frac{1}{j!}\right) < \frac{3}{n+1} < 1,$$

which is impossible because the second expression in this inequality is an integer.

Exercises 9.6

1. (a) Converges. **(b)** Diverges. **(c)** Converges. **(d)** Converges. **(e)** Converges.
(f) Diverges. **(g)** Converges. **(h)**, **(i)**, and **(j)** All converge. **(k)** Diverges.
(l) Diverges.
Hints: In (f), use the fact that $(x - 1)/\log x \to 1$ as $x \to 1$. In (g), use the
fact that $\log(\sin x)/\log x \to 1$ as $x \to 0+$. In (h), (i), and (j), observe that
$(\log \log \log x)^{\log x} = \exp[(\log x)(\log \log \log \log x)]$. Now for x sufficiently large
we have $\log \log \log \log x > 2$; in other words, $(\log \log \log x)^{\log x} > x^2$.

4. (a) Converges. **(b)** Diverges. **(c)** Converges. **(d)** Converges.
Hints: In (b), use the fact that $2 \sin^2 x = 1 - \cos 2x$. In (d), put $u = e^x$.

Exercises 9.9

3. $(\sqrt{\pi}/2) \exp(-y^2)$.

4. Use exercise 3. The integral is $\pi/4$. With the order of integration inverted, the
integral diverges.

5. $\frac{1}{3}$

7. Put $u = \arctan x$ in the integral evaluated in exercise 6. Then integrate twice
by parts.

11. (h) $\qquad B(\alpha, \alpha) = 2 \int_0^{\pi/2} \sin^{2\alpha-1} \theta \cos^{2\alpha-1} \theta \, d\theta = 2^{2-2\alpha} \int_0^{\pi/2} \sin^{2\alpha-1} 2\theta \, d\theta.$

Now put $u = 2\theta$ and use part (g).
(i) The integral is $\frac{1}{2}B(\frac{3}{4}, \frac{1}{4})$. Put $\alpha = \frac{1}{4}$ in part (h).

Exercises 10.7

1. The series converges for $p > 1$ and diverges for $p \le 1$.

2. (a) converges, (b) converges, (c) diverges, and (d) converges.

3. (a) diverges. For (b), evaluate a limit similar to the one evaluated in part (a),
and conclude that (b) diverges. For (c) and (d), make use of the fact that
$2 < (1 + 1/n)^n < e$, and conclude that (c) diverges and (d) converges.
(f) diverges, and (g) converges. For (h), use the fact that $(\log n)^{(\log n)} =
\exp[(\log n)(\log \log n)]$, and deduce that this is greater than n^2 for n large
enough to make $\log \log n > 2$. Therefore, (h) converges. (i) converges. For (j),
use the fact that $(\log n)^{(\log\log n)} = \exp[(\log \log n)(\log \log n)]$. Then observe
that $(\log \log n)^2/(\log n) \to 0$ as $n \to \infty$ and that therefore $(\log n)^{(\log\log n)} < n$ for n
sufficiently large. Therefore, (j) diverges. (k) converges, (l) diverges,
(m) diverges, and (n) converges.

4. Compare with $\Sigma 1/n^\alpha$. The series converges for $\alpha > 1$ and diverges for $\alpha \le 1$.

12. $\log 4$.

13. $2 - \log 4$ and $\log 4 - 1$.

14. $\log p$.

Exercises 10.9

1. (a) Converges. (b) Converges. (c) Diverges. (d) Diverges. (e) Converges.
2. (b) Converges for $\alpha < 1$ and diverges for $\alpha \geq 1$. (c) Diverges by d'Alembert's test. Use Exercise 7.16(13).
4. The series converges if $x > 2$ and diverges if $x \leq 2$. The nth term of this series converges to zero when $x > 1$.
7. Use Theorem 10.8.7 to show that the series converges if $\alpha > 2$ and diverges if $\alpha < 2$. Use Theorem 10.8.11 to show that the series diverges if $\alpha = 2$.
8. Theorem 10.8.7 shows that the series converges when $\beta - \alpha > 2$ and diverges when $\beta - \alpha < 2$. When $\beta - \alpha = 2$, the series diverges. To see this, compare with $\Sigma \, 1/\sqrt{n}$.

Exercises 10.11

1. (a), (b), and (c) converge. (d) converges if $0 < \delta \leq 1$ and diverges if $\delta > 1$. (e) converges. To determine whether or not the nth terms of (d) and (e) tend to zero as $n \to \infty$, use Theorem 10.8.9.
2. (a) converges absolutely when $\frac{1}{3} < x < 1$ and converges conditionally when $x = \frac{1}{3}$. (b) converges absolutely when $e^{-1} < x < e$ and conditionally when $x = e^{-1}$. (c) converges absolutely when $0 \leq x < 1$ and conditionally when $x = 1$. (d) converges absolutely when $|x| < \frac{4}{27}$ and conditionally when $x = -\frac{4}{27}$. (e) converges absolutely when $|x| < e$ and never converges conditionally. (f) converges absolutely when $|x| < 1/e$ and conditionally when $x = -1/e$. See Example 10.8.10(2).
3. (a) converges as long as x/π is not an even integer. (b) converges for all x. (c) diverges for all x. (d) diverges unless x/π is an integer. To establish the divergence of (c), use the identity $2\cos^2 x = 1 + \cos 2x$ and the series in part (a). Since $|\sin nx| \geq \sin^2 nx$, the divergence of (d) may be obtained similarly.
4. If α is a nonnegative integer, the series converges for all x. Otherwise, if $\alpha > 0$, the series converges for $|x| \leq 1$; if $-1 < \alpha < 0$, the series converges for $-1 < x \leq 1$; and if $\alpha \leq -1$, the series converges for $|x| < 1$.
9. Suppose that (a_n) is a decreasing sequence of positive numbers and that $\Sigma \, b_n$ converges. Suppose that $a_n \to a$ as $n \to \infty$. By Dirichlet's test the series $\Sigma(a_n - a)b_n$ converges. Now look at the identity $a_n b_n = (a_n - a)b_n + ab_n$.
11. Define $a_n = 1/2^n$ if n is even and $a_n = 1/n$ if n is odd. Define $b_n = (-1)^n$ for all n.

Exercises 10.15

2. The nth term of the Cauchy product is x^n if n is even and is 0 if n is odd. The Cauchy product is therefore $\Sigma \, x^{2n}$. Mertens's theorem as stated for these series says that if $|x| < 1$, then

$$\left(\frac{1}{1-x}\right)\left(\frac{1}{1+x}\right) = \frac{1}{1-x^2}.$$

3. The Cauchy product is $\Sigma (x + y)^n/n!$. Mertens's theorem gives $e^x e^y = e^{x+y}$.
5. Evaluate the Cauchy product of the series $\Sigma (-1)^n/(n + 1)$ with itself. Express $1/(n + 1 - r)(r + 1)$ in partial fractions, and then show that the nth term of this Cauchy product is $[(-1)^n/(n + 2)] S_n$. Use Dirichlet's test to show that the Cauchy product converges, and then use Theorem 10.14.4. Alternatively, the convergence of this Cauchy product can be deduced at once from the Hardy-Edmonds theorem.

Exercises 11.4

1. **(a)** $f_n \rightarrow 0$ boundedly but not uniformly.
   (b) $f_n \rightarrow 0$ pointwise but not boundedly.
   (c) $f_n \rightarrow 0$ boundedly but not uniformly.
   (d) $f_n \rightarrow 0$ pointwise but not boundedly.
   (e) $f_n \rightarrow 0$ uniformly.
4. For each n and each point $x \in S$, define $f_n(x) = f(x) + 1/n$.
6. For each n and each $x \in [0, 1)$, define $f_n(x) = x^n$.
8. To obtain a contradiction, assume that $f_n \rightarrow 0$ pointwise but not uniformly on S. Choose $\varepsilon > 0$ such that for each n we have $\sup f_n \geq \varepsilon$. For each n, choose a point $x_n \in S$ such that $f_n(x_n) > \varepsilon/2$. Choose a partial limit x of (x_n). Using the fact that $f_n(x) \rightarrow 0$, choose a natural N such that $f_N(x) < \varepsilon/2$. Now using the continuity of the function f_N and the fact that x is a partial limit of (x_n), choose a natural $n \geq N$ such that $f_N(x_n) < \varepsilon/2$. Then since the sequence (f_n) is decreasing, we have $f_n(x_n) \leq f_N(x_n) < \varepsilon/2$, which is impossible.

Exercises 11.7

2. The fact that the given series converges uniformly on $[-\delta, \delta]$ whenever $0 \leq \delta < 1$ follows at once from the Weierstrass comparison test. We must now show that the series does not converge uniformly on $(-1, 1)$. Given $x \in (-1, 1)$ and any natural n, define

$$f_n(x) = \sum_{j=0}^{n} \frac{(2j)!}{4^j(j!)^2} x^j \quad \text{and} \quad f(x) = \sum_{j=0}^{\infty} \frac{(2j)!}{4^j(j!)^2} x^j.$$

To obtain a contradiction, assume that $f_n \rightarrow f$ uniformly on $(0, 1)$, and choose a natural N such that whenever $n \geq N$, we have $\sup|f_n - f| < 1$. Using the fact that the series $\Sigma (2j)!/4^j(j!)^2$ diverges, choose a natural n such that $\Sigma_{j=N}^{n} (2j)!/4^j(j!)^2 > 1$; and choose a number $x \in (0, 1)$ sufficiently close to 1 to make $\Sigma_{j=N}^{n} [(2j)!/4^j(j!)^2] x^j > 1$. This gives

$$|f_N(x) - f(x)| = \sum_{j=N}^{\infty} \frac{(2j)!}{4^j(j!)^2} x^j \geq \sum_{j=N}^{n} \frac{(2j)!}{4^j(j!)^2} x^j > 1,$$

contradicting the choice of N.
6. Use Example 9.7.2(1). If $f_n \leq f_1$ for each n, then the dominated convergence theorem applies.

7. The statement is true. Prove it.

8. The statement is true.

11. Prove that if $-1 < \alpha < 0$, then the given series converges uniformly on $[-\delta, 1]$ whenever $0 \leq \delta < 1$ but does not converge uniformly on $(-1, 1]$. Prove that if $\alpha \leq -1$, then the given series converges uniformly on $[-\delta, \delta]$ whenever $0 \leq \delta < 1$ but does not converge uniformly on $(-1, 1)$.

Bibliography

(1) Tom Apostol. *Mathematical Analysis*. 2nd ed. Reading, Mass.: Addison-Wesley, 1974.

(2) Kendall E. Atkinson. *An Introduction to Numerical Analysis*. New York: Wiley, 1978.

(3) E. T. Bell. *Men of Mathematics*. London: Victor Gollancz, 1937.

(4) George Berkeley. *The Analyst, or a Discourse Addressed to an Infidel Mathematician*. Vol. III of *The Works of George Berkeley*. Edited by A. C. Fraser. Oxford: 1901.

(5) R. P. Boas. *A Primer of Real Functions*. Carus Mathematical Monographs, No. 13. New York: Wiley, 1960.

(6) B. R. Gelbaum and J. M. H. Olmsted. *Counterexamples in Analysis*. San Fransisco: Holden-Day, 1964.

(7) John L. Kelley. *General Topology*. Princeton, N.J.: Van Nostrand, 1955.

(8) Morris Kline. *Mathematical Thought from Ancient to Modern Times*. New York: Oxford University Press, 1972.

(9) Jonathan Lewin. "Automatic Continuity of Measurable Group Homomorphisms." *Proc. A.M.S.* 87(1) (January 1983): 78–82.

(10) Jonathan Lewin. "Some Applications of the Bounded Convergence Theorem for an Introductory Course in Analysis." *American Mathematical Monthly*, 94(10) (December 1987): 988–993.

(11) W. A. J. Luxemburg. "The Abstract Riemann Integral and a Theorem of G. Fichtenholz on Equality of Repeated Riemann Integrals." IA and IB. *Proc. Ned. Akad. Wetensch. Ser. A* 64 (1961):516–545; *Indag. Math.* 23 (1961).

(12) John M. H. Olmsted. *Advanced Calculus*. Englewood Cliffs, N.J.: Prentice-Hall, 1961.

(13) Walter Rudin. *Principles of Mathematical Analysis*. 3rd ed. New York: Mc-Graw-Hill, 1976.

(14) George B. Thomas and Ross L. Finney. *Calculus and Analytic Geometry*. 6th ed. Reading, Mass.: Addison-Wesley, 1984.

(15) F. M. A. Voltaire. *Letters Concerning the English Nation*. London: 1733.

Index of Symbols

π, the number pi, 10

$\Rightarrow$, implies, 12

$\forall$, for every, 13

$\exists$, there exists, 13

$\boldsymbol{R}$, system of reals, 23

$\boldsymbol{Z}^+$, system of naturals, 26

$\boldsymbol{Q}$, system of rationals, 27

$\boldsymbol{Z}$, system of integers, 27

Φ, empty set, 31

$\subseteq$, subset of 31

sup, supremum, least upper bound, 31

inf, infimum, greatest lower bound, 31

$[a, b], [a, b), (a, b], (a, b)$, 34

$(-\infty, a), (-\infty, a], (a, \infty), [a, \infty)$, 34

$-\infty, \infty$, infinity, 38

$[-\infty, \infty]$, extended real number system, 39

$\in$, belongs to, 41

$A \cup B$, union of two sets, 41

$A \cap B$, intersection of two sets, 41

$A \backslash B$, difference of two sets, 42

$\bigcup\limits_{n=1}^{\infty} A_n, \bigcap\limits_{n=1}^{\infty} A_n$, union and intersection of a sequence of sets, 42

$f: A \rightarrow B, f$ maps A into B, 43

$f[A]$, image of A under f, 43

$g \circ f$, composition g following f, 44

$A \sim B$, equivalence of sets A and B, 45

$A \sim \subseteq B$, A is equivalent to a subset of B, 45

$\pi(A)$, power set of A, 52

χ_A, characteristic function of A, 54, 173

A^B, set of all functions from B into A, 54

$\bar{A}$, closure of A, 62

$\mathscr{L}(A)$, set of real limit points of A, 65

(x_n), sequence with nth term x_n, 70

$x_n \to x$, convergence of a sequence, 77

$\lim\limits_{n \to \infty} x_n$, limit of a sequence, 77

$\liminf x_n$, lower limit of a sequence, 92

$\limsup x_n$, upper limit of a sequence, 92

$\lim\limits_{x \to a} f(x)$, $\lim\limits_{a} f$, limit of a function, 107

ρ_A, distance function to a set, 117

f', derivative of f, 134, 136

f'', second derivative of f, 144

$f^{(n)}$, nth derivative of f, 144

$\|\mathcal{P}\|$, mesh of a partition $\mathcal{P}$, 167

$\Sigma(\mathcal{P}, f)$, sum of f over $\mathcal{P}$, 168

$\int_a^b f$, integral of a function, 164, 169, 171

$\int_{-\infty}^{\infty} f$, where f is a step function, 171

$\int_E f$, integral over an elementary set, 175

$m(E)$, Lebesgue measure of an elementary set E, 176

$\underline{\int_a^b} f$, lower integral of a function, 182

$\overline{\int_a^b} f$, upper integral of a function, 183

$u(\mathcal{P}, f)$, upper function of f on $\mathcal{P}$, 192

$\ell(\mathcal{P}, f)$, lower function of f on $\mathcal{P}$, 192

$\omega(\mathcal{P}, f)$, oscillation function of f on $\mathcal{P}$, 192

$\int_a^{\to b} f$, improper integral of a function, 228

$\Gamma(x)$, gamma function at x, 240

$B(x, y)$, beta function at (x, y), 240

$\int_a^{\to b} f(\cdot, y) = \int_a^{\to b} f(x, y)\, dx$, integral with parameter y, 243

$\int_a^{\to b} f(\cdot, y)\, dy$, integral with parameter x, 243

$\sum a_n, \sum\limits_{n} a_n, \sum\limits_{n \geq N} a_n$, series with nth term a_n, 253

$\sum\limits_{n=1}^{\infty} a_n$, sum of a series, 255

Index